PRODUCTS OF RANDOM VARIABLES

PURE AND APPLIED MATHEMATICS

A Program of Monographs, Textbooks, and Lecture Notes

MONOGRAPHS AND TEXTBOOKS IN PURE AND APPLIED MATHEMATICS

1. *K. Yano,* Integral Formulas in Riemannian Geometry (1970)
2. *S. Kobayashi,* Hyperbolic Manifolds and Holomorphic Mappings (1970)
3. *V. S. Vladimirov,* Equations of Mathematical Physics (A. Jeffrey, ed.; A. Littlewood, trans.) (1970)
4. *B. N. Pshenichnyi,* Necessary Conditions for an Extremum (L. Neustadt, translation ed.; K. Makowski, trans.) (1971)
5. *L. Narici et al.,* Functional Analysis and Valuation Theory (1971)
6. *S. S. Passman,* Infinite Group Rings (1971)
7. *L. Dornhoff,* Group Representation Theory. Part A: Ordinary Representation Theory. Part B: Modular Representation Theory (1971, 1972)
8. *W. Boothby and G. L. Weiss, eds.,* Symmetric Spaces (1972)
9. *Y. Matsushima,* Differentiable Manifolds (E. T. Kobayashi, trans.) (1972)
10. *L. E. Ward, Jr.,* Topology (1972)
11. *A. Babakhanian,* Cohomological Methods in Group Theory (1972)
12. *R. Gilmer,* Multiplicative Ideal Theory (1972)
13. *J. Yeh,* Stochastic Processes and the Wiener Integral (1973)
14. *J. Barros-Neto,* Introduction to the Theory of Distributions (1973)
15. *R. Larsen,* Functional Analysis (1973)
16. *K. Yano and S. Ishihara,* Tangent and Cotangent Bundles (1973)
17. *C. Procesi,* Rings with Polynomial Identities (1973)
18. *R. Hermann,* Geometry, Physics, and Systems (1973)
19. *N. R. Wallach,* Harmonic Analysis on Homogeneous Spaces (1973)
20. *J. Dieudonné,* Introduction to the Theory of Formal Groups (1973)
21. *I. Vaisman,* Cohomology and Differential Forms (1973)
22. *B.-Y. Chen,* Geometry of Submanifolds (1973)
23. *M. Marcus,* Finite Dimensional Multilinear Algebra (in two parts) (1973, 1975)
24. *R. Larsen,* Banach Algebras (1973)
25. *R. O. Kujala and A. L. Vitter, eds.,* Value Distribution Theory: Part A; Part B: Deficit and Bezout Estimates by Wilhelm Stoll (1973)
26. *K. B. Stolarsky,* Algebraic Numbers and Diophantine Approximation (1974)
27. *A. R. Magid,* The Separable Galois Theory of Commutative Rings (1974)
28. *B. R. McDonald,* Finite Rings with Identity (1974)
29. *J. Satake,* Linear Algebra (S. Koh et al., trans.) (1975)
30. *J. S. Golan,* Localization of Noncommutative Rings (1975)
31. *G. Klambauer,* Mathematical Analysis (1975)
32. *M. K. Agoston,* Algebraic Topology (1976)
33. *K. R. Goodearl,* Ring Theory (1976)
34. *L. E. Mansfield,* Linear Algebra with Geometric Applications (1976)
35. *N. J. Pullman,* Matrix Theory and Its Applications (1976)
36. *B. R. McDonald,* Geometric Algebra Over Local Rings (1976)
37. *C. W. Groetsch,* Generalized Inverses of Linear Operators (1977)
38. *J. E. Kuczkowski and J. L. Gersting,* Abstract Algebra (1977)
39. *C. O. Christenson and W. L. Voxman,* Aspects of Topology (1977)
40. *M. Nagata,* Field Theory (1977)
41. *R. L. Long,* Algebraic Number Theory (1977)
42. *W. F. Pfeffer,* Integrals and Measures (1977)
43. *R. L. Wheeden and A. Zygmund,* Measure and Integral (1977)
44. *J. H. Curtiss,* Introduction to Functions of a Complex Variable (1978)
45. *K. Hrbacek and T. Jech,* Introduction to Set Theory (1978)
46. *W. S. Massey,* Homology and Cohomology Theory (1978)
47. *M. Marcus,* Introduction to Modern Algebra (1978)
48. *E. C. Young,* Vector and Tensor Analysis (1978)
49. *S. B. Nadler, Jr.,* Hyperspaces of Sets (1978)
50. *S. K. Segal,* Topics in Group Kings (1978)
51. *A. C. M. van Rooij,* Non-Archimedean Functional Analysis (1978)

52. *L. Corwin and R. Szczarba,* Calculus in Vector Spaces (1979)
53. *C. Sadosky,* Interpolation of Operators and Singular Integrals (1979)
54. *J. Cronin,* Differential Equations (1980)
55. *C. W. Groetsch,* Elements of Applicable Functional Analysis (1980)
56. *I. Vaisman,* Foundations of Three-Dimensional Euclidean Geometry (1980)
57. *H. I. Freedan,* Deterministic Mathematical Models in Population Ecology (1980)
58. *S. B. Chae,* Lebesgue Integration (1980)
59. *C. S. Rees et al.,* Theory and Applications of Fourier Analysis (1981)
60. *L. Nachbin,* Introduction to Functional Analysis (R. M. Aron, trans.) (1981)
61. *G. Orzech and M. Orzech,* Plane Algebraic Curves (1981)
62. *R. Johnsonbaugh and W. E. Pfaffenberger,* Foundations of Mathematical Analysis (1981)
63. *W. L. Voxman and R. H. Goetschel,* Advanced Calculus (1981)
64. *L. J. Corwin and R. H. Szczarba,* Multivariable Calculus (1982)
65. *V. I. Istrățescu,* Introduction to Linear Operator Theory (1981)
66. *R. D. Järvinen,* Finite and Infinite Dimensional Linear Spaces (1981)
67. *J. K. Beem and P. E. Ehrlich,* Global Lorentzian Geometry (1981)
68. *D. L. Armacost,* The Structure of Locally Compact Abelian Groups (1981)
69. *J. W. Brewer and M. K. Smith, eds.,* Emmy Noether: A Tribute (1981)
70. *K. H. Kim,* Boolean Matrix Theory and Applications (1982)
71. *T. W. Wieting,* The Mathematical Theory of Chromatic Plane Ornaments (1982)
72. *D. B.Gauld,* Differential Topology (1982)
73. *R. L. Faber,* Foundations of Euclidean and Non-Euclidean Geometry (1983)
74. *M. Carmeli,* Statistical Theory and Random Matrices (1983)
75. *J. H. Carruth et al.,* The Theory of Topological Semigroups (1983)
76. *R. L. Faber,* Differential Geometry and Relativity Theory (1983)
77. *S. Barnett,* Polynomials and Linear Control Systems (1983)
78. *G. Karpilovsky,* Commutative Group Algebras (1983)
79. *F. Van Oystaeyen and A. Verschoren,* Relative Invariants of Rings (1983)
80. *I. Vaisman,* A First Course in Differential Geometry (1984)
81. *G. W. Swan,* Applications of Optimal Control Theory in Biomedicine (1984)
82. *T. Petrie and J. D. Randall,* Transformation Groups on Manifolds (1984)
83. *K. Goebel and S. Reich,* Uniform Convexity, Hyperbolic Geometry, and Nonexpansive Mappings (1984)
84. *T. Albu and C. Năstăsescu,* Relative Finiteness in Module Theory (1984)
85. *K. Hrbacek and T. Jech,* Introduction to Set Theory: Second Edition (1984)
86. *F. Van Oystaeyen and A. Verschoren,* Relative Invariants of Rings (1984)
87. *B. R. McDonald,* Linear Algebra Over Commutative Rings (1984)
88. *M. Namba,* Geometry of Projective Algebraic Curves (1984)
89. *G. F. Webb,* Theory of Nonlinear Age-Dependent Population Dynamics (1985)
90. *M. R. Bremner et al.,* Tables of Dominant Weight Multiplicities for Representations of Simple Lie Algebras (1985)
91. *A. E. Fekete,* Real Linear Algebra (1985)
92. *S. B. Chae,* Holomorphy and Calculus in Normed Spaces (1985)
93. *A. J. Jerri,* Introduction to Integral Equations with Applications (1985)
94. *G. Karpilovsky,* Projective Representations of Finite Groups (1985)
95. *L. Narici and E. Beckenstein,* Topological Vector Spaces (1985)
96. *J. Weeks,* The Shape of Space (1985)
97. *P. R. Gribik and K. O. Kortanek,* Extremal Methods of Operations Research (1985)
98. *J.-A. Chao and W. A. Woyczynski, eds.,* Probability Theory and Harmonic Analysis (1986)
99. *G. D. Crown et al.,* Abstract Algebra (1986)
100. *J. H. Carruth et al.,* The Theory of Topological Semigroups, Volume 2 (1986)
101. *R. S. Doran and V. A. Belfi,* Characterizations of C*-Algebras (1986)
102. *M. W. Jeter,* Mathematical Programming (1986)
103. *M. Altman,* A Unified Theory of Nonlinear Operator and Evolution Equations with Applications (1986)
104. *A. Verschoren,* Relative Invariants of Sheaves (1987)
105. *R. A. Usmani,* Applied Linear Algebra (1987)

106. *P. Blass and J. Lang,* Zariski Surfaces and Differential Equations in Characteristic $p > 0$ (1987)
107. *J. A. Reneke et al.,* Structured Hereditary Systems (1987)
108. *H. Busemann and B. B. Phadke,* Spaces with Distinguished Geodesics (1987)
109. *R. Harte,* Invertibility and Singularity for Bounded Linear Operators (1988)
110. *G. S. Ladde et al.,* Oscillation Theory of Differential Equations with Deviating Arguments (1987)
111. *L. Dudkin et al.,* Iterative Aggregation Theory (1987)
112. *T. Okubo,* Differential Geometry (1987)
113. *D. L. Stancl and M. L. Stancl,* Real Analysis with Point-Set Topology (1987)
114. *T. C. Gard,* Introduction to Stochastic Differential Equations (1988)
115. *S. S. Abhyankar,* Enumerative Combinatorics of Young Tableaux (1988)
116. *H. Strade and R. Farnsteiner,* Modular Lie Algebras and Their Representations (1988)
117. *J. A. Huckaba,* Commutative Rings with Zero Divisors (1988)
118. *W. D. Wallis,* Combinatorial Designs (1988)
119. *W. Więsław,* Topological Fields (1988)
120. *G. Karpilovsky,* Field Theory (1988)
121. *S. Caenepeel and F. Van Oystaeyen,* Brauer Groups and the Cohomology of Graded Rings (1989)
122. *W. Kozlowski,* Modular Function Spaces (1988)
123. *E. Lowen-Colebunders,* Function Classes of Cauchy Continuous Maps (1989)
124. *M. Pavel,* Fundamentals of Pattern Recognition (1989)
125. *V. Lakshmikantham et al.,* Stability Analysis of Nonlinear Systems (1989)
126. *R. Sivaramakrishnan,* The Classical Theory of Arithmetic Functions (1989)
127. *N. A. Watson,* Parabolic Equations on an Infinite Strip (1989)
128. *K. J. Hastings,* Introduction to the Mathematics of Operations Research (1989)
129. *B. Fine,* Algebraic Theory of the Bianchi Groups (1989)
130. *D. N. Dikranjan et al.,* Topological Groups (1989)
131. *J. C. Morgan II,* Point Set Theory (1990)
132. *P. Biler and A. Witkowski,* Problems in Mathematical Analysis (1990)
133. *H. J. Sussmann,* Nonlinear Controllability and Optimal Control (1990)
134. *J.-P. Florens et al.,* Elements of Bayesian Statistics (1990)
135. *N. Shell,* Topological Fields and Near Valuations (1990)
136. *B. F. Doolin and C. F. Martin,* Introduction to Differential Geometry for Engineers (1990)
137. *S. S. Holland, Jr.,* Applied Analysis by the Hilbert Space Method (1990)
138. *J. Okní nski,* Semigroup Algebras (1990)
139. *K. Zhu,* Operator Theory in Function Spaces (1990)
140. *G. B. Price,* An Introduction to Multicomplex Spaces and Functions (1991)
141. *R. B. Darst,* Introduction to Linear Programming (1991)
142. *P. L. Sachdev,* Nonlinear Ordinary Differential Equations and Their Applications (1991)
143. *T. Husain,* Orthogonal Schauder Bases (1991)
144. *J. Foran,* Fundamentals of Real Analysis (1991)
145. *W. C. Brown,* Matrices and Vector Spaces (1991)
146. *M. M. Rao and Z. D. Ren,* Theory of Orlicz Spaces (1991)
147. *J. S. Golan and T. Head,* Modules and the Structures of Rings (1991)
148. *C. Small,* Arithmetic of Finite Fields (1991)
149. *K. Yang,* Complex Algebraic Geometry (1991)
150. *D. G. Hoffman et al.,* Coding Theory (1991)
151. *M. O. Gonzá lez,* Classical Complex Analysis (1992)
152. *M. O. Gonzá lez,* Complex Analysis (1992)
153. *L. W. Baggett,* Functional Analysis (1992)
154. *M. Sniedovich,* Dynamic Programming (1992)
155. *R. P. Agarwal,* Difference Equations and Inequalities (1992)
156. *C. Brezinski,* Biorthogonality and Its Applications to Numerical Analysis (1992)
157. *C. Swartz,* An Introduction to Functional Analysis (1992)
158. *S. B. Nadler, Jr.,* Continuum Theory (1992)
159. *M. A. Al-Gwaiz,* Theory of Distributions (1992)

160. *E. Perry*, Geometry: Axiomatic Developments with Problem Solving (1992)
161. *E. Castillo and M. R. Ruiz-Cobo*, Functional Equations and Modelling in Science and Engineering (1992)
162. *A. J. Jerri*, Integral and Discrete Transforms with Applications and Error Analysis (1992)
163. *A. Charlier et al.*, Tensors and the Clifford Algebra (1992)
164. *P. Biler and T. Nadzieja*, Problems and Examples in Differential Equations (1992)
165. *E. Hansen*, Global Optimization Using Interval Analysis (1992)
166. *S. Guerre-Delabrière*, Classical Sequences in Banach Spaces (1992)
167. *Y. C. Wong*, Introductory Theory of Topological Vector Spaces (1992)
168. *S. H. Kulkarni and B. V. Limaye*, Real Function Algebras (1992)
169. *W. C. Brown*, Matrices Over Commutative Rings (1993)
170. *J. Loustau and M. Dillon*, Linear Geometry with Computer Graphics (1993)
171. *W. V. Petryshyn*, Approximation-Solvability of Nonlinear Functional and Differential Equations (1993)
172. *E. C. Young*, Vector and Tensor Analysis: Second Edition (1993)
173. *T. A. Bick*, Elementary Boundary Value Problems (1993)
174. *M. Pavel*, Fundamentals of Pattern Recognition: Second Edition (1993)
175. *S. A. Albeverio et al.*, Noncommutative Distributions (1993)
176. *W. Fulks*, Complex Variables (1993)
177. *M. M. Rao*, Conditional Measures and Applications (1993)
178. *A. Janicki and A. Weron*, Simulation and Chaotic Behavior of α-Stable Stochastic Processes (1994)
179. *P. Neittaanmäki and D. Tiba*, Optimal Control of Nonlinear Parabolic Systems (1994)
180. *J. Cronin*, Differential Equations: Introduction and Qualitative Theory, Second Edition (1994)
181. *S. Heikkilä and V. Lakshmikantham*, Monotone Iterative Techniques for Discontinuous Nonlinear Differential Equations (1994)
182. *X. Mao*, Exponential Stability of Stochastic Differential Equations (1994)
183. *B. S. Thomson*, Symmetric Properties of Real Functions (1994)
184. *J. E. Rubio*, Optimization and Nonstandard Analysis (1994)
185. *J. L. Bueso et al.*, Compatibility, Stability, and Sheaves (1995)
186. *A. N. Michel and K. Wang*, Qualitative Theory of Dynamical Systems (1995)
187. *M. R. Darnel*, Theory of Lattice-Ordered Groups (1995)
188. *Z. Naniewicz and P. D. Panagiotopoulos*, Mathematical Theory of Hemivariational Inequalities and Applications (1995)
189. *L. J. Corwin and R. H. Szczarba*, Calculus in Vector Spaces: Second Edition (1995)
190. *L. H. Erbe et al.*, Oscillation Theory for Functional Differential Equations (1995)
191. *S. Agaian et al.*, Binary Polynomial Transforms and Nonlinear Digital Filters (1995)
192. *M. I. Gil'* , Norm Estimations for Operation-Valued Functions and Applications (1995)
193. *P. A. Grillet*, Semigroups: An Introduction to the Structure Theory (1995)
194. *S. Kichenassamy*, Nonlinear Wave Equations (1996)
195. *V. F. Krotov*, Global Methods in Optimal Control Theory (1996)
196. *K. I. Beidar et al.*, Rings with Generalized Identities (1996)
197. *V. I. Arnautov et al.*, Introduction to the Theory of Topological Rings and Modules (1996)
198. *G. Sierksma*, Linear and Integer Programming (1996)
199. *R. Lasser*, Introduction to Fourier Series (1996)
200. *V. Sima*, Algorithms for Linear-Quadratic Optimization (1996)
201. *D. Redmond*, Number Theory (1996)
202. *J. K. Beem et al.*, Global Lorentzian Geometry: Second Edition (1996)
203. *M. Fontana et al.*, Prü fer Domains (1997)
204. *H. Tanabe*, Functional Analytic Methods for Partial Differential Equations (1997)
205. *C. Q. Zhang*, Integer Flows and Cycle Covers of Graphs (1997)
206. *E. Spiegel and C. J. O' Donnell*, Incidence Algebras (1997)
207. *B. Jakubczyk and W. Respondek*, Geometry of Feedback and Optimal Control (1998)

208. *T. W. Haynes et al.*, Fundamentals of Domination in Graphs (1998)
209. *T. W. Haynes et al., eds.*, Domination in Graphs: Advanced Topics (1998)
210. *L. A. D' Alotto et al.*, A Unified Signal Algebra Approach to Two-Dimensional Parallel Digital Signal Processing (1998)
211. *F. Halter-Koch*, Ideal Systems (1998)
212. *N. K. Govil et al., eds.*, Approximation Theory (1998)
213. *R. Cross*, Multivalued Linear Operators (1998)
214. *A. A. Martynyuk*, Stability by Liapunov's Matrix Function Method with Applications (1998)
215. *A. Favini and A. Yagi*, Degenerate Differential Equations in Banach Spaces (1999)
216. *A. Illanes and S. Nadler, Jr.*, Hyperspaces: Fundamentals and Recent Advances (1999)
217. *G. Kato and D. Struppa*, Fundamentals of Algebraic Microlocal Analysis (1999)
218. *G. X.-Z. Yuan*, KKM Theory and Applications in Nonlinear Analysis (1999)
219. *D. Motreanu and N. H. Pavel*, Tangency, Flow Invariance for Differential Equations, and Optimization Problems (1999)
220. *K. Hrbacek and T. Jech*, Introduction to Set Theory, Third Edition (1999)
221. *G. E. Kolosov*, Optimal Design of Control Systems (1999)
222. *N. L. Johnson*, Subplane Covered Nets (2000)
223. *B. Fine and G. Rosenberger*, Algebraic Generalizations of Discrete Groups (1999)
224. *M. Väth*, Volterra and Integral Equations of Vector Functions (2000)
225. *S. S. Miller and P. T. Mocanu*, Differential Subordinations (2000)
226. *R. Li et al.*, Generalized Difference Methods for Differential Equations: Numerical Analysis of Finite Volume Methods (2000)
227. *H. Li and F. Van Oystaeyen*, A Primer of Algebraic Geometry (2000)
228. *R. P. Agarwal*, Difference Equations and Inequalities: Theory, Methods, and Applications, Second Edition (2000)
229. *A. B. Kharazishvili*, Strange Functions in Real Analysis (2000)
230. *J. M. Appell et al.*, Partial Integral Operators and Integro-Differential Equations (2000)
231. *A. I. Prilepko et al.*, Methods for Solving Inverse Problems in Mathematical Physics (2000)
232. *F. Van Oystaeyen*, Algebraic Geometry for Associative Algebras (2000)
233. *D. L. Jagerman*, Difference Equations with Applications to Queues (2000)
234. *D. R. Hankerson et al.*, Coding Theory and Cryptography: The Essentials, Second Edition, Revised and Expanded (2000)
235. *S. Dăscălescu et al.*, Hopf Algebras: An Introduction (2001)
236. *R. Hagen et al.*, C*-Algebras and Numerical Analysis (2001)
237. *Y. Talpaert*, Differential Geometry: With Applications to Mechanics and Physics (2001)
238. *R. H. Villarreal*, Monomial Algebras (2001)
239. *A. N. Michel et al.*, Qualitative Theory of Dynamical Systems: Second Edition (2001)
240. *A. A. Samarskii*, The Theory of Difference Schemes (2001)
241. *J. Knopfmacher and W.-B. Zhang*, Number Theory Arising from Finite Fields (2001)
242. *S. Leader*, The Kurzweil-Henstock Integral and Its Differentials (2001)
243. *M. Biliotti et al.*, Foundations of Translation Planes (2001)
244. *A. N. Kochubei*, Pseudo-Differential Equations and Stochastics over Non-Archimedean Fields (2001)
245. *G. Sierksma*, Linear and Integer Programming: Second Edition (2002)
246. *A. A. Martynyuk*, Qualitative Methods in Nonlinear Dynamics: Novel Approaches to Liapunov' s Matrix Functions (2002)
247. *B. G. Pachpatte*, Inequalities for Finite Difference Equations (2002)
248. *A. N. Michel and D. Liu*, Qualitative Analysis and Synthesis of Recurrent Neural Networks (2002)
249. *J. R. Weeks*, The Shape of Space: Second Edition (2002)
250. *M. M. Rao and Z. D. Ren*, Applications of Orlicz Spaces (2002)
251. *V. Lakshmikantham and D. Trigiante*, Theory of Difference Equations: Numerical Methods and Applications, Second Edition (2002)

252. *T. Albu*, Cogalois Theory (2003)
253. *A. Bezdek*, Discrete Geometry (2003)
254. *M. J. Corless and A. E. Frazho*, Linear Systems and Control: An Operator Perspective (2003)
255. *I. Graham and G. Kohr*, Geometric Function Theory in One and Higher Dimensions (2003)
256. *G. V. Demidenko and S. V. Uspenskii*, Partial Differential Equations and Systems Not Solvable with Respect to the Highest-Order Derivative (2003)
257. *A. Kelarev*, Graph Algebras and Automata (2003)
258. *A. H. Siddiqi*, Applied Functional Analysis: Numerical Methods, Wavelet Methods, and Image Processing (2004)
259. *F. W. Steutel and K. van Harn*, Infinite Divisibility of Probability Distributions on the Real Line (2004)
260. *G. S. Ladde and M. Sambandham*, Stochastic Versus Deterministic Systems of Differential Equations (2004)
261. *B. J. Gardner and R. Wiegandt*, Radical Theory of Rings (2004)
262. *J. Haluška*, The Mathematical Theory of Tone Systems (2004)
263. *C. Menini and F. Van Oystaeyen*, Abstract Algebra: A Comprehensive Treatment (2004)
264. *E. Hansen and G. W. Walster*, Global Optimization Using Interval Analysis: Second Edition, Revised and Expanded (2004)
265. *M. M. Rao*, Measure Theory and Integration: Second Edition, Revised and Expanded (2004)
266. *W. J. Wickless*, A First Graduate Course in Abstract Algebra (2004)
267. *R. P. Agarwal, M. Bohner, and W-T Li*, Nonoscillation and Oscillation Theory for Functional Differential Equations (2004)
268. *J. Galambos and I. Simonelli*, Products of Random Variables: Applications to Problems of Physics and to Arithmetical Functions (2004)

Additional Volumes in Preparation

PRODUCTS OF RANDOM VARIABLES

Applications to Problems of Physics and to Arithmetical Functions

JANOS GALAMBOS
Temple University
Philadelphia, Pennsylvania, U.S.A.

ITALO SIMONELLI
Texas A&M University
Commerce, Texas, U.S.A.

MARCEL
DEKKER

MARCEL DEKKER, INC.   NEW YORK • BASEL

Library of Congress Cataloging-in-Publication Data
A catalog record for this book is available from the Library of Congress.

ISBN: 0-8247-5402-6

This book is printed on acid-free paper.

Headquarters
Marcel Dekker, Inc., 270 Madison Avenue, New York, NY 10016, U.S.A.
tel: 212-696-9000; fax: 212-685-4540

Distribution and Customer Service
Marcel Dekker, Inc., Cimarron Road, Monticello, New York 12701, U.S.A.
tel: 800-228-1160; fax: 845-796-1772

World Wide Web
http://www.dekker.com

The publisher offers discounts on this book when ordered in bulk quantities. For more infor-mation, write to Special Sales/Professional Marketing at the headquarters address above.

Current printing (last digit):

10 9 8 7 6 5 4 3 2 1

PRINTED IN THE UNITED STATES OF AMERICA

Preface

We present in this book a unified theory of complete convergence of products of independent random variables. These results are then applied to diverse fields in which products of random variables play a central role. Seemingly unrelated topics such as interactive particle systems of physics, asymptotic distribution of order statistics, asymptotic properties of arithmetical functions, and symmetry of limiting distribution of products of random variables can successfully be treated by the very recently developed asymptotic theory of products of independent random variables. When the variables are dependent, a new Bonferroni-type argument on the distribution of an odd number of occurrences in a sequence of events is utilized. The theory developed in this book also leads to a new group of characterizations of probability distributions, which we included in a separate chapter.

We assume that the reader is familiar with the fundamentals of probability theory. Therefore, we do not include the development of the foundation of probability theory. Rather, in the first section of Chapter 1 we collect, mainly for notational purposes, the basic concepts and a few theorems that we apply in the book. The reader who wants to review these and other results from the foundations of probability theory can find them in a book by J. Galambos (*Advanced Probability Theory*, 2nd edition, Marcel Dekker, Inc., 1995) which we call throughout the book 'our basic reference'.

We believe that the theory of products of random variables is completely covered in the book, and have also sought to provide a comprehensive list of references. The chapters on applications, however, are not intended to cover those topics in full. In fact, there are specialized books on interactive particle systems as well as on order statistics and on arithmetical functions, which provide much more detail than our present book. Yet, we hope that our approach has contributed to those fields.

The book should be of interest to mathematicians, statisticians, and physicists. It is a research reference, but it can be used in special courses at any level of graduate study.

We started working together on products of random variables and their applications about ten years ago. At the early stages of our joint work we had the fortunate opportunity to discuss different topics with Paul Erdös and Vladimir M. Zolotarev, and our close contact with Jonas Kubilius and Imre Kátai had much influence on our work. Long before this project had started, we had learned much from Alfréd Rényi. We are grateful to them all. We wish particularly to thank Professor Kubilius for bringing us in contact with the authors of the only book available on products, Gintautas Bareikis and Jonas Šiaulys, who in turn were very kind to have sent us a copy of their book. Their book helped us in understanding several older results (and forced us to learn the Lithuanian language). We tried to keep to a minimum the overlap between their book and ours.

We thank Ms. Maria Allegra, Acquisitions Editor and Manager at Marcel Dekker Inc.. Ms. Allegra's encouragement and understanding of our need for extra time for submission of the manuscript allowed us to make this a better book.

Janos Galambos
Department of Mathematics
TU 038-16, Temple University
Philadelphia, Pennsylvania 19122
USA
Italo Simonelli
Department of Mathematics
Texas A&M University-Commerce
Commerce, Texas 75429-3011
USA

Contents

Chapter 1

Foundations and extension of a theorem of Simonelli

1.1 Basic concepts and notation

The aim of this section is to collect the concepts of probability theory and to introduce the notation used throughout the book. Since it is assumed that the reader is familiar with the foundations of probability theory, the theory itself is not developed in this book. The reader may find details in Chapters 1-4 of our basic reference [58].

1.1.1 Events and probability

Every probability statement assumes that we work in a specific probability space. It consists of a basic set Ω, a sigma field $\mathcal{A}$ of subsets of Ω, and a probability on $\mathcal{A}$. Members of $\mathcal{A}$ are called events. We use the standard set theoretical notation for operations with sets: union $A \cup B$, intersection $A \cap B$, and taking complements A^c. The probability $P(A)$ of the event A is defined by the axioms:

(A1) $0 \leq P(A) \leq 1$

(A2) $P(\Omega) = 1$

(A3) For every finite or denumerably infinite set T of positive integers, if the events A_i, $i \in T$, are mutually exclusive, then

$$P(\cup_{i \in T} A_i) = \sum_{i \in T} P(A_i).$$

From the axioms we derive some elementary but important properties of probabilities.

Since for any set A, $\Omega = A \cup A^c$, (A1) and (A3) immediately give

Theorem 1.1 $P(A^c) = 1 - P(A)$.

If A and B are arbitrary events, $B = (B \cap A) \cup (B \cap A^c)$, and (A1) and (A3) imply

Theorem 1.2 *For arbitrary A and B, $P(A \cap B^c) = P(A) - P(A \cap B)$. Furthermore, if $A \subset B$, then $P(A) \leq P(B)$.*

Similarly, by rewriting $A \cup B = A \cup (B \cap A^c)$ one obtains

Theorem 1.3 $P(A \cup B) = P(A) + P(B) - P(A \cap B)$.

From the above theorem one immediately has $P(A \cup B) \leq P(A) + P(B)$, and by induction this can be extended to any finite number of events.

Theorem 1.4 *For $n \geq 2$,*

$$P(\cup_{i=1}^{n} A_i) \leq \sum_{i=1}^{n} P(A_i).$$

A sequence of events A_n, $n \geq 2$, is said to be an *increasing* (*decreasing*) sequence if for all $n \geq 1$, $A_n \subset A_{n+1}$ ($A_{n+1} \subset A_n$). Axiom 3 with T denumerable is equivalent to the following property.

Theorem 1.5 *(Continuity of probability) Let A_n, $n \geq 2$ be an increasing (decreasing) sequence of events, and let $A = \cup_n A_n$ ($A = \cap_n A_n$). Then $\lim_{n \to +\infty} P(A_n)$ exists and*

$$\lim_{n \to +\infty} P(A_n) = P(A).$$

By using Theorem 1.5 one can extend Theorem 1.4 to infinite sequences.

Theorem 1.6 *For an arbitrary sequence A_n of events*

$$P(\cup_{n=1}^{+\infty} A_n) \leq \sum_{i=1}^{+\infty} P(A_i).$$

We say that two events A and B are *independent* if

$$P(A \cap B) = P(A)P(B).$$

If the events A and B are independent, then A and A^c are also independent, and consequently the events in each pair $\{A, B\}$, $\{A, B^c\}$, $\{A^c, B\}$ and $\{A^c, B^c\}$ are independent concurrently.

A collection of events $A_1, A_2, \cdots, A_n$ is said to be independent if for every $k \geq 1$ and every $1 \leq i_1 < i_2 < \cdots < i_k \leq n$,

$$P(A_{i_1} \cap A_{i_2} \cap \cdots \cap A_{i_k}) = P(A_{i_1})P(A_{i_2}) \cdots P(A_{i_k}).$$

A countable collection of events is said to consist of *independent* events if every finite subset of these events is independent.

1.1.2 Random variables

Given a probability space $(\Omega, \mathcal{A}, P)$ a *random variable* X is a real valued function $X = X(\omega)$, defined on Ω, such that for every real number z the set $\{\omega : X(\omega) \leq z\} = \{X \leq z\}$ belongs to $\mathcal{A}$. Let $sign(X) = -$ if $X < 0$, $sign(X) = +$ if $x > 0$ and $sign(X) = 0$ if $X = 0$. X is called *discrete* if there is a finite or denumerable collection of distinct values z_i with $p_i = P(X = z_i)$ and $\sum_i p_i = 1$.

Given a random variable X, its *distribution function* is

$$F(z) = F_X(z) = P(X \leq z) = \int_{-\infty}^{z} dF(x), \tag{1.1}$$

where the above integral is a Lebesgue-Stieltjes integral. If X is a discrete random variable (1.1) becomes

$$F(z) = \sum_{x_i \leq z} P(X = x_i),$$

whereas if $f(x) = F'(x)$ exists for almost all (Lebesgue) x, we have

$$F(z) = \int_{-\infty}^{z} f(x)\, dx.$$

The function $f(x)$ is called a *density* function of $F(x)$. If $f(x) = F'(x)$ exists for all x, then $f(x)$ is called the density function.

If g is a Borel-measurable function on the real line, then $g(X)$ is also a random variable and the *expectation* of $g(X)$ is

$$E[g(X)] = \int_{-\infty}^{+\infty} g(x)dF(x).$$

In particular if $g(x) = x$, $\mu_X = \mu = E[X]$ is called the *expected value* of X, and if $g(x) = x^n$, $n > 0$, $E[X^n]$ is called the *n-th moment* of X. The *variance* of X is defined by $V(X) = \sigma^2 = E[(X - E(X))^2]$, and the standard deviation is $\sqrt{V(X)} = \sigma$. Note that, by the linear properties of integration (expectation), another form of $V(X)$ is as follows:

$$V(X) = E\left[X^2\right] - E^2\left[X\right].$$

If X is a nonnegative random variable with finite expected value $\mu > 0$, then for any $a > 0$,

$$P(X \geq a\mu) = \int_{a\mu}^{+\infty} dF(x) \leq \frac{1}{a\mu} \int_0^{+\infty} xdF(x),$$

thus giving

Theorem 1.7 *(Markov) Let $X \geq 0$ with finite $\mu > 0$. Then for arbitrary $a > 0$,*

$$P(X \geq a\mu) \leq \frac{1}{a}.$$

Upon replacing X by $(X - \mu)^2$, Markov's inequality becomes that of Chebyshev.

Theorem 1.8 *(Chebyshev) Let X be a random variable with finite expectation μ and variance $V(X) > 0$. Then for arbitrary $a > 0$,*

$$P(|X - \mu| \geq a) \leq \frac{V(X)}{a^2}.$$

In dealing with products of random variables the following inequality is a frequent tool.

Theorem 1.9 *(Hölder) Let X and Y be arbitrary random variables. Then*

$$E(|XY|) \leq E^{1/p}(|X|^p)E^{1/q}(|Y|^q),$$

where $p > 0$, $q > 0$ and $1/p + 1/q = 1$.

The special case $p = 2$ is known as the Cauchy-Schwarz inequality.

The definition of distribution function can be extended to vectors. Let $(X_1, X_2, \cdots, X_n)$ be a vector whose components are random variables on the probability space $(\Omega, \mathcal{A}, P)$. Then the function

$$F(z_1, z_2, \cdots, z_n) = P(X_1 \leq z_1, X_2 \leq z_2, \cdots, X_n \leq z_n)$$

defined for all real vectors $(z_1, z_2, \cdots, z_n)$ is called the *multivariate distribution function* of the vector $(X_1, X_2, \cdots, X_n)$. Note that upon letting one or several z_j tend to infinity, by the continuity of the limit (expressed in Theorem 1.5) we get the lower dimensional distribution functions of the vector $(X_1, X_2, \cdots, X_n)$. In particular if $n-1$ of the variables z_j, $1 \leq j \leq n$, tend to infinity, we obtain the univariate distributions $F_t(z_t) = P(X_t \leq z_t)$, $1 \leq t \leq n$.

The random variables X_n, $n \geq 1$ are said to be *independent* if the events $\{X_i \leq z_i\}$ are independent for all real numbers z_i.

Let $X_1, X_2, \cdots,$ be random variables on $(\Omega, \mathcal{A}, P)$. We denote by

$$\sigma(X_k, X_{k+1}, \cdots, X_{k+m})$$

the smallest sigma field contained in $\mathcal{A}$ with respect to which the random variables $X_k, \cdots, X_{k+m}$ are measurable, and by $\sigma(X_k, X_{k+1}, \cdot\cdot)$ the smallest σ-field over the field of the union of the sequence $\sigma(X_k, X_{k+1}, \cdots, X_{k+m})$, $m \geq 0$, of σ-fields. The tail σ-field of the random variables X_i, $i \geq 1$, is the intersection of the σ-fields $\sigma(X_k, X_{k+1}, \cdots)$, $k \geq 1$. Random variables that are measurable with respect to the tail σ-field of X_i, $i \geq 1$, are called *tail functions* of X_i.

Theorem 1.10 *(Kolmogorov's zero-one law) Let $X_1, X_2, \cdots$ be independent random variables. If the event A is in the tail σ-field of the random variables X_i, $i \geq 1$, then $P(A) = 0$ or 1. The tail functions of X_i are degenerate.*

1.1.3 Weak convergence; characteristic functions

We say that a sequence $F_n(z)$ of distribution functions *converges weakly* to the distribution function $F(z)$ if $F_n(z) \to F(z)$ as $n \to +\infty$ at each continuity point z of F. If, in addition, $F_n(-\infty) \to F(-\infty)$ and $F_n(+\infty) \to F(+\infty)$, weak convergence becomes *complete convergence.* If F_n and F are the distribution functions of X_n and X, respectively, then X_n is said to *converge weakly* (*completely*) to X when F_n converges weakly (completely) to F.

Theorem 1.11 *(Weak compactness theorem) Every infinite sequence $F_n(x)$, $n \geq 1$, of distribution functions has a weakly convergent subsequence $F_{n_k}(x)$, $k \geq 1$.*

Let $m_{k,n}$ denote the k-th moment of X_n.

Theorem 1.12 *(Method of moments) Let F_n be a sequence of distribution functions. Assume that for every $k \geq 0$, $m_{k,n} \to m_k$ finite as $n \to +\infty$. Then the sequence $\{m_k\}$ is necessarily the sequence of moments of a distribution function F. If it is known that the moments m_k uniquely determine F, then F_n converges to F completely.*

If U and V are two random variables on $(\Omega, \mathcal{A}, P)$, then $Y = U + iV$, $i = \sqrt{-1}$, is a complex valued random variable. Let $e^{itx} = cos(tx) + isin(tx)$. The *characteristic function* of a random variable X is given by

$$\phi(t) = \phi_X(t) = \int_{-\infty}^{+\infty} e^{itx} dF(x). \tag{1.2}$$

From $\phi(t)$ one can obtain the moments of X by applying the following differentiation rules. Assume that for some integer $k > 0$, $E[X^k]$ is finite. Then

$$\phi^{(j)}(0) = i^j E\left[X^j\right] \qquad 0 < j \leq k.$$

Another important property of $\phi(t)$ is given by the following theorem.

Theorem 1.13 *The characteristic function of a random variable X uniquely determines its distribution function.*

Theorem 1.13 implies that there is a one to one correspondence between characteristic functions and distribution functions of random variables. This can be proved by deriving an inversion formula from which the distribution function $F(x)$ can be recovered from its characteristic function $\phi(t)$. For any continuity points $a < b$ of $F(x)$, this formula is given by

$$F(b) - F(a) = \lim_{U \to +\infty} \frac{1}{2\pi} \int_{-U}^{U} \frac{e^{-ita} - e^{-itb}}{it} \phi(t)\, dt. \tag{1.3}$$

The integral in (1.2) is also known as a Fourier integral. In general if $H(x)$ is any real function defined on $-\infty < x < +\infty$ such that

$$\int_{-\infty}^{+\infty} |H(x)| e^{itx} dx$$

converges for some t, then

$$\hat{H}(t) = \int_{-\infty}^{+\infty} e^{itx} H(x) dx$$

is called the Fourier transform of $H(x)$. In this case the inverse Fourier transform is given by

$$H(x) = \frac{1}{2\pi} \int_{-\infty}^{+\infty} e^{-itx} \hat{H}(t) dt. \tag{1.4}$$

If X and Y are independent random variables with distribution functions $F_X(x)$ and $F_Y(y)$, respectively, the distribution function $F(z)$ of $X + Y$, given by

$$F(z) = \int_{-\infty}^{+\infty} F_Y(z - x) dF_X(x), \tag{1.5}$$

is called the convolution of $F_X(x)$ and $F_Y(y)$.

In case of independence, the characteristic function of $X_1 + X_2$ satisfies the following important property

$$\phi_{X_1+X_2}(t) = E\left[e^{it(X_1+X_2)}\right] = E\left[e^{it(X_1)}\right] E\left[e^{it(X_2)}\right] = \phi_{X_1}(t)\phi_{X_2}(t).$$

More generally if $X_1, X_2, \cdots, X_n$ are independent random variables with characteristic functions $\phi_1(t), \phi_2(t), \cdots, \phi_n(t)$, the characteristic function of $Y = X_1 + X_2 + \cdots + X_n$ is $\phi(t) = \phi_1(t)\phi_2(t)\cdots\phi_n(t)$.

The one to one correspondence between characteristic functions and distribution functions continues to hold under some limiting processes.

Theorem 1.14 *Let $F_n(x)$, $n \geq 1$, be a sequence of distribution functions with corresponding characteristic functions $\phi_n(t)$. If $F_n(x)$ converges to $F(x)$ completely, then $\phi_n(t)$ converges to the characteristic function $\phi(t)$ of $F(x)$. Conversely, if $\phi_n(t)$ converges to a function $\phi(t)$ continuous at zero, then $\phi(t)$ is the characteristic function of a proper distribution function $F(x)$, and $F_n(x)$ converges to $F(x)$ completely.*

1.1.4 Convergence in probability and almost sure convergence

We say that a sequence X_n of random variables converges in probability to the random variable X if, for every $\epsilon > 0$, as $n \to +\infty$,

$$P(|X_n - X| \geq \epsilon) \to 0. \tag{1.6}$$

On the other hand, if

$$P\Big(\lim_{n\to+\infty} X_n = X\Big) = 1 \tag{1.7}$$

we say that X_n converges to X almost surely, which we abbreviate to a.s..

The two types of convergence defined at (1.6) and (1.7) are closely related but not identical. It easily follows that (1.7) implies (1.6) but the converse it is not true.

Example 1.1 Let $X_j \geq 2$ be integer valued i.i.d. random variables with $P(X_j = k) = 1/(k(k-1))$, $k \geq 2$. Set $S_n = X_1 + X_2 + \cdots + X_n$. Then

$$\frac{S_n}{n\log n} \to 1 \quad \text{in probability} \quad \text{as } n \to +\infty \tag{1.8}$$

but

$$P\Big(\lim_{n\to+\infty} \frac{S_n}{n\log n} = A \text{ finite}\Big) = 0. \tag{1.9}$$

For proving (1.8), truncate each X_j at $n\log n$, and let the truncated variables be $Y_{j,n}$, $1 \leq j \leq n$. That is,

$$Y_{j,n} = \begin{cases} X_j & \text{if } X_j \leq n\log n \\ 0 & \text{otherwise.} \end{cases}$$

Then $E(Y_{j,n}) = (1+\epsilon)\log n$ and $V(Y_{j,n}) \leq E(Y_{j,n}^2) \sim n \log n$. Upon applying Chebyshev's inequality (Theorem 1.8) to the sum $Y_{1,n} + Y_{2,n} + \cdots + Y_{n,n}$ and noting that

$$\begin{aligned} & P(X_1 + X_2 + \cdots + X_n \neq Y_{1,n} + Y_{2,n} + \cdots + Y_{n,n}) \\ \leq & \sum_{j=1}^{n} P(X_j \neq Y_{j,n}) = \sum_{j=1}^{n} P(X_j > n \log n) \\ \leq & \; n \frac{1}{n \log n} \to 0 \end{aligned}$$

(1.8) follows.

We turn to (1.9). Since the event on the left hand side of (1.9) is a tail event, its probability is either 0 or 1 by the zero-one law expressed in Theorem 1.10. Since by (1.8)

$$P\Big(\limsup_{n\to+\infty} \frac{S_n}{n \log n} \geq 1 - \epsilon \Big) > 0, \tag{1.10}$$

if the limit in (1.9) exists, then $A \geq 1$, and the right hand side of (1.9) would be 1. On the other hand

$$\liminf_{n\to+\infty} \frac{S_n}{n \log n} \leq \Big(\liminf_{n\to+\infty} \frac{S_n}{X_{n+1}} \Big) \Big(\limsup_{n\to+\infty} \frac{X_{n+1}}{n \log n} \Big) \to 0$$

whenever the lim sup in (1.10) is finite, because the last factor is evidently finite by our assumption and

$$\liminf_{n\to+\infty} \frac{S_n}{X_{n+1}} = 0$$

by independence and by the assumed distribution of the X_j (the details are left to the reader). This proves (1.9).

We add to our observation on the relation of the two types of convergence at (1.6) and (1.7) that convergence in probability implies complete convergence which, once again, is not reversable. However, it is remarkable that in the case of infinite sums of independent random variables these three types of convergence coincide. The following theorems are valid.

Theorem 1.15 *Let X_i, $i \geq 1$ be independent random variables and let $F_n(z)$ be the distribution function of $Y_n = \sum_{i=1}^{n} X_i$. Then $\sum_{i=1}^{+\infty} X_i$ converges almost surely to a finite random variable Y if, and only if, $F_n(z)$ converges completely to the distribution function $F(z)$ of Y.*

Theorem 1.16 *(Kolmogorov's three series theorem) Let X_i, $i \geq 1$ be an infinite sequence of random variables. Then $\sum_{i=1}^{+\infty} X_i$ converges a.s. if and only if each of the three series*

$$(i) \sum P(|X_i| \geq M) \qquad (ii) \sum E(X_{i,M}) \qquad (iii) \sum V(X_{i,M})$$

converges, where $0 < M < +\infty$ is arbitrary and $X_{i,M}$ is X_i truncated at M.

For the next theorem we need the following definition.

The real number m is a *median* of the random variable X, or of its distribution function $F(x)$, if, for every $\epsilon > 0$,

$$P(X - \epsilon \leq m) = F(m + \epsilon) \geq \frac{1}{2}$$

and

$$P(X + \epsilon \leq m) = F(m - \epsilon) \leq \frac{1}{2}.$$

A distribution function $F(x)$ can have several medians. Yet, its concept plays a significant role in the convergence of sums.

Theorem 1.17 *Let X_j, $j \geq 1$, be independent random variables. There are constants a_j^*, $j \geq 1$, such that the infinite series $\sum(X_j - a_j^*)$ converges a.s. if, and only if, there are constants a_n, $a_n \geq 1$, such that the distribution function of $\sum_{j=1}^{n} X_j - a_n$ converges completely to a proper distribution function $F(x)$. When the above constants exist, one can always choose $a_j^* = m_j + E[(X_j - m_j)_M]$ and $a_n = a_1^* + a_2^* + \cdots + a_n^*$, where m_j is a median of X_j and $(X_j - m_j)_M$ is $X_j - m_j$ truncated at $M > 0$.*

Theorem 1.18 *(Two-series theorem) With the notation and assumptions of the preceding theorem, there are constants a_j^* such that the infinite sum $\sum(X_j - a_j^*)$ converges a.s. if, and only if, with arbitrary constant $M > 0$, the two series $\sum P(|X_j - m_j| \geq M)$ and $\sum V[(X_j - m_j)_M]$ converge. When such constants a_j^* exist, they can be chosen as in Theorem 1.17.*

In Chapter 2, we shall present significant extensions of these theorems. Theorems 1.15 through 1.18 add a new light to our example on (1.6) and(1.7), Example 1.1. These theorems tell that for independent random variables X_j with $S_n = \sum_{j=1}^n X_j$, the complete convergence of S_n or $S_n - a_n$ implies a.s. convergence. However, this implication is no longer true when a scale normalization S_n/b_n, $b_n > 0$, $b_n \to +\infty$ with n, is involved.

1.2 Elementary formulas for products of random variables

One faces products of random variables in the very early stages of studying measurements associated with random variables. Indeed, given two random variables X and Y, the variance

$$\begin{aligned} V(X+Y) &= E\left[(X+Y-E(X+Y))^2\right] \\ &= E\left[(X-E(X))^2\right] + E\left[(Y-E(Y))^2\right] \\ &+ 2E\left[(X-E(X))(Y-E(Y))\right] \end{aligned}$$

which, if one wants to directly compute the last term on the right hand side, requires the computation of the distribution function of XY.

First, we discuss two cases: (i) both X and Y are discrete, and (ii) the general case with $P(X=0)=0$.

We start with the discrete case. Let X take the values x_j, $j \geq 1$, with distribution p_j, $j \geq 1$, and let Y take the values y_t, $t \geq 1$, with distribution r_t, $t \geq 1$. Let the joint distribution $P(X = x_j, Y = y_t) = u_{j,t}$, $j \geq 1$, $t \geq 1$. Then XY takes the values $x_j\, y_t$, and if $s = x_j\, y_t$ for several pairs (j,t), then

$$P(XY = s) = \sum u_{jt} \tag{1.11}$$

where the summation is over all pairs (j,t) such that $s = x_j\, y_t$. No considerable simplification obtains if X and Y are independent. The formula (1.11) remains unchanged except that $u_{j,t} = p_j\, r_t$ for all (j,t) in (1.11).

Example 1.2 Let X take the values +1 and -1, with probabilities $p_1 > 0$ and $p_2 = 1 - p_1 > 0$, respectively, and let Y take these same values with probabilities $r_1 > 0$ and $r_2 = 1 - r_1 > 0$. Then XY also takes +1 and -1, and

$$P(XY = 1) = P(X = 1, Y = 1) + P(X = -1, Y = -1) \qquad (1.12)$$

which, in the case of independence, becomes $P(XY = 1) = p_1 r_1 + (1 - p_1)(1 - r_1)$. Observe that if $p_1 = 1/2$ then, regardless of r_1, $P(XY = 1) = P(XY = -1) = 1/2$.

We shall later call X in the above example symmetric if $p_1 = 1/2\,(= p_2)$. We thus have that XY also is symmetric if X is, regardless of Y. In the case of Example 1.2 the opposite is also true: if neither X nor Y is symmetric then XY cannot be symmetric (a direct proof is easy based on (1.12), and we encourage the reader to carry out the computations). As we shall see in Section 1.4, this last property does not extend to arbitrary independent random variables. Note that if X and Y are not independent then (1.12) yields that XY can be symmetric even if neither X nor Y is symmetric.

Example 1.3 Let $\Omega = (0, 1)$, and let P be Lebesgue measure. Define $X = 1$ on $(0, 3/4)$ and $X = -1$ otherwise, and let $Y = 1$ on $(0, 1/4)$, and $Y = -1$ otherwise. Then $P(X = -1, Y = 1) = 1/4$, $P(X = -1, Y = -1) = 1/2$, and $P(X = 1, Y = 0) = 0$; so, by (1.12), $P(XY = 1) = P(XY = -1) = 1/2$, but $P(X = 1) = P(Y = -1) = 3/4$.

Next, let $F(x, y) = P(X \le x, Y \le y)$. Then

$$P(XY \le z) = \iint_{xy \le z} dF(x, y).$$

If $P(X = 0) = P(Y = 0) = 0$, then $x \le z/y$ if $y > 0$ and $x \ge z/y$ if $y < 0$, so

$$\begin{aligned} P(XY \le z) &= \int_{-\infty}^{0} \left(\int_{z/y}^{+\infty} dF(x, y) \right) \\ &\quad + \int_{0}^{+\infty} \left(\int_{-\infty}^{z/y} dF(x, y) \right) \end{aligned} \qquad (1.13)$$

where in both iterated integrals in 1.13 we first integrate in x, and then in y.

This time, if X and Y are independent, that is,

$$F(x,y) = P(X \leq x)P(Y \leq y) = F_X(x)F_Y(y),$$

then we can simplify (1.13). We get

$$\begin{aligned} P(XY \leq z) &= \int_{-\infty}^{0} \left(1 - F_X\left(\frac{z}{y} - 0\right)\right) dF_Y(y) \\ &\quad + \int_{0}^{+\infty} F_X\left(\frac{z}{y}\right) dF_Y(y). \end{aligned} \tag{1.14}$$

Furthermore, if $g(y) = F'_Y(y)$ exists then $dF_Y(y) = g(y)dy$.

Example 1.4 Assume X and Y are independent random variables with Pareto distributions $F_X(x) = 1 - 1/x$ and $F_Y(y) = 1 - 1/y$, $x \geq 1$ and $y \geq 1$, respectively. Then the first integral in (1.14) is zero, and, in the second integral $z/y \geq 1$ as well as $y \geq 1$. Since $g(y) = F'_Y(y) = 1/y^2$ if $y \geq 1$ and $g(y) = 0$ otherwise, by (1.14), for $z \geq 1$,

$$P(XY \leq z) = \int_1^z \left(1 - \frac{y}{z}\right)\frac{1}{y^2}dy = 1 - \frac{1}{z} - \frac{1}{z}\ln(z) = 1 - \frac{1 + \ln(z)}{z}.$$

Evidently, $P(XY \leq 1) = 0$, so $P(XY \leq z) = 0$ if $z \leq 1$.

Next we consider the general case, where no assumption is made on $P(X = 0)$ and $P(Y = 0)$. Obviously, if one of $P(X = 0)$ or $P(Y = 0)$ equals one, XY has distribution function degenerate at zero. So let us assume that $P(X = 0)$ and $P(Y = 0)$ are both less than 1. In this case we introduce the following random quantity.

For any arbitrary random variable V, with $0 \leq P(V = 0) < 1$, we define $\tilde{V}$ to be the random variable with distribution function

$$\tilde{F}(z) = P(\tilde{V} \leq z) = P(V \leq z \mid V \neq 0). \tag{1.15}$$

We shall assume that $\tilde{V} \overset{d}{=} 1$ if $P(V = 0) = 1$.

Hence

$$P(XY \leq z) = P(\tilde{X}\tilde{Y} \leq z)P(XY \neq 0) + \delta_0(z)\, P(XY = 0), \tag{1.16}$$

where $\delta_0(z)$ denotes a distribution function degenerate at zero. Since $P(\tilde{X} = 0) = P(\tilde{Y} = 0) = 0$, formula (1.13), and if $\tilde{X}$ and $\tilde{Y}$ are independent formula (1.14), can be used to compute $P(\tilde{X}\tilde{Y} \leq z)$ in (1.16).

By utilizing (1.16) the distribution of XY could be reduced to the distribution of $\ln|\tilde{X}| + \ln|\tilde{Y}|$. Indeed one can write

$$\tilde{X}\tilde{Y} = |\tilde{X}\tilde{Y}| sign(\tilde{X}\tilde{Y}) = sign(\tilde{X}\tilde{Y}) e^{(\ln|\tilde{X}|+\ln|\tilde{Y}|)} \tag{1.17}$$

and thus we can treat $sign(\tilde{X}\tilde{Y})$ and $e^{(\ln|\tilde{X}|+\ln|\tilde{Y}|)}$ separately, and the distribution problem is reduced to the distribution of a sum. This approach makes the problem simpler on some occasions, while it may make the distribution problem more complicated because the summation of $\ln|\tilde{X}|$ and $\ln|\tilde{Y}|$ is not necessarily simple. In addition, if we treat a product of a large number of factors, the sign of such product is not easily computed. We shall return to these problems in subsequent sections.

One example when the above procedure leading to (1.17) is simple is the case of independent lognormal distributions. By definition, X is lognormal if $X \stackrel{d}{=} \tilde{X} > 0$, and $\ln X$ is normal. So, by utilizing that the sum of independent normal variates is normal, we get from (1.17) that the product of independent lognormal variables is lognormal.

In order to apply (1.17) to distributions other than lognormal we introduce the concept of logarithmically $F(x)$-distributed random variables. We say that X is *logarithmically $F(x)$-distributed* if the distribution function of $Y = \ln|X|$ is $F(x)$. When $F(x)$ is normal then X is logarithmically normal variable. This concept fundamentally differs from the traditional lognormal variables because we allow X to be negative since we define the relation to $F(x)$ by $\ln|X|$. However, if $X > 0$, we continue to use both terms "logarithmically normal" and "lognormal". We have from the theory of summation of independent random variables that the logarithmically Poisson and Cauchy variables are closed families under multiplication (see (1.17)), and so are all logarithmically (sum) stable distributions. But when our interest is different from the closedness of distributions under multiplication then other distributions can come into focus. Note that the logarithmically exponential variable X, $P(\ln X \leq y) = 1 - e^{-y}$, $y \geq 0$ is the Pareto variable $P(X \leq z) = 1 - 1/z$, $z \geq 1$, which we discussed in an earlier example. This is not a closed family under multiplication but its significance will be seen later. Subsequent chapters will bring other important distributions to light.

1.3 Mellin transform: the uniqueness theorem

It was mentioned in Section 1.1 that the characteristic function of the sum of independent random variables can be expressed as the product of their characteristic functions. This property has made the Fourier transform an essential tool in studying the distribution of sums of independent random variables. This property however does not carry over to products, where another transform, the Mellin transform, is often preferable. In this section we introduce this transform and give some applications.

Let Y be a nonnegative random variable with distribution function $F_Y(y)$. The *Mellin transform* of Y or $F(y)$ is defined as

$$\mathcal{M}_Y(s) = \int_{-\infty}^{+\infty} y^s dF_Y(y),$$

where s is a complex number and $\mathcal{M}_Y(s)$ is assumed finite. We like to stress that in the definition above we use $0^s = 0$ for all s (including $s = 0$).

To extend the definition of this transform to arbitrary random variables X we use the following procedure. Let X^+ and X^- denote the nonnegative and negative part of X, respectively. That is,

$$X^+(\omega) = \begin{cases} X(\omega) & \text{if } X(\omega) \geq 0, \\ 0 & \text{otherwise,} \end{cases}$$

and

$$X^-(\omega) = \begin{cases} -X(\omega) & \text{if } X(\omega) < 0, \\ 0 & \text{otherwise.} \end{cases}$$

Since both X^+ and X^- are nonnegative their Mellin transforms, denoted respectively by $\mathcal{M}_{X^+}(s)$ and $\mathcal{M}_{X^-}(s)$, are well defined. Let

γ be a formal indeterminate with $\gamma\gamma = \gamma^2 = 1$. We define the *Mellin transform* of X or $F_X(x)$ as

$$\mathcal{M}_X(s) = \mathcal{M}_{X^+}(s) + \gamma \mathcal{M}_{X^-}(s).$$

We refer to $\mathcal{M}_{X^+}(s)$ and $\mathcal{M}_{X^-}(s)$ as the components of $\mathcal{M}_X(s)$.

Since

$$|\mathcal{M}_{X^\eta}(s)| \leq \int_{-\infty}^{+\infty} |x|^{Re(s)} dF(x),$$

where $\eta = +$ or $-$ and $Re(s)$ denotes the real part of s, $\mathcal{M}_X(s)$ exists for all values of s in some complex strip $D = \{s : a \leq Re(s) \leq b\}$ which contains the imaginary axis, and possibly equals to this axis (see Zolotarev [158]). Since at times there is no advantage in keeping such generality, whenever it suffices we restrict our discussion to $s = it$, where t is real. In such cases we write $\mathcal{M}_X(t)$ which is $\mathcal{M}_X(s)$ as above with $s = it$.

Next we are going to show that, as in the case of characteristic functions, there is a one-to-one correspondence between distribution functions and their Mellin transforms $\mathcal{M}_X(t)$, t real ($s = it$).

Theorem 1.19 *The Mellin transform $\mathcal{M}_X(t)$ of a random variable X uniquely determines its distribution function.*

PROOF. We start by assuming that $P(X > 0) = 1$. In this case

$$\mathcal{M}_X(it) = \mathcal{M}_X(t) = \int_0^{+\infty} x^{it} dF(x) = \int_0^{+\infty} e^{it\ln(x)} dF(x), \quad (1.18)$$

and (1.18) can be viewed as the characteristic function of the random variable $\ln(X)$. This, Theorem 1.13, the equivalence of $P(X \leq c)$ and $P(\ln(X) \leq \ln(c))$ when $c > 0$, and (1.3) immediately give that

$$F(b) - F(a) = \lim_{U\to+\infty} \frac{1}{2\pi} \int_{-U}^{U} \frac{e^{-it\ln(a)} - e^{-it\ln(b)}}{it} \mathcal{M}_X(t)\, dt, \quad (1.19)$$

for all continuity points $a > 0$ and $b > 0$ of $F(x)$. Hence $\mathcal{M}_X(t)$ uniquely determines the distribution function of X.

If $P(X \geq 0) = 1$ but $0 < P(X = 0) < 1$, then we define the random variable Y with distribution function

$$F_Y(y) = \begin{cases} 0 & \text{if } y < 0, \\ \frac{F(y)-F(0)}{1-F(0)} & \text{if } y \geq 0. \end{cases}$$

Y is nonnegative, continuous at zero, with

$$\mathcal{M}_X(t) = (1 - F(0))\mathcal{M}_Y(t),$$

and the validity of the theorem now follows from the validity of the previous case. An appeal to the property that $P(X = 0) = 1$ if and only if $\mathcal{M}_X(t)$ is identically zero completes the proof of the theorem in the case X is nonnegative. Since the case $P(X \leq 0) = 1$ can be recovered by our previous discussion, it is left to consider the case $0 < P(X \leq 0) < 1$. In this case we define two nonnegative random variables, $Y^{(+)}$ and $Y^{(-)}$, with respective distribution functions

$$F^+(y) = \begin{cases} \frac{F(y)-F(0)}{1-F(0)} & \text{if } y \geq 0, \\ 0 & \text{otherwise,} \end{cases}$$

and

$$F^-(y) = \begin{cases} \frac{F(0-)-F(-y)}{F(0-)} & \text{if } y > 0, \\ 0 & \text{otherwise.} \end{cases}$$

Since

$$F(x) = \begin{cases} F(0-)(1 - F^-(-x)) & \text{if } x < 0, \\ (1 - F(0))F^+(x) + F(0) & \text{if } x \geq 0, \end{cases} \tag{1.20}$$

the first part of the proof immediately gives that $\mathcal{M}_{Y^{(+)}}(t)$ and $\mathcal{M}_{Y^{(-)}}(t)$, and consequently $\mathcal{M}_X(t)$, uniquely determine $F(x)$. The proof of the theorem is now complete. ∎

It is important to notice that from the proof of Theorem 1.19 one obtains an explicit inversion formula for the Mellin transform of an arbitrary distribution function (simply apply (1.19) with $\mathcal{M}_X(t)$

replaced by $\mathcal{M}_{X^+}(t)$ and $\mathcal{M}_{X^-}(t)$ in (1.20)). If $F'(x) = f(x)$ exists, an inversion formula for $f(x)$ can be obtained by using the following procedure. Write $f(x) = f_+(x) + f_-(x)$, where $f_+(x)$ and $f_-(x)$ denote $f(x)$ restricted on $[0, +\infty)$ and $(-\infty, 0)$, respectively. Then from (1.19) one obtains

$$f_\eta(x) = \frac{1}{2\pi} \int_{-\infty}^{+\infty} |x|^{-it-1} \mathcal{M}_{X^\eta}(t)\, dt, \tag{1.21}$$

$\eta = +$ or $-$.

Next we prove the analog of Theorem 1.14 for Mellin transforms. To this end we say that a sequence of Mellin transforms $\mathcal{M}_{X_n}(t)$ converges to a function $\mathcal{M}(t) = \mathcal{M}_1(t) + \gamma\mathcal{M}_2(t)$ if and only if the components of $\mathcal{M}_{X_n}(t)$ converge to the corresponding components of $\mathcal{M}(t)$.

Theorem 1.20 *Let F_n, $n \geq 1$ be a sequence of distribution functions of the random variables X_n, with corresponding Mellin transforms $\mathcal{M}_{X_n}(t)$.*
(i) If $F_n \to F$ completely and $F_n(0) \to F(0)$ and $F_n(0-) \to F(0-)$ as $n \to +\infty$, then $\mathcal{M}_{X_n}(t)$ converges to the Mellin transform of F (as $n \to +\infty$).
(ii) Conversely, if $\mathcal{M}_{X_n}(t)$ converges to a function $\mathcal{M}(t) = \mathcal{M}_1(t) + \gamma\mathcal{M}_2(t)$ (as $n \to +\infty$) with components continuous at zero, then $\mathcal{M}(t)$ is the Mellin transform of a distribution function F and F_n converges to F completely. Moreover $F_n(0) \to F(0)$ and $F_n(0-) \to F(0-)$ as $n \to +\infty$.

PROOF. We start by proving (i). Let us assume $F_n \to F$ completely, $F_n(0) \to F(0)$, and $F_n(0-) \to F(0-)$ as $n \to +\infty$. Let X be a random variable with distribution function $F(x)$. If $P(X \geq 0) = 1$ and $0 \leq P(X = 0) < 1$, then for n sufficiently large we have $P(X_n = 0) < 1$. Hence for such n's we can define a positive random variable $Y_n^{(+)}$ with distribution function

$$F_n^+(x) = \begin{cases} \frac{F_n(x) - F_n(0)}{1 - F_n(0)} & \text{if x} \geq 0, \\ 0 & \text{otherwise.} \end{cases} \tag{1.22}$$

From (1.22) we obtain

$$\mathcal{M}_{X_n^+}(t) = (1 - F_n(0))\mathcal{M}_{Y_n^{(+)}}(t) = (1 - F_n(0))\phi_{\ln(Y_n^{(+)})}(t), \tag{1.23}$$

where $\phi_{\ln(Y_n^{(+)})}(t)$ is the characteristic function of $\ln(Y_n^{(+)})$, and the convergence of $\mathcal{M}_{X_n^+}(t)$ to $\mathcal{M}_{X^+}(t) = \mathcal{M}_X(t)$ is now guaranteed by Theorem 1.14 and our initial assumptions. The validity of (i) now follows from

$$|\mathcal{M}_{X_n^-}(t)| \leq P(X_n < 0) \to 0 \qquad \text{as } n \to +\infty.$$

If $P(X = 0) = 1$,

$$|\mathcal{M}_{X_n^\eta(t)}| \leq P(X_n \neq 0) \to 0 \qquad \text{as } n \to +\infty,$$

where $\eta =$ + or -, and the theorem certainly holds. Since the case $P(X \leq 0) = 1$ can be recovered from our previous discussion, it is left to consider the case $0 < P(X \leq 0) < 1$. In this case, when n is sufficiently large, we can define two random variables $Y_n^{(+)}$ and $Y_n^{(-)}$ with distribution functions F_n^+, defined as in (1.22), and

$$F_n^-(x) = \begin{cases} \frac{F_n(0-) - F_n(-x)}{F_n(0-)} & \text{if } x > 0, \\ 0 & \text{otherwise.} \end{cases} \tag{1.24}$$

Now $\mathcal{M}_{X_n^+}(t)$ is still given by (1.23) whereas

$$\mathcal{M}_{X_n^-}(t) = (F_n(0-))\mathcal{M}_{Y_n^{(-)}}(t) = (F_n(0-))\phi_{\ln(Y_n^{(-)})}(t).$$

Theorem 1.14 and our initial assumptions on the convergence of $F_n(0)$ and $F_n(0-)$ now imply the convergence of $\mathcal{M}_{X_n}(t)$. It is easy to see that these limits coincide with the components of the Mellin transform of X. The proof of (i) is now complete.

Next we turn to the proof of (ii) Let us assume that $\mathcal{M}_{X_n}(t)$ converges to a function $\mathcal{M}(t) = \mathcal{M}_1(t) + \gamma\mathcal{M}_2(t)$ continuous at zero. Note that this immediately implies that

$$\lim_{n\to+\infty} F_n(0-) = \lim_{n\to+\infty} \mathcal{M}_{X_n^-}(0) = \mathcal{M}_2(0), \tag{1.25}$$

and

$$\lim_{n\to+\infty} F_n(0) = \lim_{n\to+\infty} (1 - \mathcal{M}_{X_n^+}(0)) = 1 - \mathcal{M}_1(0). \tag{1.26}$$

The proof consists of considering several cases depending on whether or not the components of $\mathcal{M}(t)$ are identically zero. If both components of $\mathcal{M}(t)$ are zero, then from (1.25) and (1.26) we immediately have that F_n converges completely to a distribution function degenerate at 0, and clearly $\mathcal{M}(t)$ is its Mellin transform. Next we assume that $\mathcal{M}_2(t)$ is identically zero but $\mathcal{M}_1(t)$ is not. In this case if n is sufficiently large we can define a nonnegative random variable $Y_n^{(+)}$ with distribution function F_n^+ defined as in (1.22). Since

$$(\frac{1}{1-F_n(0)})\mathcal{M}_{X_n^+}(t) = \mathcal{M}_{Y_n^{(+)}}(t) = \phi_{\ln(Y_n^{(+)})}(t) \tag{1.27}$$

the convergence of the left hand side of (1.27) implies that $F_n^+ \to F^+$ completely, as $n \to +\infty$, and from (1.22), (1.25) and (1.26) this further implies that for all continuity points x of F^+, $F_n(x)$ converges completely to

$$F(x) = \begin{cases} \mathcal{M}_1(0)F^+(x) + (1 - \mathcal{M}_1(0)) & \text{if } x \geq 0, \\ 0 & \text{otherwise.} \end{cases}$$

By taking the limit as $n \to +\infty$ on both sides of (1.29) we also obtain that $\mathcal{M}(t)$ is the Mellin transform of F. The case $\mathcal{M}_1(t)$ is identically zero but $\mathcal{M}_2(t)$ is not can be treated similarly. It is left to consider the case in which neither $\mathcal{M}_1(t)$ nor $\mathcal{M}_2(t)$ is identically zero. In this case when n is sufficiently large we can define two random variables $Y_n^{(+)}$ and $Y_n^{(-)}$ with distribution functions F_n^+ and F_n^- defined as in (1.22) and (1.26), respectively. Now (1.27) continues to hold, and

$$\frac{\mathcal{M}_{X_n^-}(t)}{F_n(0-)} = \mathcal{M}_{Y_n^{(-)}}(t) = \phi_{\ln(Y_n^{(-)})}(t).$$

The rest of the proof now follows the lines of the previous case, and the details are left to the reader. The proof of the theorem is complete. ∎

The above discussion shows the connection between Mellin transforms and characteristic functions. This connection motivates the following example.

Example 1.5 Let X and Y be independent random variables with respective characteristic functions $\phi_X(t)$ and $\phi_Y(t)$. Let $\nu = -, +$, $P(\nu = -) = p$, ν independent of X and Y, and put

$$V = e^X 1_{\{\nu=+\}} - e^Y 1_{\{\nu=-\}}.$$

Then

$$\mathcal{M}_V(t) = \mathcal{M}_{V^+}(t) + \gamma \mathcal{M}_{V^-}(t) = (1-p)\phi_X(t) + \gamma p \phi_Y(t).$$

The above steps can be reversed. That is, if $a > 0$, $b > 0$, $a + b = 1$, and $\phi_1(t)$ and $\phi_2(t)$ are two arbitrary characteristic functions, then there exists a random variable X with Mellin transform

$$\mathcal{M}_X(t) = a\,\phi_1(t) + \gamma\, b\, \phi_2(t).$$

The above example will be utilized in Examples 1.16 and 1.17 to construct Mellin transforms with desired properties. For relations between Mellin transform and other integral transforms see Zolotarev [158].

We end this section with two applications of the Mellin transform. The first deals with the computation of the moments of X, and as in the case of the Fourier transform, this can be accomplished without first computing the inverse transform. We start by considering the special case X is a nonnegative random variable. Assume a positive integer k is in the domain of $\mathcal{M}_X(s) = \mathcal{M}_{X^+}(s)$. Then

$$E[X^j] = \mathcal{M}_X(j), \qquad 0 < j \leq k.$$

Example 1.6 Let X be a uniform random variable on the interval (a, b), $a \geq 0$, with density function

$$f(x) = \begin{cases} \frac{1}{b-a} & \text{if } a < x < b, \\ 0 & \text{otherwise.} \end{cases}$$

Then

$$\mathcal{M}_X(s) = \int_a^b x^s\, f(x)\, dx = \frac{b^{s+1} - a^{s+1}}{(b-a)(s+1)},$$

where $\mathrm{Re}(s) > \;1$. Hence if $s = k$, a positive integer,

$$\mathcal{M}_X(k) = E[X^k] = \frac{b^{k+1} - a^{k+1}}{(b-a)(k+1)}.$$

The special cases $k = 1, 2$ could be further exploited to obtain

$$\begin{aligned} V(X) &= \mathcal{M}_X(2) - (\mathcal{M}_X(1))^2 \\ &= \frac{b^3 - a^3}{3(b-a)} - \left(\frac{b^2 - a^2}{2(b-a)}\right)^2 = \frac{(b-a)^2}{12} \end{aligned}$$

Example 1.7 Let X be a lognormal random variable with parameters m and $\sigma^2 > 0$ ($\log N(m, \sigma^2)$), and density function

$$f(x) = \begin{cases} \frac{1}{x\sqrt{\sigma^2 2\pi}} e^{-\frac{(\log(x)-m)^2}{2\sigma^2}} & \text{if } x > 0, \\ 0 & \text{otherwise.} \end{cases}$$

Then

$$\mathcal{M}_X(s) = \int_0^{+\infty} x^s f(x)\, dx = e^{sm} e^{+\frac{s^2\sigma^2}{2}}.$$

The special case $s = k$, a positive integer, gives $E[X^k] = e^{km + \frac{k^2\sigma^2}{2}}$.

Next we turn to the general case. Let X be an arbitrary random variable with Mellin transform $\mathcal{M}_X(s) = \mathcal{M}_{X^+}(s) + \gamma\mathcal{M}_{X^-}(s)$. Assume a positive integer k is in the domain of $\mathcal{M}_X(s)$. Then

$$E[X^j] = \mathcal{M}_{X^+}(j) + (-1)^j \mathcal{M}_{X^-}(j), \qquad 0 < j \leq k.$$

Example 1.8 Let X be a uniform random variable on the interval (a, b), $a < 0 < b$. Then

$$\mathcal{M}_X(s) = \mathcal{M}_{X^+}(s) + \gamma\mathcal{M}_{X^-}(s) = \frac{b^{s+1}}{(b-a)(s+1)} + \gamma\frac{|a|^{s+1}}{(b-a)(s+1)},$$

where $\text{Re}(s) > -1$. Hence if k is a positive integer,

$$E\left[X^k\right] = \frac{b^{k+1}}{(k+1)(b-a)} + (-1)^k \frac{|a|^{k+1}}{(k+1)(b-a)}.$$

Example 1.9 Let X be a normal random variable (standard normal) with density function

$$f(x) = \frac{1}{\sqrt{2\pi}} e^{-\frac{x^2}{2}} \quad -\infty < x < +\infty.$$

We start by computing $\mathcal{M}_{X^+}(s)$.

$$\begin{aligned}
\mathcal{M}_{X^+}(s) &= \int_0^{+\infty} x^s f(x) \\
&= \frac{2^{\frac{s-1}{2}}}{\sqrt{2\pi}} \int_0^{+\infty} u^{(s+1)/2-1} e^{-u}\, du = \frac{2^{\frac{s-1}{2}}}{\sqrt{2\pi}} \Gamma\left(\frac{s+1}{2}\right),
\end{aligned}$$

where $\mathrm{Re}(s) > -1$, and $\Gamma(u)$ is the gamma function. This and the symmetry of $f(x)$ immediately give that

$$\mathcal{M}_X(s) = \frac{2^{\frac{s-1}{2}}}{\sqrt{2\pi}} \Gamma\left(\frac{s-1}{2}\right) + \gamma \frac{2^{\frac{s+1}{2}}}{\sqrt{2\pi}} \Gamma\left(\frac{s+1}{2}\right),$$

where $\mathrm{Re}(s) > -1$. By exploiting the definition of $\mathcal{M}_X(s)$ and the property that for $u > 0$, $\Gamma(u+1) = u\Gamma(u)$ and $\Gamma(1/2) = \sqrt{\pi}$, one immediately has,

$$E\left[X^k\right] = \begin{cases} 0 & \text{if } k \text{ is odd,} \\ 2\mathcal{M}_X(k) = \frac{2^{k/2}}{\sqrt{\pi}} \Gamma(\frac{k+1}{2}) = \prod_{i=1}^{k/2} (2i-1) & \text{if } k \text{ is even.} \end{cases}$$

The next application deals with products. The importance of the Mellin transform in the study of products of independent random variables is based on the property that the Mellin transform of XY can be expressed as the product of the Mellin transforms of X and Y. In fact if X and Y are independent, and s is in the domain of both $\mathcal{M}_X(s)$ and $\mathcal{M}_Y(s)$,

$$\begin{aligned}
\mathcal{M}_{XY}(s) &= \mathcal{M}_{(XY)^+}(s) + \gamma \mathcal{M}_{(XY)^-}(s) \\
&= E\left[((XY)^+)^s\right] + \gamma E\left[((XY)^-)^s\right] \\
&= E\left[(X^+)^s\right] E\left[(Y^+)^s\right] + E\left[(X^-)^s\right] E\left[(Y^-)^s\right] \\
&\quad + \gamma E\left[(X^+)^s\right] E\left[(Y^-)^s\right] + \gamma E\left[(X^-)^s\right] E\left[(Y^+)^s\right] \\
&= \left(E\left[(X^+)^s\right] + \gamma E\left[(X^-)^s\right]\right)\left(E\left[(Y^+)^s\right] + \gamma E\left[(Y^-)^s\right]\right) \\
&= \mathcal{M}_X(s) \mathcal{M}_Y(s). \qquad (1.28)
\end{aligned}$$

The above result easily generalizes to any number of independent random variable, and it will be assumed to hold with no further reference throughout the rest of the book.

Next we show how Mellin transforms can be used to compute the density function of products of random variables. The general procedure consists of computing the Mellin transform of a product by using (1.28), and then trying to recover its density function by computing the inverse Mellin transform. Note that if one considers $\mathcal{M}_{X^\eta}(s)$, rather than $\mathcal{M}_{X^\eta}(t)$, then (1.21) becomes

$$f_\eta(x) = \frac{1}{2i\pi} \lim_{U\to+\infty} \int_{c-iU}^{c+iU} |x|^{-s-1} \mathcal{M}_{X^\eta}(s)\, ds,$$

where the path of integration, given by the line $(c - i\infty, c + i\infty)$, lies within the strip of analyticity of $\mathcal{M}_{X^\eta}(s)$. We do not discuss the computational difficulties that one faces in computing the inverse Mellin transforms (see Springer [141] for a discussion of this problem and for references), and simply describe the procedure with few examples.

Example 1.10 Let $Y = X_1 X_2 \cdots X_n$, where the X_i's are independent uniform random variables on the interval $(0, b)$. By utilizing Example 1 we have

$$\mathcal{M}_Y(s) = \prod_{i=1}^{n} \mathcal{M}_{X_i}(s) = \left(\frac{b^{s+1}}{b(s+1)} \right)^n,$$

where Re$(s) > -1$. By the above inversion formula the density function of Y is given by

$$h(y) = \lim_{U\to+\infty} \frac{1}{2i\pi} \int_{c-iU}^{c+iU} y^{-s-1} \frac{b^{ns}}{(s+1)^n}\, ds.$$

The above integral can be evaluated by changing the path of integration and then by applying the residue theorem, which give, for $0 \le y \le b^n$,

$$h(y) = \frac{1}{(n-1)!} H^{(n-1)}(-1) = \frac{1}{b^n (n-1)!} \left(\ln \left(\frac{b^n}{y} \right) \right)^{n-1},$$

where

$$H^{(n-1)}(s) = \frac{d^{n-1}}{ds^{n-1}}\left((s+1)^n y^{-s-1}\frac{b^{ns}}{(s+1)^n}\right).$$

In the next examples we use the simple facts that $\gamma^n = \gamma$ if n is odd and $\gamma^n = 1$ for even n. hence, $(1+\gamma)^n = 2^{n-1} + \gamma\, 2^{n-1}$.

Example 1.11 Let $Y = X_1X_2\cdots X_n$ where the X_i's are independent random variables, $X_i = (-1)^{\nu_i}V_i$, where ν_i and V_i are independent, $\nu_i = 0, 1$, $P(\nu_i = 0) = P(\nu_i = 1) = 1/2$, and V_i is $\log N(m, \sigma^2)$. By utilizing Example 1.7 we have

$$\begin{aligned}
\mathcal{M}_Y(s) &= \prod_{i=1}^{n}\mathcal{M}_{X_i}(s) = \left(\frac{1}{2}e^{sm}e^{\frac{s^2\sigma^2}{2}} + \gamma\frac{1}{2}e^{sm}e^{\frac{s^2\sigma^2}{2}}\right)^n \\
&= \frac{2^{n-1}}{2^n}e^{smn}e^{\frac{s^2n\sigma^2}{2}} + \gamma\,\frac{2^{n-1}}{2^n}e^{smn}e^{\frac{s^2n\sigma^2}{2}} \\
&= \frac{1}{2}e^{smn}e^{\frac{s^2n\sigma^2}{2}} + \gamma\frac{1}{2}e^{smn}e^{\frac{s^2n\sigma^2}{2}}.
\end{aligned}$$

This, Example 1.7 and Theorem 1.19 imply $Y = (-1)^{\nu}V$, where ν and V are independent, $\nu = 0, 1$, $P(\nu = 0) = P(\nu = 1) = 1/2$, and V is $\log N(nm, n\sigma^2)$.

Example 1.12 Let $Y = X_1X_2\cdots X_n$, where the X_i's are independent uniform random variables on the interval $(-b, b)$, $b > 0$. Then by utilizing Example 1.6 we have

$$\begin{aligned}
\mathcal{M}_Y(s) &= \prod_{i=1}^{n}\mathcal{M}_{X_i}(s) = \left(\frac{b^{s+1}}{2b(s+1)} + \gamma\,\frac{b^{s+1}}{2b(s+1)}\right)^n \\
&= 2^{n-1}\frac{b^{ns}}{2^n(s+1)^n} + \gamma\, 2^{n-1}\frac{b^{ns}}{2^n(s+1)^n} \\
&= \frac{b^{ns}}{2(s+1)^n} + \gamma\,\frac{b^{ns}}{2(s+1)^n},
\end{aligned}$$

where Re$(s)> -1$. Because of symmetry, the density function of Y is simply given by

$$\begin{aligned} h(y) &= \lim_{U\to+\infty} \frac{1}{2i\pi} \int_{c-iU}^{c+iU} y^{-s-1} \frac{b^{ns}}{2(s+1)^n} \, ds, \\ &= \frac{1}{2b^n(n-1)!} \left[\ln \left(\frac{b^n}{|y|} \right) \right]^{n-1} \qquad 0 < |y|^n \le b^n, \end{aligned}$$

where the last equation follows from Example 1.10, and $h(y) = 0$ otherwise.

Example 1.13 Let $Y = X_1 X_2$ where the X_1 and X_2 are independent standard normal random variables. Then by Example 1.9

$$\begin{aligned} \mathcal{M}_Y(s) &= \mathcal{M}_{X_1}(s)\mathcal{M}_{X_2}(s) \\ &= \left(\frac{2^{\frac{s-1}{2}}}{\sqrt{2\pi}} \Gamma\left(\frac{s+1}{2}\right) + \gamma \frac{2^{\frac{s+1}{2}}}{\sqrt{2\pi}} \Gamma\left(\frac{s+1}{2}\right) \right)^2 \\ &= \frac{2^{s-1}}{\pi} \Gamma\left(\frac{s+1}{2}\right)^2 + \gamma \frac{2^{s-1}}{\pi} \Gamma\left(\frac{s+1}{2}\right)^2, \end{aligned}$$

where Re$(s)> -1$. The symmetry of $\mathcal{M}_Y(s)$ simplifies the computation of the density function $h(y)$ of Y, given by

$$\begin{aligned} h(y) &= \lim_{U\to+\infty} \frac{1}{2i\pi} \int_{c-iU}^{c+iU} y^{-s-1} \frac{2^{s-1}}{\pi} \Gamma\left(\frac{s+1}{2}\right)^2 ds, \quad -\infty < y < +\infty \\ &= \frac{1}{\pi} K_o(y), \quad -\infty < y < +\infty, \end{aligned}$$

where $K_o(y)$ is Bessel's function of the second kind with a purely imaginary argument of order zero (see Epstein [44]).

1.4 Symmetric distributions

In this section we investigate symmetry of products of independent random variables.

A random variable X, or equivalently its distribution function $F(x)$, is said to be *symmetric* about zero, or simply symmetric, if X and $-X$ have the same distribution.

It is important to stress that symmetry and $P(X < 0) = 1/2$ are not equivalent.

Example 1.14 Let $X = Y$ with probability 1/2, and -1 with probability 1/2, where Y is unit exponential. Clearly, $P(X < 0) = 1/2$ but X is not symmetric. We take this one step further, and consider the product $Z = XV$, where X and V are independent, X as before, and V is unit exponential again. Then Z has the property that $P(Z < 0) = 1/2$, but Z is not symmetric.

An immediate consequence of the definition is that X is symmetric if, and only if,

$$\phi_X(t) = \phi_{-X}(t) = \phi_X(-t) = \overline{\phi_X(t)}, \tag{1.29}$$

where $\overline{\phi_X(t)}$ denotes the complex conjugate of $\phi_X(t)$, the characteristic function of X. That is, X is symmetric if, and only if, its characteristic function is real.

Let X and Y be two independent random variables, and let $W = X+Y$. From (1.29) and the property that the characteristic function of W is the product of the characteristic functions of X and Y, one immediately has that the sum of two symmetric random variables is always symmetric. On the other hand, if only X is symmetric, then W is symmetric if, and only if, the support of the imaginary part of the characteristic function of Y does not overlap with the support of the characteristic function of X. Surprisingly, as we are going to show next, it is possible for W to be symmetric even if X and Y are not symmetric and identically distributed. From our previous discussion this can only happen if the supports of the real and imaginary parts of their common characteristic function are disjoint. Such a random variable will be obtained by constructing a density function $f(x) = f_1(x) + f_2(x)$, where the Fourier transform $\phi_1(t)$ of $f_1(x)$ is real, the Fourier transform $\phi_2(t)$ of $f_2(x)$ is imaginary, and $\phi_1(t)\phi_2(t) = 0$ for all t. The following explicit example, based on a result due to Freedman and Diaconis [51], is due to Romano and Siegel [132].

Example 1.15 Let $V = (V_1 + V_2)/2$, where V_1 and V_2 are independent identically distributed random variables, uniformly distributed on the interval (-1,1), and let $h(x)$ be the density function of V. Since

$$\phi_{V_1}(t) = \int_{-1}^{1} \frac{e^{itx}}{2}\, dx = \frac{sin(t)}{t},$$

we immediately have that

$$\phi_V(t) = \int_{-1}^{1} e^{itx} h(x)\, dx = \frac{4sin(t/2)^2}{t^2}.$$

Let $f_1(x) = \phi_V(x)/(2\pi)$. Note that $f_1(x)$ is an even function, and thus its Fourier transform and its inverse Fourier transform coincide. Consequently, the Fourier transform $\phi_1(t)$ of $f_1(t)$ equals 2π times the inverse Fourier transform of $\phi_V(x)$, that is,

$$\phi_1(t) = \begin{cases} 1 - |t| & \text{if } -1 < t < 1, \\ 0 & \text{otherwise,} \end{cases}$$

the density function of V. Next we define $\phi(t)$ to be the convolution (defined for arbitrary functions just as at 1.5 in Section 1.1.1) between $\phi_1(t)$ and $g(t) = i[\delta_{-5}(t) - \delta_5(t)]/2$, where $\delta_a(t) = 1$ if $t = a$, 0 otherwise. By definition $\phi(t)$ is imaginary with support $[-6,-4] \cup [4,6]$. Let $f_2(x)$ be the Fourier transform of $\phi(t)/(2\pi)$. Since the Fourier transform of $g(t)$ is $sin(5x)$, and the Fourier transform of $\phi(t)$ is the product of the Fourier transforms of $\phi_1(t)$ and $g(t)$, we have that

$$f_2(x) = f_1(x) sin(5x).$$

Note that the Fourier transform of $f_2(x)$, $\phi_2(t)$, equals $-\phi(t)$. Moreover

$$f(x) = f_1(x)(1 + sin(5x)) \geq 0 \;\text{ for all } x$$

and

$$\int_{-\infty}^{+\infty} f(x)\, dx = 1,$$

making $f(x)$ a density function. Hence, if we let X and Y be two independent random variables with common density $f(x)$, then the Fourier transform of $f(x)$ is $\phi_1(t) + \phi_2(t)$, and from the above discussion we immediately have that $X + Y$ is symmetric, even if X (and consequently Y) is not.

Next we investigate symmetry of products of random variables. We are going to derive necessary conditions for the product to be symmetric in the case the factors are independent, and necessary and sufficient if the factors are independent and identically distributed (i.i.d.).

Let X be a random variable with Mellin transform

$$\mathcal{M}_X(s) = \mathcal{M}_{X^+}(s) + \gamma \mathcal{M}_{X^-}(s).$$

Since $\mathcal{M}_{X^+}(s)$ and $\mathcal{M}_{X^-}(s)$ uniquely determine the distribution of X^+ and X^-, respectively, (see the proof of Theorem (1.19), then X is symmetric if, and only if, $\mathcal{M}_{X^+}(s) = \mathcal{M}_{X^-}(s)$ for all s in some strip containing the imaginary axis. In fact it is enough for this equality to hold for $s = it$, t real. In order to see the need for this requirement we look at the following example.

Example 1.16 Let U be a random variable with density function

$$g(x) = \frac{1}{\pi}\frac{2sin^2(x/2)}{x^2},$$

and let Y be a discrete random variable taking the values $0, \pm(2k+1)\pi$, $k \geq 0$, with distribution $P(Y = 0) = 1/2$ and $P(Y = (2k+1)\pi) = P(Y = -(2k+1)\pi) = 2/(\pi^2(2k+1)^2)$. Then $\phi_U(t) = 1 - |t|$, $|t| \leq 1$, $\phi_U(t) = 0$ otherwise, (see Example 1.15), and $\phi_Y(t) = \phi_U(t)$, for $-1 \leq t \leq 1$, and $\phi_Y(t) = \phi_Y(2+t)$ (see Exercises 7 and 8 on pp. 117-118 together with their solutions on p. 426, in [58]). Then by Example 1.5 there exists a random variable X such that

$$\mathcal{M}_X(t) = \frac{1}{2}\phi_U(t) + \gamma\frac{1}{2}\phi_Y(t).$$

By construction,

$$\mathcal{M}_{X^+}(t) = \frac{1}{2}\phi_U(t) = \frac{1}{2}\phi_Y(t) = \mathcal{M}_{X^-}(t), \quad |t| \leq 1,$$

X^+ is absolutely continuous, while X^- is discrete (i.e., X cannot be symmetric).

The simple criterium we derived for symmetry of products turns out to be quite useful. We start by considering symmetry of finite products, when the factors are i.i.d., and we show Example 1.15 does not extend to products.

Lemma 1.1 *The product of any finite number of independent identically distributed random variables is symmetric if, and only if, the random variables in the product are symmetric.*

PROOF. We start by considering the product of two i.i.d. random variables Y and X. Let $Z = XY$, and let $\mathcal{M}_X(s)$ and $\mathcal{M}_Y(s)$ be the Mellin transforms of X and Y, respectively. Then

$$\begin{aligned}\mathcal{M}_Z(s) &= \mathcal{M}_{Z^+}(s) + \gamma\mathcal{M}_{Z^-}(s)\\ &= (\mathcal{M}_{X^+}(s) + \gamma\mathcal{M}_{X^-}(s))(\mathcal{M}_{Y^+}(s) + \gamma\mathcal{M}_{Y^-}(s))\\ &= \mathcal{M}_{X^+}(s)\mathcal{M}_{Y^+}(s) + \mathcal{M}_{X^-}(s)\mathcal{M}_{Y^-}(s)\\ &\quad +\gamma(\mathcal{M}_{X^+}(s)\mathcal{M}_{Y^-}(s) + \mathcal{M}_{X^-}(s)\mathcal{M}_{Y^+}(s)).\end{aligned}$$

Hence Z is symmetric if, and only if, for all s in a strip containing the imaginary axis,

$$\begin{aligned}0 &= \mathcal{M}_{Z^+}(s) - \mathcal{M}_{Z^-}(s)\\ &= (\mathcal{M}_{X^+}(s)\mathcal{M}_{Y^+}(s) + \mathcal{M}_{X^-}(s)\mathcal{M}_{Y^-}(s))\\ &- (\mathcal{M}_{X^+}(s)\mathcal{M}_{Y^-}(s) + \mathcal{M}_{X^-}(s)\mathcal{M}_{Y^+}(s))\\ &= (\mathcal{M}_{X^+}(s) - \mathcal{M}_{X^-}(s))(\mathcal{M}_{Y^+}(s) - \mathcal{M}_{Y^-}(s))\\ &= (\mathcal{M}_{X^+}(s) - \mathcal{M}_{X^-}(s))^2\end{aligned}$$

which holds if, and only if, X and Y are symmetric. The validity of the general case now follows from a simple induction argument. The details are left to the reader. ∎

The above result does not extend to infinite sequences of i.i.d. random variables (see Theorems 1.22 and 1.23).

If one relaxes the assumption that the factors have the same distribution, then in general Lemma 1.1 does not hold. That is, XY can be symmetric even if X and Y are not. This possibility was first recognized by Zolotarev [159]. Here we provide an explicit counterexample.

Example 1.17 Let X and Y be as in Example 1.16, U_1 an independent copy of U, and $0 < a < 1$. Then by Example 1.5 there exist independent random variables X and V such that

$$\begin{aligned}\mathcal{M}_X(t) &= \frac{1}{2}\phi_U(t) + \gamma\frac{1}{2}\phi_Y(t) \qquad \text{and} \\ \mathcal{M}_V(t) &= a\phi_{U_1}(t) + \gamma(1-a)\phi_{U_1}(t).\end{aligned}$$

Clearly V is not symmetric if $a \neq 1/2$, and recall that neither is X. However

$$\mathcal{M}_{XV}(t) = \mathcal{M}_X(t)\mathcal{M}_V(t) = \frac{1}{2}\phi_U(t)^2 + \gamma\frac{1}{2}\phi_U(t)^2,$$

where in the derivation of the above equation we used $\phi_U(t)\phi_Y(t) = \phi_U(t)^2$ (see Example 1.16). Hence $\mathcal{M}_{(XV)^+}(t) = \mathcal{M}_{(XV)^-}(t)$, and XV is symmetric.

Examples 1.16 and 1.17 are very important because they show that in general it is not possible to weaken the conditions for symmetry we developed above, i.e., $\mathcal{M}_{X^+}(t) = \mathcal{M}_{X^-}(t)$ for t real. However, alternative conditions can be obtained by imposing additional assumptions on the random variables. For example, assume X and Y are independent. By utilizing the calculation in Lemma 1.1 we immediately get that XY is symmetric iff

$$(\mathcal{M}_{(X^+)}(s) - \mathcal{M}_{X^-}(s))(\mathcal{M}_{Y^+}(s) - \mathcal{M}_{Y^-}(s)) = 0$$

for all s in some strip containing the imaginary axis. In particular one of the two differences must be zero for all s in a set containing a limit point. If the corresponding components of the Mellin transform were analytic, then by the Identity Theorem for analytic functions, see Marsden and Hoffman [111], for example, we would further have that this factor must be identically zero, i.e., the corresponding random variable is symmetric. Hence we have the following.

Lemma 1.2 *Let X and Y be independent random variables, and assume that the Mellin transforms of both the positive and negative part of X or Y are analytic. Then XY is symmetric if, and only if, X or Y is symmetric.*

As we will see in Chapter 3, the normal, the Poisson and other important random variables satisfy Lemma 1.2.

In the next example we consider another important collection of random variables for which it is simple to characterize symmetry for products. Those functions play an important role in probabilistic number theory, and will further be considered in Chapter 5.

Example 1.18 Let $p_1, p_2, \cdots$ denote the prime integers, and let e_{p_j}, $j = 1, 2, \cdots$, be independent random variables taking the values $0, 1, 2, \cdots$ with distribution

$$P(e_{p_j} \geq k) = \frac{1}{p_j^k}.$$

Let $g(m)$ be a real valued function defined on the integers, and let $X_j = g(p_j^{e_{p_j}})$, $j = 1, 2, \cdots$. Now

$$\begin{aligned} \left|\mathcal{M}_{X_j^+}(t) - \mathcal{M}_{X_j^-}(t)\right| &\geq 1 - \frac{1}{p_j} - \sum_{k=1}^{+\infty} \left||g(p_j^k)|^{it}\right| \left(\frac{1}{p_j^k} - \frac{1}{p_j^{k+1}}\right) \\ &\geq 1 - \frac{1}{p_j} - \sum_{k=1}^{+\infty} \left(\frac{1}{p_j^k} - \frac{1}{p_j^{k+1}}\right) = 1 - \frac{2}{p_j} \geq 0, \end{aligned}$$

and $1 - 2/p_j > 0$ except if $p_j = 2$. Hence $\mathcal{M}_{X_j^+}(t) - \mathcal{M}_{X_j^-}(t) = 0$ for some t only if $X_j = g(2^{e_2})$, and $g(2^k) < 0$ for $k \geq 1$. Note that in the estimates above we have equations only if $g(2^k) = -1$ for all $k \geq 1$. So in this case the difference above is zero for all t. This shows that a finite product of the X_js is symmetric if, and only if, one of the factors in the product is $g(2^{e_2})$, and $g(2^k) = -1$ for $k \geq 1$.

Next we turn to symmetry of infinite products. We first consider convergence and symmetry of the special product $sign(X_1 X_2 \cdots)$. The next result is due to Simonelli [139].

Theorem 1.21 *Let $X_1, X_2, \cdots$ be independent random variables with $P(X_j = 0) = 0$ for all $j \geq 1$.*

(i) $\lim_{n \to +\infty} P(sign(X_1 X_2 \cdots X_n) = -)$ *exists if, and only if, either*

$$\lim_{n \to +\infty} P(X_n < 0) = 0, \quad \text{or} \quad P(X_i < 0) = \frac{1}{2} \text{ for some } i, \quad \text{or}$$

$$\sum_{i=1}^{+\infty} \min\{P(X_i < 0), P(X_i > 0)\} = +\infty.$$

(ii) $$\lim_{n \to +\infty} P(sign(X_1 X_2 \cdots X_n) = -) = \frac{1}{2}$$ *if, and only if, either*

$$P(X_i < 0) = \frac{1}{2} \textit{ for some } i, \quad \textit{or}$$

$$\sum_{i=1}^{+\infty} \min\{P(X_i < 0), P(X_i > 0)\} = +\infty.$$

PROOF. We start by proving part (ii) of the theorem. Let $\nu_n = sign(X_1 X_2 \cdots X_n)$. By the total probability rule and the independence of the X_i's

$$\begin{aligned} P(\nu_n = -) &= P(\nu_n = -|X_n < 0)P(X_n < 0) \\ &\quad + P(\nu_n = -|X_n > 0)P(X_n > 0) \\ &= P(\nu_{n-1} = +)P(X_n < 0) + P(\nu_{n-1} = -)P(X_n > 0). \end{aligned} \tag{1.30}$$

Similarly

$$P(\nu_n = +) = P(\nu_{n-1} = -)P(X_n < 0) + P(\nu_{n-1} = +)P(X_n > 0).$$

Subtracting the last two equations yields

$$\begin{aligned} &P(\nu_n = -) - P(\nu_n = +) \\ &= (P(\nu_{n-1} = -) - P(\nu_{n-1} = +))(1 - 2P(X_n < 0)). \end{aligned}$$

This is a formula on absolute values because, if you multiply by -1, the identity is similar. So, by induction

$$\begin{aligned} |P(\nu_n = -) - P(\nu_n = +)| &= |\prod_{i=1}^{n}(1 - 2P(X_i < 0))| \\ &= \prod_{i=1}^{n}(1 - 2\min\{P(X_i < 0), P(X_i > 0)\}), \end{aligned}$$

which immediately gives

$$\lim_{n\to+\infty} P(\nu_n- = -1) = \frac{1}{2} \iff P(X_i < 0) = \frac{1}{2} \text{ for some } i$$

$$\text{or} \quad \sum_{i=1}^{+\infty} \min\{P(X_i < 0), P(X_i > 0)\} = +\infty.$$

This proves (ii) and part of (i). To complete the proof of (i) we assume that $P(X_i < 0) \neq 1/2$ for any i and $\sum_{i=1}^{+\infty} \min\{P(X_i < 0), P(X_i > 0)\} < +\infty$. We show that $P(\nu_n = -) \to \alpha$, $\alpha \neq 1/2$, if, and only if, $P(X_n < 0) \to 0$.

Upon rearranging the terms in (1.30), we have

$$\begin{aligned} &|P(\nu_n = -) - P(\nu_{n-1} = -)| \\ &= |P(\nu_{n-1} = -) - P(\nu_{n-1} = +)| P(X_n < 0). \end{aligned}$$

So, if $P(\nu_n = -) \to \alpha$, $\alpha \neq 1/2$, the left hand side above converges to zero, and thus on the right hand side we must have $P(X_n < 0) \to 0$.

Conversely, if $P(X_n < 0) \to 0$, then $\min\{P(X_n < 0), P(X_n > 0)\}$ is ultimately $P(X_n < 0)$, so the assumed convergence of the sum $\sum_{i=1}^{+\infty} P(X_i < 0)$ and the estimates

$$\begin{aligned} &|P(\nu_n = -) - P(\nu_{n+k} = -)| \\ &\leq \sum_{i=1}^{k} |P(\nu_{n+i-1} = -) - P(\nu_{n+i} = -)| \\ &= \sum_{i=1}^{k} |2P(\nu_{n+i-1} = -) - 1|\, P(X_{n+i} < 0) \\ &\leq \sum_{i=n+1}^{+\infty} P(X_i < 0) \longrightarrow 0 \quad \text{as } n \longrightarrow +\infty, \end{aligned}$$

imply the convergence of $P(\nu_n = -1)$ to a limit different from $1/2$. The proof of the theorem is complete. ∎

Next we extend Theorem 1.21 to arbitrary independent random variables.

Theorem 1.22 *Let $X_1, X_2, \cdots$ be independent random variables with $P(X_i = 0) < 1$ for all $i \geq 1$, and let $\nu_n = sign(X_1 X_2 \cdots X_n)$.*

(i) $\lim_{n\to+\infty} P(\nu_n = 0) = 1$ *if, and only if,*

$$\sum_{i=1}^{+\infty} P(X_i = 0) = +\infty.$$

(ii) *Assume* $\sum_{i=1}^{+\infty} P(X_i = 0) < +\infty$. *Then for* $\eta = -, 0, +$

$$\lim_{n\to+\infty} P\big(\nu_n = \eta\big)$$

exist if, and only if, either

$\lim_{n\to+\infty} P(X_n < 0) = 0$ *or* $P(X_j < 0 | X_j \neq 0) = 1/2$ *for some* j,

or

$$\sum_{i=1}^{+\infty} \min\{P(X_i < 0), P(X_i > 0)\} = +\infty.$$

(iii) $\lim_{n\to+\infty} P(\nu_n = -) = \lim_{n\to+\infty} P(\nu_n = +)$ *if, and only if, either* $P(X_i < 0 | X_i \neq 0) = \frac{1}{2}$ *for some* i, *or*

$$\sum_{i=1}^{+\infty} \min\{P(X_i < 0), P(X_i > 0)\} = +\infty.$$

PROOF. We start our proof by proving (i). By turning to complements and then using the independence of the X_i we obtain

$$\begin{aligned} P(\nu_n = 0) &= P(X_1 X_2 \cdots X_n = 0) \\ &= 1 - P(X_1 X_2 \cdots X_n \neq 0) \\ &= 1 - \prod_{i=1}^{n} P(X_i \neq 0) \\ &= 1 - \prod_{i=1}^{n} (1 - P(X_i = 0)) \to 1 \end{aligned}$$

if, and only if, $\prod_{i=1}^{n}(1 - P(X_i = 0)) \to 0$, which can happen if, and only if, $\sum_{i=1}^{+\infty} P(X_i = 0) = +\infty$. This proves (i).

Next we assume $\sum_{i=1}^{+\infty} P(X_i = 0) < +\infty$, and turn to the proof of (ii) and (iii). The estimate

$$\begin{aligned}
&\left|P(\nu_n = 0) - P(\nu_{n+k} = 0)\right| \\
&= \left|P(\nu_n = 0) - P(\nu_n = 0 \text{ or } X_{n+1} \cdots X_{n+k} = 0)\right| \\
&\leq 2 \sum_{i=n+1}^{+\infty} P(X_i = 0) \to 0,
\end{aligned}$$

as $n \to +\infty$, implies the convergence of $P(\nu_n = 0)$ to a limit different from 1. By using the notation introduced in (1.15) we write

$$P(\nu_n = -) = P(sign(\tilde{X}_1 \tilde{X}_2 \cdots \tilde{X}_n) = -)\, P(X_1 X_2 \cdots X_n \neq 0).$$

Since $P(X_1 X_2 \cdots X_n \neq 0)$ converges, to complete the proof of the theorem it suffices to show that the conditions for convergence and symmetry of $sign(\tilde{X}_1 \tilde{X}_2 \cdots \tilde{X}_n)$ given by Theorem 1.21 are equivalent to the conditions in part (ii) and (iii) of Theorem 1.22. Our assumption on the convergence of $\sum_{i=1}^{+\infty} P(X_i = 0)$ implies $P(X_n \neq 0) \to 1$, as $n \to= \infty$, which immediately gives

$$\lim_{n \to +\infty} P(\tilde{X}_n < 0) = 0 \quad \text{if, and only if,} \quad \lim_{n \to +\infty} P(X_n < 0) = 0,$$

and

$$\sum_{i=1}^{+\infty} \min\{P(X_i < 0), P(X_i > 0)\} = +\infty$$

if, and only if,

$$\sum_{i=1}^{+\infty} \min\left\{P(\tilde{X}_i < 0), P(\tilde{X}_i > 0)\right\} = +\infty.$$

The proof of the theorem is now complete. ■

During the proof of Theorem 1.21 we have proved that

$$P(sign(X_1 X_2 \cdots X_n) = \frac{1}{2}$$

if, and only if, $P(X_i < 0) = 1/2$ for some i, $1 \leq i \leq n$. This generalizes to arbitrary random variables, thus implying the following result for symmetry of finite products.

Lemma 1.3 *Let $X_1, X_2, \cdots X_n$ be independent random variables, and assume that $P(X_i < 0) \neq P(X_i > 0)$, $i = 1, 2, \cdots, n$. Then $X_1 X_2 \cdots X_n$ is not symmetric. In fact,*

$$P\left(sign(X_1 X_2 \cdots X_n) = -\big|X_1 X_2 \cdots X_n \neq 0\right) \neq 1/2$$

under the preceding condition.

Next we consider symmetry of arbitrary infinite products of random variables. The following theorem was motivated by a result of Simonelli [139]. We take this opportunity to correct some of the statements made in that paper.

Remark 1.1 In dealing with symmetry Simonelli [139] only required for $\mathcal{M}_{X^+}(t) = \mathcal{M}_{X^-}(t)$ for t in some interval containing zero, and this statement has some effects on the necessity parts of his Theorems 2.2 and 2.3. However all results of Simonelli [139], including his Theorems 2.2 and 2.3, are correct under some mild additional assumptions on the random variables. For example, they are valid for the random variables covered by Lemmas 1.1, 1.2, 1.3, and Example 1.18. The correct formulation of these theorems is given in Theorems 1.23 and 1.27.

Theorem 1.23 *Let $X_1, X_2, \cdots$ be a sequence of independent random variables, and assume that for some sequence of real numbers $\{a_n\}$, $(\prod_{i=1}^{n} X_i)e^{-a_n}$ converges completely to a random variable W, $P(W = 0) < 1$. Then W is symmetric if either $X_1 X_2 \cdots X_n e^{-a_n}$ is symmetric for some n, or*

$$\sum_{i=1}^{+\infty} \min\left\{P\left(X_i \leq 0\right), P\left(X_i > 0\right)\right\} = +\infty. \tag{1.31}$$

Furthermore, if $P(X_i < 0 \big| X_i \neq 0) \neq 1/2$ for any i, then (1.31) is also necessary for the symmetry of W.

PROOF. From Theorem 1.22, $P(W = 0) < 1$ implies that

$$\sum_{i=1}^{+\infty} P(X_i = 0) < +\infty,$$

which implies that

$$\lim_{n \to +\infty} P(X_1 X_2 \cdots X_n \neq 0) \quad \text{exists.}$$

This and the convergence of $(\prod_{i=1}^{n} X_i)e^{-a_n}$ imply the convergence of $(\prod_{i=1}^{n} \tilde{X}_i)e^{-a_n}$, where $\tilde{X}_i$ is as in (1.15), and hence the convergence of $\sum_{j=1}^{n} \ln|\tilde{X}_j| - a_n$. Denote by Y the limit of $(\prod_{i=1}^{n} \tilde{X}_i)e^{-a_n}$. By Theorem 1.17 there exist constants a_j^* such that $\sum_{j=1}^{+\infty}(\ln|\tilde{X}_j| - a_j^*)$, and consequently $\prod_{j=1}^{n} |\tilde{X}_j e^{-a_j^*}|$, converges almost surely. Put $Y_j = \tilde{X}_j e^{-a_j^*}$. We choose a fixed n and write

$$Y = \Big(\prod_{j=1}^{n} Y_j\Big)\Big(\prod_{j=n+1}^{+\infty} Y_j\Big) = U_1 U_2, \qquad \text{say.} \tag{1.32}$$

If n is such that U_1 is symmetric then so is the product $Y = U_1 U_2$ (it is implicit in the proof of Lemma 1.1).

Let us assume that $P(X_i < 0 \big| X_i \neq 0) \neq 1/2$ for any i. Then $P(sign(U_1) = - \big| U_1 \neq 0) \neq 1/2$ whatever n (see Lemma 1.3). Consequently, if Y is symmetric then, by (1.32), $P(sign(U_2) = - \big| U_2 \neq 0) = 1/2$. With an appeal to Theorem 1.22 (iii) we get that (1.31) must hold.

Next we prove sufficiency of the conditions. If $X_1 X_2 \cdots X_n e^{-a_n}$ is symmetric for some n, then $(\prod_{i=1}^{t} X_i)e^{-a_t}$ is symmetric for all $t > n$, and there is nothing to prove. So let us assume that $X_1 X_2 \cdots X_n e^{-a_n}$ is not symmetric for any n, and (1.31) holds. Let $V_n = (\prod_{i=1}^{n} |\tilde{X}_i|)e^{-a_n}$, and for arbitrary positive integers i and j, $i < j$, let $\nu_{i,j}^- = -1$ if $sign(X_i \cdots X_j) = -1$, $\nu_{i,j}^- = 1$ otherwise. For arbitrary continuity point a of Y let

$$J = \lim_{k \to +\infty} \lim_{n \to +\infty} E\left[1_{\{V_k \leq a\}} 1_{\{\nu_{n+k}^- = -1\}}\right].$$

Then

$$\begin{aligned} J &= \lim_{k\to+\infty}\lim_{n\to+\infty}\left(E\left[1_{\{\nu^-_{k+1,n+k}=1\}}\right]E\left[1_{\{V_k\le a\}}1_{\{\nu^-_k=-1\}}\right]\right. \\ &\quad \left.+E\left[1_{\{\nu^-_{k+1,n+k}=-1\}}\right]E\left[1_{\{V_k\le a\}}1_{\{\nu^-_k=1\}}\right]\right) \\ &= \lim_{k\to+\infty}\left(\frac{1}{2}E\left[1_{\{V_k\le a\}}\,1_{\{\nu^-_k=-1\}}\right]+\frac{1}{2}E\left[1_{\{V_k\le a\}}\,1_{\{\nu^-_k=1\}}\right]\right) \\ &= \lim_{k\to+\infty}\frac{1}{2}\,E\left[1_{\{V_k\le a\}}\right]=\frac{1}{2}E\left[1_{\{|Y|\le a\}}\right], \end{aligned}$$

where the first equation follows from the independence of the $\tilde{X}_i$'s, and the second equation follows from (1.31) and Theorem 1.22. This and

$$\begin{aligned} &\lim_{k\to+\infty}\limsup_{n\to+\infty}\left|E\left[1_{\{V_k\le a\}}\,1_{\{\nu^-_{k+n}=1\}}\right]-E\left[1_{\{V_{k+n}\le a\}}1_{\{\nu^-_{k+n}=-1\}}\right]\right| \\ &\le \lim_{k\to+\infty}\limsup_{n\to+\infty}E\left[\left|1_{\{V_k\le a\}}-1_{\{V_{k+n}\le a\}}\right|\right]=0, \end{aligned}$$

which follows from the dominated convergence theorem (recall that V_n converges almost surely), immediately give

$$\lim_{n\to+\infty}E\left[1_{\{|X_1\cdots X_n|e^{-a_n}\le a\}}\,1_{\{\nu^-_n=-1\}}\right]=\frac{1}{2}E\left[1_{\{|Y|\le a\}}\right],$$

which concludes the proof. ∎

We conclude this section by the following observation. It is established in the course of the proof of the sufficiency part of Theorem 1.23 that the validity of (1.31) and the convergence of $V_n = (\prod_{i=1}^n |\tilde{X}_i|)e^{-a_n}$ imply the convergence of $(\prod_{i=1}^n \tilde{X}_i)e^{-a_n}$. This fact will be used in the proof of Theorem 1.27.

1.5 A new Bonferroni-type identity and inequalities for an odd number of occurrences among n events

For finding the probability that the product of n random variables X_j, $1 \le j \le n$, is negative one faces the problem of evaluating the probability that, for an odd number of j, $1 \leq j \leq n$, $X_j < 0$. Thus, if we put $A_j = \{X_j < 0\}$, and m_n for the number of A_j which occur we want

$$P(m_n = \text{ odd }) = \sum_{\substack{1 \leq t \leq n \\ t \text{ odd}}} P(m_n = t). \tag{1.33}$$

Since problems leading to (1.33) come up in the sequel for events A_j other than just $\{X_j < 0\}$, we treat (1.33) for arbitrary events on an arbitrary probability space.

Set $S_0 = S_{0,n} = 1$, and for arbitrary $k \geq 1$,

$$S_k = S_{k,n} = E\left[\binom{m_n}{k}\right] = \sum P(A_{i_1} A_{i_2} \cdots A_{i_k}),$$

where $A_j A_l = A_j \cap A_l$, and $\sum$ is summation over all integers $1 \leq i_1 < i_2 < \cdots < i_k \leq n$. By K. Jordan's identity (see Galambos and Simonelli [62]),

$$P(m_n = t) = \sum_{k=0}^{n-t} (-1)^k \binom{k+t}{t} S_{k+t} = \sum_{u=t}^{n} (-1)^{u-t} \binom{u}{t} S_u.$$

Thus, upon noting that, for $t =$ odd, $(-1)^{u-t} = -(-1)^u = (-1)^{u-1}$, $u \geq 1$, and proceeding as in Galambos and Simonelli [64],

$$\begin{aligned} P(m_n = \text{ odd }) &= \sum_{\substack{1 \leq t \leq n \\ t \text{ odd}}} \sum_{u=t}^{n} (-1)^{u-1} \binom{u}{t} S_u \\ &= \sum_{u=1}^{n} (-1)^{u-1} S_u \sum_{\substack{1 \leq t \leq n \\ t \text{ odd}}} \binom{u}{t}. \end{aligned} \tag{1.34}$$

But, $2^u = (1+1)^u = \sum_{t=0}^{u} \binom{u}{t}$ and

$$0 = (1-1)^u = \sum_{t=0}^{u} (-1)^t \binom{u}{t} = \sum_{\substack{1 \leq t \leq n \\ t \text{ even}}} \binom{u}{t} - \sum_{\substack{1 \leq t \leq n \\ t \text{ odd}}} \binom{u}{t},$$

therefore

$$\sum_{\substack{1\le t\le n\\ t \text{ odd}}} \binom{u}{t} = \frac{1}{2}2^u = 2^{u-1}. \tag{1.35}$$

We thus have from (1.34) and (1.35)

$$P(m_n = \text{ odd }) = \sum_{u=1}^{n}(-2)^{u-1}S_u. \tag{1.36}$$

In the special case when the A_i are independent and $P(A_i) = p$, $i = 1, 2, \cdots, n$, (1.36) gives

$$\begin{aligned} P(m_n = \text{ odd }) &= \sum_{u=1}^{n}(-2)^{u-1}\binom{n}{u}p^u \\ &= -\frac{1}{2}\Big[\sum_{u=1}^{n}(-2)^u\binom{n}{u}p^u + 1 - 1\Big] \\ &= \frac{1}{2}\Big[1-(1-2p)^n\Big]. \end{aligned} \tag{1.37}$$

Inequalities similar to those on $P(m_n = t)$ are also valid. However, such inequalities do not follow from the classical inequalities by utilizing (1.33), and their derivation and validity require separate proofs. Since $\{m_n = \text{odd}\}$ can be written in terms of Boolean functions of the A_i, the method of indicators can be used in proving inequalities for $P(m_n = \text{ odd })$. It is interesting to note that the method of polynomials applies as well (see [64]).

Theorem 1.24 *The validity of the Bonferroni-type inequalities*

$$a\,P(m_n = \ odd\ ) + \sum_{k=0}^{n} a_k S_k \ge 0 \tag{1.38}$$

is equivalent to the validity of the following collection of polynomial inequalities

$$a\frac{1}{2}\Big[1-(1-2p)^n\Big] + \sum_{k=0}^{n} a_k\binom{n}{k}p^k \ge 0, \tag{1.39}$$

for all n and all $p \in [0,1]$.

PROOF. The validity of (1.38) clearly implies the validity of (1.39), and therefore only the converse needs proof. By the method of indicators (1.38) is equivalent to

$$a\delta_{0,t} + \sum_{k=0}^{t} a_k \binom{t}{k} \geq 0 \tag{1.40}$$

for $0 \leq t \leq n$, and $\delta_{0,t} = 1$ if t is odd, and $\delta_{0,t} = 0$ if t is even. Hence it suffices to show that (1.39) implies (1.40). Let us assume that (1.39) holds. Then for any t

$$a\frac{1}{2}\left[1 - (1-2p)^t\right] + \sum_{k=0}^{t} a_k \binom{n}{k} p^k \geq 0$$

for all $0 \leq p \leq 1$. If we let $p \to 1$ in the above inequality we get

$$1 + \sum_{k=0}^{t} a_k \binom{t}{k} \geq 0, \qquad \text{if } t \text{ is odd,}$$

and

$$0 + \sum_{k=0}^{t} a_k \binom{t}{k} \geq 0, \qquad \text{if } t \text{ is even.}$$

Hence (1.40) holds, and the proof is complete. ∎

Next we use Theorem 1.24 to derive the following result, due to Galambos and Simonelli [64].

Theorem 1.25 *For arbitrary integers r and d, $r \geq 1$ and $d \geq 0$,*

$$\sum_{k=1}^{2r} (-2)^{k-1} S_k \leq P(m_n = \text{ odd }) \leq \sum_{k=1}^{2d+1} (-2)^{k-1} S_k.$$

When $2r \geq n$ and $2d + 1 \geq n$, the preceding inequalities become identities.

PROOF. By Theorem 1.24 it suffices to consider the case $P(m_n =$ odd) is given by (1.37). By applying Taylor formula with remainder to $(1-2p)^n$ we get

$$(1-2p)^n = \sum_{k=0}^{t}(-1)^k\binom{n}{k}2^k p^k + E_t(p),$$

where

$$E_t(p) = \int_o^p \frac{(p-x)^t}{t!}\frac{d^{t+1}}{dx^{t+1}}\Big[(1-2x)^n\Big]\,dx.$$

Clearly if $t > n$, $E_t(p) = 0$, and the theorem holds. Next we assume that $t \leq n$, and observe that

$$\frac{d^{t+1}}{dx^{t+1}}(1-2x)^n = (-2)^{t+1}(t+1)!\binom{n}{t+1}(1-2x)^{n-t-1}.$$

Hence if $0 \leq p \leq 1/2$, $E_t(p) \leq 0$ if t is even, and $E_t(p) \geq 0$ if t is odd, thus giving

$$\begin{aligned}(1-2p)^n &\leq \sum_{k=0}^{2r}(-2)^k\binom{n}{k}p^k \\ &\geq \sum_{k=0}^{2d+1}(-2)^k\binom{n}{k}p^k,\end{aligned}$$

$r \geq 1$ and $d \geq 0$. The validity of the above inequalities for arbitrary $1/2 < p \leq 1$ follows by observing that if $1/2 < p \leq 1$,

$$\big|E_t(1/2)\big| \geq \big|E_t(p) - E_t(1/2)\big|.$$

Hence for all $0 \leq p \leq 1$, and r and d as above,

$$\begin{aligned}\frac{1}{2}\Big[1-(1-2p)^n\Big] &\geq \frac{1}{2}\Big[1-\sum_{k=0}^{2r}(-2)^k\binom{n}{k}p^k\Big] \\ &= \sum_{k=1}^{2r}(-2)^{k-1}\binom{n}{k}p^k\end{aligned}$$

and

$$\begin{aligned}\frac{1}{2}\Big[1-(1-2p)^n\Big] &\le \frac{1}{2}\Big[1-\sum_{k=0}^{2d+1}(-2)^k\binom{n}{k}p^k\Big]\\ &= \sum_{k=1}^{2d+1}(-2)^{k-1}\binom{n}{k}p^k.\end{aligned}$$

The proof is now complete. ■

We end this section with the following examples. The first is an application of (1.37).

Example 1.19 Consider double indexed sequences X_{jn} of random variables, where $1 \le j \le k(n)$, $n \ge 1$ and $k(n) \to +\infty$ with n. For each n, the X_{jn} are i.i.d. with $P(A_{jn}) = P(X_{jn} < 0) = p_n$. Hence from (1.37) we obtain the following

$$\lim_{n\to+\infty} P(m_{k(n)} = \text{ odd }) = \begin{cases} 0 & \text{if } n\,p_n \to 0,\\ \frac{1}{2}(1-e^{-\alpha}) & \text{if } 2n\,p_n \to \alpha,\\ \frac{1}{2} & \text{if } n\,p_n \to +\infty.\end{cases}$$

The next example investigates some interesting monotonicity property of $P(m_n = \text{ odd })$.

Example 1.20 Let $A_1, A_2, \cdots, A_n$ be independent events, and assume that $P(m_n = \text{odd}) < 1/2$. Let A be an event independent of A_i, $i = 1, 2, \cdots, n$, and put $m_{n+1} = m_n + 1_A$. Then by using independence we get

$$\begin{aligned} P(m_{n+1} = \text{odd}) &= P(A^c)\,P(m_n = \text{odd}) + P(A)\,P(m_n = \text{ even })\\ P(m_{n+1} = \text{even}) &= P(A)\,P(m_n = \text{odd}) + P(A^c)\,P(m_n = \text{even}),\end{aligned}$$

which give

$$\begin{aligned}& P(m_{n+1} = \text{even}) - P(m_{n+1} = \text{ odd })\\ =\; & \Big(P(m_n = \text{even}) - P(m_n = \text{odd})\Big)\,\big(P(A^c) - P(A)\big).\end{aligned}$$

Hence if $P(m_n = \text{odd}) < 1/2$,

$$\begin{aligned} P(m_{n+1} = \text{odd}) &< \tfrac{1}{2} \quad \text{if } P(A) < \frac{1}{2}, \\ P(m_{n+1} = \text{odd}) &= \tfrac{1}{2} \quad \text{if } P(A) = \frac{1}{2}, \\ P(m_{n+1} = \text{odd}) &> \tfrac{1}{2} \quad \text{if } P(A) > \frac{1}{2}. \end{aligned}$$

The above discussion also shows that $P(m_n = \text{odd}) < 1/2$, whenever $P(A_i) < 1/2$, $i = 1, 2, \cdots, n$.

1.6 Limiting distribution of products of independent random variables

In this section we investigate weak convergence of infinite products of independent random variables, with and without normalizing constants. Throughout this section, X_i, $i \geq 1$, will denote an infinite sequence of independent random variables. We first consider convergence of $(\prod_{i=1}^{n} X_i)\, e^{-a_n}$, where a_n is an arbitrary sequence of real numbers, and for convenience we consider the case $P(X_i = 0) = 0$, for all i, and the arbitrary case, $P(X_i = 0) < 1$, separately. The next theorem corrects a theorem of Simonelli [139] (see Remark 1.1).

Theorem 1.26 *Let $X_1, X_2, \cdots$ be independent random variables with $P(X_i = 0) = 0$, $i \geq 1$.*

(i) If there exist constants a_n such that $(\prod_{i=1}^{n} X_i) e^{-a_n}$ converges completely to a nonsymmetric random variable Y, $P(Y = 0) = 0$, then

$$\left| \prod_{i=1}^{n} X_i \right| e^{-a_n} \longrightarrow |Y| \qquad \textit{completely},$$

and $\lim_{n \to +\infty} P(X_n < 0) = 0$.

(ii) For the complete convergence of $(\prod_{i=1}^{n} X_i) e^{-a_n}$ to a random variable Y, $P(Y = 0) = 0$, where the a_n are constants, it is sufficient that there exists a random variable V such that

$$\left| \prod_{i=1}^{n} X_i \right| e^{-a_n} \longrightarrow V \qquad \textit{completely}, \qquad P(V = 0) = 0,$$

and either $\lim_{n\to+\infty} P(X_n \le 0) = 0$, *or* $X_1X_2\cdots X_n$ *is symmetric for some* n, *or*

$$\sum_{i=1}^{+\infty} \min\{P(X_i \le 0), P(X_i > 0)\} = +\infty.$$

(iii) There exist constants a_n *and a random variable* Y *such that*

$$\left(\prod_{i=1}^{n} X_i\right) e^{-a_n} \longrightarrow Y \quad \textit{completely}, \qquad P(Y = 0) > 0,$$

if, and only if,

$$\left(\prod_{i=1}^{n} X_i\right) e^{-a_n} \longrightarrow 0 \qquad \textit{almost surely}.$$

Proof. Let $V_n = (\prod_{i=1}^n |X_i|)e^{-a_n}$, $\nu_n = sign(X_1X_2\cdots X_n)$, and $\eta = -,+$. We start by proving (i). Clearly, the complete convergence of $(\prod_{i=1}^n X_i)\, e^{-a_n}$ to a random variable Y immediately implies the complete convergence of V_n to $V \stackrel{d}{=} |Y|$. Since $P(V = 0) = 0$, by first turning to logarithms, and then applying Theorem 1.15, we also obtain that V_n converges to V almost surely. Since Y is not symmetric, by Theorem 1.24 we have that

$$\sum_{j=1}^{+\infty} \min\{P(X_j < 0),\ P(X_j > 0)\} < +\infty.$$

Let us assume that the theorem does not hold. Then we can find a subsequence $\{X_{n_k}\}$ with $P(X_{n_k} \le 0) \to 1$, as $n_k \to +\infty$, and real numbers a_1, $P(Y = a_1) = 0$, $\epsilon_1 > 0$, $M_1 = M_1(\epsilon_1, a_1)$, such that whenever $n > M_1$,

$$\left| E\left[1_{\{V_n \le a_1\}} 1_{\{\nu_n^- = \eta\}}\right] - E\left[1_{\{V_n \le a_1\}} 1_{\{\nu_n^- = -\eta\}}\right]\right| \ge \epsilon_1.$$

Moreover, weak convergence of $(\prod_{i=1}^n X_i)\, e^{-a_n}$ and almost sure convergence of V_n imply the existence of $M_2 = M_2(a_1, \epsilon_1/3)$ such that whenever j, t, k, l are arbitrary positive integers and j, t, k are greater than M_2,

$$\left|E\left[1_{\{V_j\leq a_1\}}1_{\{\nu_l=\eta\}}\right]-E\left[1_{\{V_t\leq a_1\}}1_{\{\nu_l=\eta\}}\right]\right|\leq\frac{\epsilon_1}{3},$$

and

$$\left|E\left[1_{\{V_k\leq a_1\}}1_{\{\nu_k=\eta\}}\right]-E\left[1_{\{V_{k+1}\leq a_1\}}1_{\{\nu_{k+1}=\eta\}}\right]\right|\leq\frac{\epsilon_1}{3}.$$

Then if $n_k > \max\{M_1, M_2\} + 1$,

$$\begin{aligned}
\frac{\epsilon_1}{3} &\geq \left|E\left[1_{\{V_{n_k-1}\leq a_1\}}1_{\{\nu_{n_k-1}=\eta\}}\right]-E\left[1_{\{V_{n_k}\leq a_1\}}1_{\{\nu_{n_k}=\eta\}}\right]\right|.\\
&\geq \left|E\left[1_{\{V_{n_k-1}\leq a_1\}}1_{\{\nu_{n_k-1}=\eta\}}\right]-E\left[1_{\{V_{n_k-1}\leq a_1\}}1_{\{\nu_{n_k}=\eta\}}\right]\right|-\frac{\epsilon_1}{3}\\
&= \left|E\left[1_{\{V_{n_k-1}\leq a_1\}}1_{\{\nu_{n_k-1}=\eta\}}\right]-E\left[1_{\{V_{n_k-1}\leq a_1\}}1_{\{\nu_{n_k-1}=-\eta\}}\right]\right|\\
&\quad\cdot P(X_{n_k}\leq 0)-\frac{\epsilon_1}{3}\\
&\geq \epsilon_1 P(X_{n_k}\leq 0)-\frac{\epsilon_1}{3}\longrightarrow\frac{2}{3}\epsilon_1
\end{aligned}$$

as $n_k \to +\infty$, which is not possible. Hence under our initial assumptions the theorem must hold.

Next we turn to the proof of part (ii). Since the validity of the theorem is obvious in the case $X_1X_2\cdots X_n$ is symmetric for some n, and it also follows from the proof of Theorem 1.24 that if

$$\sum_{i=1}^{+\infty}\min\{P(X_i\leq 0), P(X_i>0)\}=+\infty$$

then Y exists with a symmetric distribution (see last paragraph of Section 1.4), it suffices to prove the convergence of $\left(\prod_{i=1}^{n}X_i\right)e^{-a_n}$ under the assumption that V_n converges and $\sum_{i=1}^{+\infty}P(X_i\leq 0)<+\infty$. Let a, $P(V=a)=0$, and $\epsilon>0$ be given. Then our assumptions imply the existence of $N_1=N_1(\epsilon/2, a)$ such that whenever $n>N_1$

and k is arbitrary

$$\begin{aligned}
D_{n,k} &= \left|E\left[1_{\{V_n\le a\}}1_{\{\nu_n=\eta\}}\right]-E\left[1_{\{V_{n+k}\le a\}}1_{\{\nu_{n+k}=\eta\}}\right]\right| \\
&\le \left|E\left[1_{\{V_n\le a\}}1_{\{\nu_n=\eta\}}\right]-E\left[1_{\{V_n\le a\}}1_{\{\nu_{n+k}=\eta\}}\right]\right|+\frac{\epsilon}{2} \\
&\le \sum_{i=1}^{k}\left|E\left[1_{\{V_n\le a\}}1_{\{\nu_{n+i-1}=\eta\}}\right]-E\left[1_{\{V_n\le a\}}1_{\{\nu_{n+i}=\eta\}}\right]\right|+\frac{\epsilon}{2} \\
&= \sum_{i=1}^{k}\left(\left|E\left[1_{\{V_n\le a\}}1_{\{\nu_{n+i-1}=\eta\}}\right]\right.\right. \\
&\qquad \left.\left.-E\left[1_{\{V_n\le a\}}1_{\{\nu_{n+i-1}=-\eta\}}\right]\right|P(X_{n+i}\le 0)\right)+\frac{\epsilon}{2} \\
&\le \sum_{i=n+1}^{+\infty}P(X_i\le 0)+\frac{\epsilon}{2}\le\epsilon,
\end{aligned}$$

which implies convergence of $\{E[1_{\{V_n\le a\}}1_{\{\nu_n=\eta\}}]\}$. The proof of part (ii) is now complete.

Finally, we turn to the proof of (iii). Let $\mathcal{B}=\{\omega|\,(\prod_{i=1}^{n}X_i)\,e^{-a_n}\longrightarrow 0\}$. Since $P\,(X_i=0)=0$ for all i, $\mathcal{B}$ is a tail event, and from this the validity of (iii) immediately follows. ■

Next we extend Theorem 1.26 to arbitrary random variables.

Theorem 1.27 *Let $X_1, X_2, \cdots$ be independent random variables with $P(X_j=0)<1$, $j\ge 1$.*

(i) $\lim_{n\to+\infty} P(X_1X_2\cdots=0)=L$ always exists and $L=1$ if, and only if,

$$\sum_{j=1}^{+\infty}P(X_j=0)=+\infty.$$

Assume that $L<1$.

(ii) If there exist constants a_n such that $(\prod_{i=1}^{n}X_i)e^{-a_n}$ converges completely to a nonsymmetric random variable Y, $P(Y=0)<1$, then

$$\left|\prod_{i=1}^{n}X_i\right|e^{-a_n}\longrightarrow |Y| \quad \textit{completely,}$$

and $\lim_{n\to+\infty} P(X_n < 0) = 0$.

(iii) For the complete convergence of $(\prod_{i=1}^n X_i)e^{-a_n}$ *to a random variable* Y, $P(Y = 0) < 1$, *where the* a_n *are constants, it is sufficient that there exists a random variable* V *such that*

$$\left|\prod_{i=1}^n X_i\right| e^{-a_n} \longrightarrow V \quad \textit{completely,} \qquad P(V = 0) < 1,$$

and either $\lim_{n\to+\infty} P(X_n \le 0) = 0$, *or* $X_1X_2\cdots X_n$ *is symmetric for some* n, *or*

$$\sum_{i=1}^{+\infty} \min\{P(X_i \le 0), P(X_i > 0)\} = +\infty.$$

Whenever the above limits exist, then there exists a random variable W, $P(W = 0) = 0$, *such that for arbitrary continuity point* a *of* Y,

$$P(Y \le a) = P(W \le a)(1 - L) + L\,\delta_0(a)$$

where $\delta_0(z)$ *represents a distribution function degenerate at 0.*

(iv) There exist constants a_n *and a random variable* Y *such that*

$$\left(\prod_{i=1}^n X_i\right) e^{-a_n} \longrightarrow Y \quad \textit{completely,} \qquad P(Y = 0) > L,$$

if, and only if,

$$\left(\prod_{i=1}^n X_i\right) e^{-a_n} \longrightarrow 0 \qquad \textit{almost surely.}$$

PROOF. The proof of (i) follows from the first part of the proof of Theorem 1.20 after noticing that $\{P(X_1X_2\cdots X_n = 0)\}$ is a nondecreasing bounded sequence. To complete the proof of the theorem we first write

$$\begin{aligned} P\Big(\big(\prod_{i=1}^n X_i\big)e^{-a_n} \le a\Big) &= P\Big(\big(\prod_{i=1}^n \tilde{X}_i\big)e^{-a_n} \le a\Big)\, P(X_1X_2\cdots X_n \ne 0) \\ &\quad + P(X_1X_2\cdots X_n \ne 0)\,\delta_0(a). \end{aligned}$$

This, part (i), and the equivalence of parts (ii)-(iv) with parts (i)-(iii) of Theorem 1.26 applied to $\prod_{i=1}^n \tilde{X}_i e^{-a_n}$, immediately give the validity of the remaining parts of Theorem 1.27. The proof is complete. ■

Next we investigate convergence of infinite products of random variables with normalization

$$\left|\prod_{i=1}^n X_i\right|^{\frac{1}{b_n}} e^{-a_n} sign(X_1 \cdots X_n),$$

with $b_n \to +\infty$ as $n \to +\infty$. The following result was first proved by Abramov [3]. To get the full extent of the convergence criterium, the reader should combine Theorems 1.22 and 1.28 below. In this generality the case $p = 1/2$ is due to Simonelli [139], and the case $p \neq 1/2$ is new here. We take this opportunity to point out that in Theorem 2.4 of [139], $(-1)^{\nu^-}$ should of course be ν^-.

Theorem 1.28 *Let $\nu_n = sign(X_1 X_2 \cdots X_n)$, and assume that ν_n converges completely to a random variable ν. For arbitrary constants a_n and b_n, with $b_n \to +\infty$ as $n \to +\infty$, there exists a random variable Y such that*

$$\left|\prod_{i=1}^n X_i\right|^{\frac{1}{b_n}} e^{-a_n} \nu_n \to Y \quad \textit{completely,} \tag{1.41}$$

if, and only if, there exists a random variable W such that

$$\left|\prod_{i=1}^n X_i\right|^{\frac{1}{b_n}} e^{-a_n} \longrightarrow W \quad \textit{completely.} \tag{1.42}$$

Whenever the above limits exist, $W \stackrel{d}{=} |Y|$, and Y is always of the form $Y \stackrel{d}{=} \nu W$, with ν and W independent.

PROOF. Clearly (1.41) implies (1.42). The converse will be proved by showing the convergence of the Mellin transform of the normalized product. Let us assume that (1.42) holds. Let $p = P(\nu = -)$, $q = P(\nu = +)$, and for arbitrary integers k, l, $1 \le k < l$, let

$$\mathcal{M}_n^\eta(t; k, l) = E\left[|X_k \cdots X_l|^{\frac{it}{b_n}} e^{-ita_n} 1_{\{\nu_{k,l}=\eta\}}\right],$$

where $\eta = -, +$, and $\nu_{k,l} = sign(X_k \cdots X_l)$. Let $\epsilon > 0$ be given. We are going to show that

$$\limsup_{n\to+\infty} \left|\mathcal{M}_n^-(t; 1, n) - pE[\prod_{j=1}^n |X_j|^{\frac{it}{b_n}} e^{-ita_n}]\right| < \epsilon. \tag{1.43}$$

Since ν_n converges, then by Theorem 1.21 we can find a $N_1 = N_1(\epsilon)$ such that whenever $k \ge N_1$,

$$|E[1_{\{\nu_{1,k}=-\}}] - p| < \frac{\epsilon}{3} \qquad \text{and} \qquad \sum_{j=k+1}^{+\infty} P(X_j \le 0) < \frac{\epsilon}{3}, \tag{1.44}$$

if $p \ne q$, and

$$\Big|P(\nu_n = -) - P(\nu_n = +)| \le \frac{\epsilon}{3} \tag{1.45}$$

if $p = q$. Moreover, since $b_n \to +\infty$, for arbitrary finite l the dominated convergence theorem gives

$$\lim_{n\to+\infty} E[\prod_{j=1}^l |X_j|^{\frac{it}{b_n}} 1_{\{\nu_{1,l}=\eta\}}] \quad = \quad E[1_{\{\nu_{1,l}=\eta\}}] \tag{1.46}$$

and

$$\lim_{n\to+\infty} E[\prod_{j=l}^n |X_j|^{\frac{it}{b_n}} e^{-ita_n}] \quad = \quad \lim_{n\to+\infty} E[\prod_{j=1}^n |X_j|^{\frac{it}{b_n}} e^{-ita_n}]. \tag{1.47}$$

This and the independence of the X_j's immediately give that if $k \geq N_1$,

$$
\begin{aligned}
&\limsup_{n\to+\infty}\left|\mathcal{M}_n^-(t;1,n) - pE[\prod_{j=1}^n |X_j|^{\frac{it}{b_n}} e^{-ita_n}]\right| \\
= &\limsup_{n\to+\infty}\left|E[\prod_{j=1}^k |X_j|^{\frac{it}{b_n}} 1_{\{\nu_{1,k}=-\}}]\mathcal{M}_n^+(t;k+1,n)\right. \\
&+E[\prod_{j=1}^k |X_j|^{\frac{it}{b_n}} 1_{\{\nu_{1,k}=+\}}]\mathcal{M}_n^-(t;k+1,n) \\
&\left.-pE[\prod_{j=1}^n |X_j|^{\frac{it}{b_n}} e^{-ita_n}]\right| \\
= &\limsup_{n\to+\infty}\left|E[\prod_{j=1}^k |X_j|^{\frac{it}{b_n}} 1_{\{\nu_{1,k}=-\}}]E[\prod_{j=k+1}^n |X_j|^{\frac{it}{b_n}} e^{-ita_n}]\right. \\
&-pE[\prod_{j=1}^n |X_j|^{\frac{it}{b_n}} e^{-ita_n}] \\
&\left.+\mathcal{M}_n^-(t;k+1,n)(E[\prod_{j=1}^k |X_j|^{\frac{it}{b_n}} 1_{\{\nu_{1,k}=+\}}] - E[\prod_{j=1}^k |X_j|^{\frac{it}{b_n}} 1_{\{\nu_{1,k}=-\}}])\right| \\
\leq &\limsup_{n\to+\infty}\left(\left|(1-2E[1_{\{\nu_{1,k}=-\}}])\mathcal{M}_n^-(t;k+1,n)\right|\right. && (1.48) \\
&\left.+\left|E[1_{\{\nu_{1,k}=-\}}]E[\prod_{j=1}^n |X_j|^{\frac{it}{b_n}} e^{-ita_n}] - pE[\prod_{j=1}^n |X_j|^{\frac{it}{b_n}} e^{-ita_n}]\right|\right). && (1.49)
\end{aligned}
$$

Now if $p \neq q$,

$$\left|(1-2E[1_{\{\nu_{1,k}=-\}}])\mathcal{M}_n^-(t;k+1,n)\right| \leq \sum_{j=k+1}^{+\infty} P(\tilde{X}_j \leq 0) < \frac{\epsilon}{3}$$

by (1.44), and if $p = q$,

$$\left|(1-2E[1_{\{\nu_{1,k}=-\}}])\mathcal{M}_n^-(t;k+1,n)\right| \leq |1-2P(\nu_k=-)| \leq \frac{\epsilon}{3}$$

by (1.45). This, the convergence of ν_n, and (1.46) and (1.47) give that the sum of the absolute values in (1.48) and (1.49) is less than or equal to $2\epsilon/3$. Hence (1.43) holds, and since ϵ was arbitrary the validity of the theorem follows from Theorem 1.20 (and our assumption that 1.42 holds). ∎

Chapter 2

General limit theorems for products

Triangular arrays

This chapter is devoted to products with emphasis on normalizations. Just as in the theory of sums we deal with two kinds of normalization process. One group is when we have an infinite sequence of independent random variables and we normalize them by sequences of constants in order to achieve that each term should stay close to one in absolute value. The second group is when one faces double indexed sequences X_{jn} of independent random variables where, for fixed n, j is finite, $1 \leq j \leq k_n$, and n goes to infinity. Here the normalization is implicit in the distribution of X_{jn} which once again is aimed at keeping the absolute value of each X_{jn} close to one, this time in probability. Precise definitions will be given in the appropriate sections. Since the sum of $\log|X_{jn}|$, conditioned on $X_{jn} \neq 0$, plays an important role in several limit theorems that follow in the present chapter, we frequently give only outlines of proofs concerning sums, whose details can be found in our basic reference [58].

2.1 Logarithmically Poisson distributions

Let X be a discrete random variable taking the value zero with probability $p_o = P(X = 0)$, $0 \leq p_o < 1$. Then X is called a logarithmically Poisson random variable if, given that $X \neq 0$, $\log|X|$ is a Poisson random variable. That is, X takes the values $0, \pm 1, \pm e, \pm e^2, \cdots, \pm e^k, \cdot\cdot,$

and

$$P(X = e^k | X \neq 0) + P(X = -e^k | X \neq 0) = \frac{\lambda^k e^{-\lambda}}{k!}, \quad k \geq 0, \tag{2.1}$$

where $\lambda > 0$ is a fixed parameter. We set

$$p_k^{(+)} = P(X = e^k | X \neq 0), \qquad p_k^{(-)} = P(X = -e^k | X \neq 0) \tag{2.2}$$

and

$$p^{(+)} = P(X > 0), \qquad p^{(-)} = P(X < 0). \tag{2.3}$$

Clearly,

$$p^{(+)} = (1 - p_o) \sum_{k=0}^{+\infty} p_k^{(+)}, \qquad p^{(-)} = (1 - p_o) \sum_{k=0}^{+\infty} p_k^{(-)} \tag{2.4}$$

and

$$p_o + p^{(+)} + p^{(-)} = 1. \tag{2.5}$$

For the Mellin transforms we have that, under the condition $X \neq 0$,

$$\mathcal{M}_{\log|X|}(t) = e^{\lambda(e^{it} - 1)} \tag{2.6}$$

and

$$\mathcal{M}_{X^+}(t) = \sum_{k=0}^{+\infty} p_k^{(+)} e^{ikt}, \qquad \mathcal{M}_{X^-}(t) = \sum_{k=0}^{+\infty} p_k^{(-)} e^{ikt}. \tag{2.7}$$

Because the definition of a logarithmically Poisson variable involves only (the conditional) distribution expressed by (2.6), the multiplicative character of Mellin transforms for products and the uniqueness theorem yield the following

Theorem 2.1 *Let X and Y be independent logarithmically Poisson variables. Then XY is logarithmically Poisson as well.*

Definition. Let X be a logarithmically Poisson variable, and let $a > 0$ and $b > 0$ be real constants. Then $Y_{a,b} = a|X|^b sign(X)$ is called a logarithmically Poisson-type variable.

Note that, for $h > 1$ and $b = \log h$, $Y_{1,h} = Y_h = |X|^b sign(X)$ takes the values 0 and $\pm h^k$, $k \geq 0$, if X is logarithmically Poisson.

Let us look at some examples for logarithmically Poisson X.

Example 2.1 Let $0 \le p_o < 1$ and $p_k^{(+)} = p_k^{(-)}$ for all $k \ge 0$. Then, by (2.1) and (2.2), X is symmetric.

Example 2.2 Let $p_o = 0$, $\lambda = \log 2$, $p_o^{(+)} = 1/2$ and $p_k^{(+)} = 0$ for all $k \ge 1$. By (2.1), $p_o^{(-)} = 0$, and $p_k^{(-)} = \lambda^k e^{-\lambda}/k!$, $k \ge 1$ ($\lambda = \log 2$). Hence, by (2.3) and (2.4), $p^{(+)} = p^{(-)} = 1/2$. Yet, X is not symmetric.

Example 2.3 Let $p_o = 0.2$, and $p_k^{(+)} = 0.3\lambda^k e^{-\lambda}/k!$, $k \ge 0$. By (2.1), $p_k^{(-)} = 0.7\lambda^k e^{-\lambda}/k!$, $k \ge 0$. Clearly, $p^{(+)} \ne p^{(-)}$. Note, however, that $sign(X)$ and $|X|$ are independent (as they are in the symmetric case of Example 2.1).

Example 2.4 Let $h > 1$ and $b = \log h$, and we describe $Y_{1,h} = Y_h$ for the following logarithmically Poisson X. We introduce two additional parameters $0 \le \mu < \lambda$ and α for the distribution of X. Let the values in (2.1) and (2.2) be as follows ($P(X = 0) = p_o$)

$$P(X = e^k) = \frac{1}{2}(1 - p_o)\frac{\lambda^k e^{-\lambda}}{k!} + \frac{1}{2}\alpha\, e^{-\lambda+\mu}\frac{(\lambda-\mu)^k}{k!} \tag{2.8}$$

and

$$P(X = -e^k) = \frac{1}{2}(1 - p_o)\frac{\lambda^k e^{-\lambda}}{k!} - \frac{1}{2}\alpha\, e^{-\lambda+\mu}\frac{(\lambda-\mu)^k}{k!}. \tag{2.9}$$

Upon writing $\alpha = (1-p_o)\,\alpha/(1-p_o)$, we see that in order that (2.8) and (2.9) be probabilities we must have $|\alpha| \le (1-p_o)\,e^{-\mu}$ (first, take $k = 0$). The distribution of Y_h also is given by (2.8) and (2.9), except that the values of Y_h are 0, and $\pm h^k$, $k \ge 0$, in place of $\pm e^k$.

Note that Example 2.4 becomes symmetric if $\alpha = 0$, and $|X|$ and $sign(X)$ are independent if $\mu = 0$. Example 2.4 does not become a case like Example 2.2.

The following simple limit theorem provides a further example.

Theorem 2.2 *Let X_{jn}, $1 \le j \le n$, be independent and identically distributed random variables taking the values -2 and 1 with distribution r_n^* and $1 - r_n^*$, respectively. Assume that $n r_n^* \to \lambda$, $0 < \lambda < +\infty$, as $n \to +\infty$. Then $Y_n = \prod_{j=1}^n X_{jn}$ is asymptotically logarithmically Poisson-type with values $(-2)^k$, $k \ge 0$ and distribution $p_k^* = \lambda^k e^{-\lambda}/k!$, $k \ge 0$.*

PROOF. Clearly, Y_n takes the values $(-2)^k$ with distribution

$$P(Y_n = (-2)^k) = \binom{n}{k} (r_n^*)^k \, (1 - r_n^*)^{n-k},$$

that is, its distribution is binomial. Under our condition, its limiting distribution is Poisson.

Note that, in the limit, $p_k^{(+)} + p_k^{(-)} = \lambda^k e^{-\lambda}/k!$, and $p_k^{(+)} = 0$ if k is odd, while $p_k^{(-)} = 0$ if k is even. Here $p_o = 0$. ∎

The significant role of logarithmically Poisson-type variables will be seen in connection with limit theorems for triangular arrays.

2.2 The class of $\mathcal{L}$-distributions for products

Let $X_1, X_2, \cdots$ be an infinite sequence of independent random variables with distribution functions $F_k(x)$, $k \geq 1$. We assume that $F_k(x)$ is continuous at zero for each k. In Theorem 1.28 we studied the complete convergence of the normalized variables

$$U_n = U_n(a_n, b_n) = \left\{ \prod_{j=1}^{n} \left|X_j\right|^{\frac{1}{b_n}} sign(X_j) \right\} e^{-a_n}$$

where a_n and $b_n > 0$ are sequences of constants such that $b_n \to +\infty$ with n. In this section our emphasis is on the actual form of the limiting distribution.

Definition. The distribution function $F(x)$ is said to belong to the class L of distributions if there are independent random variables $Y_1, Y_2, \cdots$ and constants A_n and B_n such that $F(x)$ is the limiting distribution of the normalized sum

$$\frac{1}{B_n} \left(Y_1 + \cdots + Y_n \right) - A_n.$$

An equivalent definition is that $F(x)$ is an L-function if its characteristic function $\varphi(t)$ satisfies the equation

$$\varphi(t) = \varphi(at)\, \psi_a(t), \quad 0 < a < 1,$$

where $\psi_a(t)$ is a characteristic function.

We extend this to the class of $\mathcal{L}$ of distributions as follows.

Definition. We say that $G(x)$ belongs to class $\mathcal{L}$ of distributions if $G(x)$ is the limiting distribution of

$$U_n = U_n(a_n, b_n) = \left\{ \prod_{j=1}^{n} \left| X_j \right|^{\frac{1}{b_n}} sign(X_j) \right\} e^{-a_n} \tag{2.10}$$

where the factors are independent random variables, a_n and b_n are sequences of real numbers, $b_n \to +\infty$ as $n \to +\infty$.

Let X be a random variable with distribution function $G(x)$ in $\mathcal{L}$. Then, by definition, there exists a sequence of independent random variables X_j, $j = 1, 2, \cdots$ and real sequences a_n and b_n, $b_n \to +\infty$ with n, such that

$$\left\{ \prod_{j=1}^{n} \left| X_j \right|^{\frac{1}{b_n}} sign(X_j) \right\} e^{-a_n} \to X \quad \text{completely.}$$

By Theorem 1.28, $X = \nu|X|$, $\nu = sign(X)$, ν and $|X|$ are independent, and

$$\prod_{j=1}^{n} \left| X_j \right|^{\frac{1}{b_n}} e^{-a_n} \to |X| \quad \text{completely.}$$

Therefore $\ln|\tilde{X}|$ (recall (1.15)) is of class L, and we have the following equivalent definition of class $\mathcal{L}$.

Definition. A distribution function $G(x)$ is a member of class $\mathcal{L}$ if $G(x)$ is the distribution function of a random variable X with $X \stackrel{d}{=} \nu W$, $\nu = -, 0, +$, $W > 0$, ν and W independent, and $\ln W$ is of class L.

The above definition can be used to derive the following result, which explicitly expresses distribution functions in $\mathcal{L}$ in terms of distribution functions in L. Precisely, if X is a random variable with distribution function $G(x) \in \mathcal{L}$, then conditioning on $X \neq 0$, $G(x)$ is logarithmically-$F(x)$, $F(x) \in L$, and

$$G(x) = \begin{cases} P(\nu = -)(1 - F(\ln|x|-)) & \text{if } x < 0 \\ P(\nu \leq 0) & \text{if } x = 0 \\ P(\nu \leq 0) + P(\nu = +)F(\ln x) & \text{if } x > 0. \end{cases}$$

It is worthwhile to note that the distribution of ν is completely determined by the values $P(\nu = -) = p$ and $P(\nu = +) = q$. Clearly the converse is also true. That is, any nonnegative real numbers p, q, $0 \leq p + q \leq 1$ uniquely determine a random variable $\nu = -, 0, +$ by defining $p = P(\nu = -)$, $q = P(\nu = +)$, $1 - p - q = P(\nu = 0)$. Hence the following holds.

Theorem 2.3 *A distribution function $G(x)$ belongs to class $\mathcal{L}$ of distributions if , and only if, there exist nonnegative real numbers p, q, $0 \leq p + q \leq 1$, and a distribution function $F(x)$ in L such that*

$$G(x) = \begin{cases} p(1 - F(\ln|x|-)) & \textit{if } x < 0 \\ (1 - q) & \textit{if } x = 0 \\ (1 - q) + qF(\ln x) & \textit{if } x > 0. \end{cases} \tag{2.11}$$

Theorem 2.3 reduces the study of distributions of class $\mathcal{L}$ and their domain of attractions to those of elements of class L. However the full power of Theorem 2.3 is obtained by combining it with Theorem 1.22, which gives very simple criteria for the convergence of $sign(X_1 X_2 \cdots X_n)$.

We end this section by considering members of class $\mathcal{L}$ which are logarithmically normal.

Let Y be a normal random variable with density function

$$f(y) = \frac{1}{\sigma\sqrt{2\pi}} e^{-\frac{y^2}{2\sigma^2}} \qquad -\infty < y < +\infty,$$

and let $N(y)$ denote its distribution function. Then $G(x) \in \mathcal{L}$ is logarithmically normal if, and only if, $G(x)$ is given by (2.11) with $F(\cdot)$ replaced by $N(\cdot)$. Moreover the following holds.

Theorem 2.4 *Let $X_1, X_2, \cdots$ be a sequence of independent random variables with distribution function $F_j(x)$, $j \geq 1$. We assume that*

$E(\ln|\tilde{X}_j|) = \mu_j$ and $V(\ln|\tilde{X}_j|) = \sigma_j$ are finite, and let $c_n = \sum_{j=1}^n \mu_j$. Then there exists a sequence of real numbers b_n such that

$$\frac{1}{b_n} \max_{1 \leq j \leq n} \sigma_j \to 0 \qquad \text{as } n \to +\infty \tag{2.12}$$

and the distribution function of

$$U\left(\frac{c_n}{b_n}, b_n\right) = \left\{ \prod_{j=1}^{n} \left|X_j\right|^{\frac{1}{b_n}} sign(X_j) \right\} e^{-\frac{c_n}{b_n}}$$

converges completely to a logarithmically normal distribution function if, and only if,

(a) the X_is satisfy the conditions of part (ii) of Theorem 1.22,

(b) for every $\epsilon > 0$

$$\frac{1}{b_n^2} \sum_{j=1}^{n} \int_{|x| \geq \epsilon b_n} x^2 \, d\tilde{F}_j(x) \to 0$$

as $n \to +\infty$.

Proof. We appeal to Theorem 1.28, which states that $U(\frac{c_n}{b_n}, b_n)$ converges completely to a random variable X, $P(X = 0) \neq 1$, if and only if, $sign(X_1 X_2 \cdots X_n)$ converges to a random variable nondegenerate at zero, and

$$\left| U\left(\frac{c_n}{b_n}, b_n\right) \right| \to |X|$$

completely. The distribution function of $|X|$ is logarithmically normal if, and only if, the distribution function of

$$\frac{\sum_{j=1}^{n} \ln|\tilde{X}_j| - c_n}{b_n}$$

converges completely to a normal distribution, and from classical results for sums of independent random variables this and (2.12) both hold if, and only if, condition (b) of the theorem is satisfied (see Theorem 9, page 175, in [58]). ∎

The convergence result of Theorem 2.4 was first proved by Bakstys [16] and Abramov [3]. However these authors do not provide necessary and sufficient conditions for the convergence of $sign(X_1 \cdot\cdot X_n)$, and in their results part (a) of Theorem 2.4 is replaced by the requirement of convergence of $sign(X_1 X_2 \cdots X_n)$.

Guided by the characteristic property of $\mathcal{L}$, we end this section by introducing an expanded family of distributions, which we call $\mathcal{L}^*$, defined by the property that $sign(X)$ is independent of $|X|$. The importance and properties of this family of distributions will be discussed in Sections 2.5 and 2.6.

2.3 The Two Series and the Three Series Theorems for products

Recall from Chapter 1, Theorem 1.15 through Theorem 1.18, expressing convergence of infinite sums of independent random variables. A related theorem for products is Theorem 1.21 which essentially reduces the convergence of infinite products to their absolute values. Hence, by taking the logarithm of such products we can reformulate both the Two Series Theorem and the Three Series Theorem for products. We wish to reemphasize the fact that convergence completely to a distribution nondegenerate at zero is equivalent to almost sure convergence.

We now give new variants of all these theorems, applicable to both sums and products.

Theorem 2.5 *Let $X_1, X_2, \cdots$ be independent random variables with $P(X_j = 0) = 0$ and $m_j = 1$, $j \geq 1$, where m_j is a median of $|X_j|$. We further assume either the existence of real numbers $0 < a < b$, such that*

$$\lim_{n \to +\infty} P\Big(a < \big|X_1 \cdots X_n\big| \leq b\Big) = P \tag{2.13}$$

exists and is positive, or the existence of a real number $c > 0$ such that

$$\lim_{n \to +\infty} P\Big(\big|X_1 \cdots X_n\big| \leq c\Big) = P' \tag{2.14}$$

exists, is positive and less than one. Then $\prod_{j=1}^{n} |X_j| e^{-E_{j,M}}$ *converges almost surely, where* $E_{j,M} = E\big[(\ln |X_j|)_M\big]$, *and* $(\ln |X_j|)_M$ *is* $\ln |X_j|$ *truncated at* M.

PROOF. Let $X_1, X_2, \cdots$ be a sequence of independent random variables satisfying the conditions of the theorem. By turning to logarithms these conditions give that 0 is a median of $\log |X_j|$, $j \geq 1$, and either the existence of real numbers $0 < a < b$ such that

$$\lim_{n \to +\infty} P\Big(\log a < \log |X_1| + \cdots + \log |X_n| \leq \log b \Big) = P \qquad (2.15)$$

exists and is positive, or the existence of a real number $c > 0$ such that

$$\lim_{n \to +\infty} P\Big(\log |X_1| + \cdots + \log |X_n| \leq \log c \Big) = P' \qquad (2.16)$$

exists and $0 < P' < 1$. Let $\varphi_j(t)$ be the characteristic function of $\log |X_j|$, and $\varphi_{S_n}(t)$ the characteristic function of $S_n = \log |X_1| + \log |X_2| + \cdots + \log |X_n|$. Then

$$\varphi_{S_n}(t) = \prod_{j=1}^{n} \varphi_j(t).$$

Since $|\varphi_j(t)|^2 \leq 1$, $\prod_{j=1}^{n} |\varphi_j(t)|^2$ converges for all t, possibly to zero. We claim that under our assumptions this limit cannot be identically zero. To prove this we assume that this is not the case. That is, we assume that $\prod_{j=1}^{n} |\varphi_j(t)|^2$ converges to zero, except perhaps on a set of Lebesgue measure zero. If we first rewrite

$$\prod_{j=1}^{n} |\varphi_j(t)|^2 = \left(\prod_{j=1}^{n} \varphi_j(t) \right) \left(\prod_{j=1}^{n} \varphi_j(-t) \right)$$

and then consider arbitrary convergent subsequences of the right hand side above, we immediately obtain that $\prod_{j=1}^{n} \varphi_j(t)$ also converges and the limit must be identically zero. Let $F_n(x) = P(S_n \leq x)$. By Lemma 3 in [58], page 154, the convergence of $\prod_{j=1}^{n} \varphi_j(t)$ guarantees the existence of a distribution function $F(x)$, possibly not proper, such that $F_n(x)$ converges weakly to $F(x)$. Even in the case $F(x)$ is not proper, the inversion formula (1.3) continues to hold, which implies $F(z)-F(y) = 0$ for all continuity points $y < z$ of $F(x)$. The monotonicity and right continuity of $F(x)$ further imply that we can find continuity points of $F(x)$ $x_1 \leq a < b \leq x_2$ such that

$$\begin{aligned}\lim_{n\to+\infty} (F_n(b) - F_n(a)) &\leq \lim_{n\to+\infty} (F_n(x_2) - F_n(x_1)) \\ &= F(x_2) - F(x_1) = 0.\end{aligned}$$

Hence no real numbers $0 < a < b$ satisfying (2.15) with $P > 0$ can exist. So let us assume that there exists a real number $c > 0$ satisfying (2.16) with $0 < P' < 1$. By using again the inversion formula (1.3) with continuity points $x < c < y$, we obtain

$$\begin{aligned}P' &= \lim_{n\to+\infty} F_n(c) \leq \lim_{n\to+\infty} F_n(y) \\ &= \lim_{x\to-\infty} (F_n(y) - F_n(x)) + P(X = -\infty) \\ &= P(X = -\infty),\end{aligned}$$

where X is a random variable with distribution function $F(x)$. Hence $P(X = -\infty) \geq P' > 0$. This implies

$$\begin{aligned}&\lim_{n\to+\infty} P\left(\log|X_1| + \log|X_2| + \cdots \log|X_n| = -\infty\right) \\ = &\lim_{n\to+\infty} P\left(|X_1 X_2 \cdots X_n| = 0\right) > 0,\end{aligned}$$

which by Theorem 1.26 is equivalent to

$$\lim_{n\to+\infty} P\left(|X_1 X_2 \cdots X_n| = 0\right) = 1,$$

thus contradicting our assumption that $P' < 1$. This completes the proof of our claim, and $\prod_{j=1}^{n} |\varphi_j(t)|^2$ converges to a nonzero limit on a set of positive Lebesgue measure. Theorem 1 in [58], page 152, and Theorem 1.17 now imply that if $\ln|Y_j|$ is an independent copy of $\ln|X_j|$, then $\sum_{j=1}^{+\infty}(\ln|X_j| - \ln|Y_j|)$ converges almost surely. By Lemma 5 in [58], page 157, we further have that there are constants a_j^* such that $\sum_{j=1}^{+\infty}(\log|X_j| - a_j^*)$ converges a.s.. Moreover the a_j^* can be chosen to be $E\big[(\ln|X_j|)_M\big] = E_{j,M}$. Hence $\prod_{j=1}^{n} |X_j| e^{-E_{j,M}}$ converges a.s.. The proof of the theorem is now complete. ∎

Theorem 2.5 shows that the condition expressed in (2.13) or (2.14), together with the assumption that 1 is a median of $|X_j|$, $j \geq 1$, are enough to imply the convergence of $|X_1X_2\cdots X_n|e^{-a_n}$. Next we investigate the importance of these two conditions with respect to the almost sure convergence of $|X_1X_2\cdots X_n|$, and consequently to the almost sure convergence of $\log|X_1|+\log|X_2|+\cdots+\log|X_n|$. Before stating our next result we define the spectrum of a distribution function, and related properties.

Let $F(x)$ be a distribution function. The *spectrum* of $F(x)$, denoted by S_F, is the set of all points of increase of $F(x)$.

Clearly S_F is a closed nonempty set. Let X_1 and X_2 be two random variables with distribution functions $F_1(x)$ and $F_2(x)$, respectively, and let $F(x)$ denote the distribution function of X_1+X_2. Then

$$S_F = \overline{\{\omega : \omega = x_1 + x_2,\ x_1 \in S_{F_1},\ x_2 \in S_{F_2}\}}, \tag{2.17}$$

where $\overline{\{\cdot\}}$ denotes the closure of the set $\{.\}$ (a proof of (2.17) can be found in Lukacs [105], page 57).

Theorem 2.6 *Let $X_1, X_2, \cdots$ be independent random variables with $P(X_j = 0) = 0$ and $m_j = 1$, $j \geq 1$, where m_j is a median of $|X_j|$. We further assume the existence of a real number $c > 0$ such that*

$$\lim_{n\to+\infty} P\Big(|X_1\cdots X_n| \leq c\Big) = P'$$

exists, is positive and less than one. Then $\prod_{j=1}^{n}|X_j|$ converges almost surely.

PROOF. We appeal to Theorem 2.5, which says that under the conditions of Theorem 2.6, $\sum_{j=1}^{+\infty}\big(\log|X_j| - E_{j,m}\big)$ converges almost surely. This and Theorem 4 in [58], page 152, give that for arbitrary constants $M > 0$ the two series

$$\sum_{j=1}^{+\infty} P\Big(\big|\log|X_j|\big| > M\Big) \qquad \text{and} \qquad \sum_{j=1}^{+\infty} V\Big[\big(\log|X_j|\big)_M\Big] \tag{2.18}$$

converge, where $\big(\log|X_j|\big)_M$ is $\log|X_j|$ truncated at M.

Since $\sum_{j=1}^{+\infty}\left(\log|X_j| - E_{j,m}\right)$ converges almost surely, we immediately have that $\sum_{j=1}^{N(r)} \log|X_j|$ is an almost sure convergent subsequence of $\sum_{j=1}^{N} \log|X_j|$ if, and only if, $\sum_{j=1}^{N(r)} E_{j,M}$ is a convergent subsequence of $\sum_{j=1}^{N} E_{j,M}$. Let

$$\limsup_{N\to+\infty}\sum_{j=1}^{N} E_{j,M} = \lim_{t\to+\infty}\sum_{j=1}^{N_t} E_{j,M} = \overline{E} \tag{2.19}$$

and

$$\liminf_{N\to+\infty}\sum_{j=1}^{N} E_{j,M} = \lim_{k\to+\infty}\sum_{j=1}^{T_k} E_{j,M} = \underline{E}. \tag{2.20}$$

Our initial assumptions and the convergence of $\sum_{j=1}^{+\infty}(\log|X_j| - E_{j,M})$ imply that

$$\begin{aligned} P' &= \lim_{t\to+\infty} P\left(\sum_{j=1}^{N_t}\log|X_j| < \log c\right) \\ &= \lim_{t\to+\infty} P\left(\sum_{j=1}^{N_t}(\log|X_j| - E_{j,M}) < \log c - \sum_{j=1}^{N_t} E_{j,M}\right) \\ &= \lim_{k\to+\infty} P\left(\sum_{j=1}^{T_k}\log|X_j| < \log c\right) \\ &= \lim_{k\to+\infty} P\left(\sum_{j=1}^{T_k}(\log|X_j| - E_{j,M}) < \log c - \sum_{j=1}^{T_k} E_{j,M}\right), \end{aligned}$$

and since $0 < P' < 1$, we immediately have that $\overline{E}$ and $\underline{E}$ are both finite. We claim that $\overline{E} = \underline{E}$. Assume that this is not the case. Then the above calculation gives that

$$\begin{aligned} P' &= P\left(\sum_{j=1}^{+\infty}(\log|X_j| - E_{j,M}) \le \log c - \overline{E}\right) \\ &\le P\left(\sum_{j=1}^{+\infty}(\log|X_j| - E_{j,M}) \le \log c - \underline{E}\right) = P'. \end{aligned}$$

That is, if $\overline{E} \neq \underline{E}$, the interval $(\log c - \overline{E}, \log c - \underline{E})$ is not in the spectrum of the distribution function of $\sum_{j=1}^{+\infty}(\log|X_j| - E_{j,M})$.

Let $F_j(x)$ be the distribution function of $\log|X_j|$. Since we have that $|\overline{E}|$ and $|\underline{E}|$ are both finite, then $\lim_{j\to+\infty} E_{j,M} = 0$, as $j \to +\infty$. Furthermore, 0 being a median of $\log|X_j|$ implies that for any $\log|X_j|$ the following elements of S_{F_j} are well defined:

$$r_{j,1} = \inf\{x : S_{F_j} \cap [0, +\infty)\} \quad \text{and} \quad r_{j,2} = \sup\{x : S_{F_j} \cap (-\infty, 0]\}.$$

In the above definition we allow $r_{j,1} = r_{j,2}$. Since $(\log|X_j| - E_{j,M}) \to 0$ a.s., we further have that $r_{j,1}$ and $r_{j,2}$ approach zero as $j \to +\infty$.

Let $d = \overline{E} - \underline{E}$, $d > 0$. Let $\{N_i\}_{i=1}^{+\infty}$ and $\{T_i\}_{i=1}^{+\infty}$ be the two subsequences used in (2.19) and (2.20), respectively, and consider the two corresponding convergent subsequences of $\sum_{j=1}^{N} \log|X_j|$, $\sum_{j=1}^{N_i} \log|X_j|$ and $\sum_{j=1}^{T_i} \log|X_j|$. Then if $T_i > N_i$,

$$\begin{aligned}
\lim_{i\to+\infty} \left[\sum_{j=1}^{T_i} \log|X_j| - \sum_{j=1}^{N_i} \log|X_j|\right] &= \lim_{i\to+\infty} \sum_{j=N_i+1}^{T_i} \log|X_j| \\
= \lim_{i\to+\infty} &\left[\sum_{j=N_i+1}^{T_i} (\log|X_j| - E_{j,M}) + \sum_{j=1}^{T_i} E_{j,M} - \sum_{j=1}^{N_i} E_{j,M}\right] \\
= -d.
\end{aligned}$$

Hence the difference $\sum_{j=1}^{T_i} \log|X_j| - \sum_{j=1}^{N_i} \log|X_j|$, which we denote by Y_i, converges almost surely to $-d$. This implies that for every $\epsilon > 0$,

$$P\left(\cap_{n=1}^{+\infty} \cup_{i=n}^{+\infty} \{|Y_i + d| \geq \epsilon\}\right) = 0.$$

By taking the complement we have that there is an n, such that, for all $i \geq n$, $Y_i = \sum_{j=N_i+1}^{T_i} \log|X_j|$ is about $-d$ on a set of probability close to one. Similarly, if $\{N_i\}_{i=1}^{+\infty}$ and $\{T_i\}_{i=1}^{+\infty}$ are as before, but now $N_i > T_i$, then there is a n such that for all $i \geq n$, $V_i = \sum_{j=T_{i+1}}^{N_i} \log|X_j|$ is about d on a set of probability close to one. Let F_{Y_i} and F_{V_i} be the distribution function of Y_i and V_i, respectively. Then for every $d > \epsilon > 0$, the following elements of the spectrum of $S_{F_{Y_i}}$ and $S_{F_{V_i}}$ are well defined:

$$\begin{aligned}
c_{j,3} &= \min\{x : x \in S_{F_{Y_i}} \cap [-d-\epsilon, -d+\epsilon]\} \\
c_{j,4} &= \max\{x : x \in S_{F_{V_i}} \cap [d-\epsilon, d+\epsilon]\}.
\end{aligned}$$

From (2.17), $c_{j,3}$ and $c_{j,4}$ can be written as a sum of elements in the spectrum of the F_j involved in the definition of Y_i and V_i. Say, $c_{j,3} = \sum_{j=N_i+1}^{T_i} r_{j,3}$ and $c_{j,4} = \sum_{j=T_{i+1}}^{N_i} r_{j,4}$.

The above quantities and related properties give that by choosing i appropriately, say i' and i'', we can find elements in the spectrum of the X_j such that

$$\sum_{j=1}^{+\infty} r_{j,i'} = +\infty \quad \text{and} \quad \sum_{j=1}^{+\infty} r_{j,i''} = -\infty,$$

with $r_{j,i'}$ and $r_{j,i''}$ going to zero as j goes to infinity. This, and our result on $|\overline{E}|$ and $|\underline{E}|$ being bounded, give

$$\sum_{j=1}^{+\infty} (r_{j,i'} - E_{j,M}) = +\infty \quad \text{and} \quad \sum_{j=1}^{+\infty} (r_{j,i''} - E_{j,M}) = -\infty,$$

with $(r_{j,i'} - E_{j,M})$ and $(r_{j,i''} - E_{j,M})$ going to zero as j goes to infinity. Hence by properly choosing the value of i, every real number x can be written as a conditionally convergent sum

$$x = \sum_{j=1}^{+\infty} (r'_{j,i} - E_{M,j}),$$

showing that the spectrum of the distribution function of $\sum_{j=1}^{+\infty} (\log |X_j| - E_{j,M})$ is the whole real line (see (2.17)), thus reaching a contradiction. Hence $\overline{E} = \underline{E}$, and $\sum_{j=1}^{+\infty} E_{j,M}$ converges. This and the convergence of the two series in (2.18) imply $\sum_{j=1}^{+\infty} \log |X_j|$ converges a. s. (see Theorem 1.16). Hence $\prod_{j=1}^{n} |X_j|$ converges a.s., and the proof is completed. ∎

From the proof of Theorem 2.6 it is clear that once Theorem 2.5 applies, the property $P' < 1$ is only needed to assure that $\underline{E}$ is finite. Hence $P' < 1$ can be relaxed by combining (2.13) and (2.14) in the following way.

Theorem 2.7 *Let $X_1, X_2, \cdots$ be independent random variables with $P(X_j = 0) = 0$ and $m_j = 1$, $j \geq 1$, where m_j is a median of $|X_j|$.*

We further assume the existence of real numbers $0 < a_1 < a_2$ *such that*

$$\lim_{n\to+\infty} P\Big(\big|X_1\cdots X_n\big| \le a_i\Big) = P_i, \quad i = 1,2 \tag{2.21}$$

exist and $P_2 - P_1 > 0$. *Then* $\prod_{j=1}^{n} |X_j|$ *converges almost surely.*

PROOF. The validity of (2.21) with $P_2 - P_1 > 0$ implies the validity of (2.13) with $P = P_2 - P_1 > 0$. Hence by Theorem 2.5 we have that $\sum_{j=1}^{+\infty}(\log|X_j| - E_{j,M})$ converges almost surely, and as in the proof of Theorem 2.6 this implies the convergence of the two sums in (2.18). Let $\overline{E}$, $\underline{E}$, $\sum_{j=1}^{N_t} E_{j,M}$ and $\sum_{j=1}^{T_k} E_{j,M}$ be as in the proof of Theorem 2.6. Then, with $P = P_2 - P_1$,

$$\begin{aligned}
P &= \lim_{t\to+\infty}\left(P\Big(\sum_{j=1}^{N_t}\log|X_j| \le \log a_2\Big) - P\Big(\sum_{j=1}^{N_t}\log|X_j| \le \log a_1\Big)\right)\\
&= \lim_{t\to+\infty}\left(P\Big(\sum_{j=1}^{N_t}(\log|X_j| - E_{j,M}) \le \log a_2 - \sum_{j=1}^{N_t} E_{j,M}\Big)\right.\\
&\qquad\left. -P\Big(\sum_{j=1}^{N_t}(\log|X_j| - E_{j,M}) \le \log a_1 - \sum_{j=1}^{N_t} E_{j,M}\Big)\right)\\
&= \lim_{t\to+\infty}\left(P\Big(\sum_{j=1}^{T_k}\log|X_j| \le \log a_2\Big) - P\Big(\sum_{j=1}^{T_k}\log|X_j| \le \log a_1\Big)\right)\\
&= \lim_{t\to+\infty}\left(P\Big(\sum_{j=1}^{T_k}(\log|X_j| - E_{j,M}) \le \log a_2 - \sum_{j=1}^{T_k} E_{j,M}\Big)\right.\\
&\qquad\left. -P\Big(\sum_{j=1}^{T_k}(\log|X_j| - E_{j,M}) \le \log a_1 - \sum_{j=1}^{T_k} E_{j,M}\Big)\right),
\end{aligned}$$

which together with our initial assumption that $P > 0$ immediately give that $\overline{E}$ and $\underline{E}$ are both finite. The validity of the theorem now follows from the proof of Theorem 2.6 (see the paragraph preceding Theorem 2.7). ∎

For related results in connection with sums see also Hildebrand [81] and K. F. Kong and Q. Tang [92]. In their work the assumption that 0 is a median of $\log|X_j|$ is replaced by different assumptions, which can be expressed in terms of Lévy's metric.

2.4 Conditional distributions for products

We return to the model of Section 2.2. That is, $X_1, X_2, \cdots$ are independent random variables with distribution functions $F_j(x)$, $j \geq 1$. We assume that each $F_j(x)$ is continuous at zero, and that the sequence $F_j(0)$, $j \geq 1$, is such that $P(sign(\prod_{j=1}^{n} X_j) < 0)$ converges (very simple properties guarantees this, see Theorem 1.21). Consequently, the normalized product

$$U_n = e^{-a_n} \prod_{j=1}^{n} |X_j|^{\frac{1}{b_n}} sign(X_j),$$

where a_n and $b_n > 0$ are sequences of constants and $b_n \to +\infty$ with n, converges completely if $|U_n|$ does. Our aim is to extend such complete convergence theorem to convergence under conditioning.

We prove the following theorems.

Theorem 2.8 *Assume that $|U_n|$ converges completely to a nondegenerate distribution function $G(x)$. Then, for any event B with $P(B) > 0$,*

$$P\Big(|U_n| \leq x \big| B\Big) \to G(x) \qquad \textit{completely.}$$

Theorem 2.9 *If $|U_n|$ converges completely to a nondegenerate distribution function $G(x)$, then U_n does not converge in probability (and thus almost surely) to any random variable.*

Remark 2.1 Theorem 2.9 shows a deep contrast to the results of the preceding section when complete convergence and almost sure convergence coincide. In particular, the classical central limit theorems leading to asymptotic normality for sums, and lognormality for products, are not replaceable with almost sure convergence or convergence in probability.

Remark 2.2 When one conditions on events B, the underlying variables X_j are no longer independent. Hence, Theorem 2.8 is a dependent version of earlier limit theorems.

We turn to the proofs. The method for proof that follows is due to A. Rényi [128], although similar arguments appear in functional analysis; see W. Schmeidler [134]. The results of the present section are new here.

First, we prove Theorem 2.8 for special choices of B.

Theorem 2.8a *With the notation and assumptions of Theorem 2.8,*

$$P\Big(|U_n| \le x \Big| |U_k| \le x\Big) \to G(x) \quad \textit{completely} \tag{2.22}$$

for any fixed $k \ge 1$ *as* $n \to +\infty$.

PROOF. We show that, under the condition $|U_k| \le x$,

$$V_{k,n} = \prod_{j=1}^{k} |X_j|^{1/b_n} \text{ converges to 1 in probability.} \tag{2.23}$$

Indeed, under the condition $|U_k| \le x$,

$$\prod_{j=1}^{k} |X_j| \le (e^{a_k}x)^{b_k},$$

that is, since $b_n \to +\infty$ with n,

$$V_{k,n} \le (e^{a_k}x)^{b_k/b_n} \le 1 + \epsilon \text{ for large } n, \tag{2.24}$$

where $\epsilon > 0$ is arbitrary. Conversely, since each $F_j(x)$ is continuous at zero, and k is fixed, we have for arbitrary $\delta > 0$ and some $\eta > 0$,

$$P\left(\prod_{j=1}^{k} |X_j| \ge \eta\right) \ge 1 - \delta.$$

Hence, for large n,

$$1 - \delta \le P\big(V_{k,n} > \eta^{1/b_n}\big) \le P\big(V_{k,n} \ge 1 - \epsilon\big). \tag{2.25}$$

The estimates in (2.24) and (2.25) imply (2.23). We utilize (2.23) in the form: if Z_n is a random variable converging completely to a nondegenerate distribution function $H(x)$ then

$$Z_n V_{k,n} \text{ converges completely to } H(x) \text{ as well.} \tag{2.26}$$

(Although (2.26) is an elementary and widely applied rule, the reader can find a proof in [58] pp. 126 and 427.) So, by (2.23) and (2.26), if the first limit below exists then

$$\lim_{n\to+\infty} P\big(|U_n| \le x \big| |U_k| \le x\big) \tag{2.27}$$

$$= \lim_{n\to+\infty} P\Big(e^{-a_n}\big(\prod_{j=k+1}^{n} |X_j|^{1/b_n}\big)V_{k,n} \le x \big| |U_k| \le x\Big)$$

$$= \lim_{n\to+\infty} P\Big(e^{-a_n} \prod_{j=k+1}^{n} |X_j|^{1/b_n} \le x\Big) \tag{2.28}$$

where in the last expression we dropped the condition $|U_k| \le x$ after the application of (2.26) since the X_j are independent. Once more applying (2.26) in

$$\lim_{n\to+\infty} P\big(|U_n| \le x\big) = \lim_{n\to+\infty} P\Big(e^{-a_n}\big(\prod_{j=k+1}^{n} |X_j|^{1/b_n}\big)V_{k,n} \le x\Big)$$

the assumption of the theorem on the left hand side gives $G(x)$ while on the right hand side, on account of (2.26), (2.28) appears. That is, (2.27) is $G(x)$, and thus (2.22) obtains. The proof is complete. ∎

For showing that Theorem 2.8a implies Theorem 2.8 in its full generality we use some properties of square integrable random variables and a general theorem from the theory of Hilbert spaces. These tools are formulated below as Lemmas. Their proofs are delayed to Chapter 6.

We denote by L_2 the set of all random variables X on a given probability space $(\Omega, \mathcal{A}, P)$ for which $\int_\Omega X^2 dP < +\infty$. For X, $Y \in L_2$, we put

$$(X, Y)_{L_2} = \int_\Omega XY\, dP$$

and call it the inner-product of X and Y. The value $(X,X)_{L_2}^{1/2}$ is called the norm of X, and denoted $\| X \|_{L_2} = \| X \|$.

Note that $(X,Y)_{L_2} = (Y,X)_{L_2}$. Furthermore

$$\| X \| \geq 0 \text{ and } \| X \| = 0 \text{ if, and only if, } X = 0 \; a.s., \tag{2.29}$$

and the triangular inequality

$$\| X + Y \| \leq \| X \| + \| Y \| \tag{2.30}$$

holds. We also have

$$\big|(X,Y)_{L_2}\big| \leq \| X \| \cdot \| Y \| \tag{2.31}$$

which is known as the Cauchy-Schwarz inequality.

We say that a sequence $X_n \in L_2$ converges to zero in L_2 if $\| X_n \| \to 0$. Furthermore, $X_n \to X \in L_2$ in L_2 if $\| X_n - X \| \to 0$.

Lemma 2.1 *If $X_n \in L_2$ and $\| X_n - X_m \| \to 0$ as $n, m \to +\infty$, then there is an $X \in L_2$ such that $X_n \to X$ in L_2.*

It should be emphasized that if $X_n \to X$ a.s., it does not necessarily follow that $X_n \to X$ in L_2. Take the simple example of $\Omega = [0,1]$, $\mathcal{A}$ the set of the Borel subsets of $[0,1]$ and Lebesgue measure as P, then $X_n = \sqrt{n}$ on $[0,1/n)$ and $X_n = 0$ otherwise, does converge to $X \equiv 0$ a.s., but $\| X_n - X \| = \| X_n \| = 1$ for all n, so $X_n \not\to X$ in L_2. However, if each X_n is bounded by K, then $X_n \to X$ a.s. implies $X_n \to X$ in L_2 by the dominated convergence theorem. Note also that if $X_n \to X$ in L_2 and $Y \in L_2$, then, by (2.31), as $n \to +\infty$,

$$\begin{aligned}
\big|(X_n,Y)_{L_2} - (X,Y)_{L_2}\big| &= \Big|\int_\Omega X_n Y \, dP - \int_\Omega XY \, dP\Big| \\
&= \Big|\int_\Omega (X_n - X)\, Y dP\Big| \\
&\leq \Big(\int_\Omega (X_n - X)^2 dP\Big)^{1/2}\Big(\int_\Omega Y^2 dP\Big)^{1/2} \to 0
\end{aligned}$$

that is,

$$\int_\Omega X_n Y dP \to \int_\Omega XY \, dP, \qquad n \to +\infty. \tag{2.32}$$

If the variables X_n are bounded in norm then (2.32) is still valid without the assumption of $X_n \to X$ in L_2.

Indeed, in the following lemma, we formulate a convergence of L_2 products without assuming that the underlying sequence of variables converges in L_2 norm.

Lemma 2.2 *Let X_n, $n \geq 0$ be a sequence of random variables belonging to L_2 and such that $X_o \equiv 1$ (a.s.), and $E(X_n^2) = \| X_n \|^2$ is bounded. Assume that, for every fixed $k \geq 0$, as $n \to +\infty$,*

$$\int_\Omega X_n X_k \, dP \to 0. \tag{2.33}$$

Then, for every $Y \in L_2$, (2.32) holds with $X = 0$ a.s.

In Chapter 6, when we prove Lemma 2.2, we argue with spaces with elements more general than L_2, and, as side results, further properties of L_2 itself will be developed.

An easy consequence of Lemma 2.2 is the following result of Rényi [128].

Theorem 2.10 *Given a probability space $(\Omega, \mathcal{A}, P)$, let $A_0 = \Omega$, and A_n, $n \geq 1$, be a sequence of events satisfying $P(A_k) > 0$ for each k, and*

$$\lim_{n \to +\infty} P(A_n | A_k) = c \text{ for every fixed } k \geq 0, \quad 0 < c < 1. \tag{2.34}$$

Then, for any event B with $P(B) > 0$, $P(A_n | B) \to c$.

Proof. Let I_n be the indicator of the event A_n, that is, $I_n = 1$ if A_n occurs, and 0 otherwise. In order to satisfy (2.33) with $k = 0$, we modify I_n to consider the random variables $I_n - c = X_n$. We rewrite (2.34) as, for $k \geq 0$ fixed, as $n \to +\infty$,

$$P(A_n) \to c \; (k = 0), \quad \text{and} \quad P(A_n \cap A_k) = P(A_m | A_k) P(A_k) \to c P(A_k).$$

Then we get

$$\begin{aligned} \int_\Omega X_n X_k \, dP &= E(X_n X_k) \\ &= P(A_n \cap A_k) - c\,P(A_n) - c\,P(A_k) + c^2 \to 0, \end{aligned}$$

that is, (2.33) is valid. Hence, by Lemma 2.2, for $Y = I_B$, the indicator of B,

$$\int_\Omega X_n I_B \, dP = P(A_n \cap B) - c\,P(B) \to 0,$$

which is equivalent to $P(A_n|B) \to c$. The proof is complete. ∎

PROOF OF THEOREM 2.8. Put $A_n = \{|U_n| \le x\}$. Then we proved in Theorem 2.8a that (2.34) holds with $c = G(x)$. Thus Theorem 2.10 completes the proof. ∎

PROOF OF THEOREM 2.9. Let Y be an arbitrary random variable and set $B = \{Y \le z\}$. Then, by Theorem 2.8,

$$\begin{aligned} &P\big(|U_n| \le x, Y \le z\big) \\ &= P\big(|U_n| \le x \big| Y \le z\big)\, P(Y \le z) \to G(x)P(Y \le z). \end{aligned} \quad (2.35)$$

In particular, if $|U_n|$ converges in probability to a random variable Y then, by writing $|U_n| = (|U_n| - Y) + Y$, since $|U_n| - Y$ converges to zero in probability, an elementary rule similar to (2.26) implies that the limiting distribution $G(x)$ of $|U_n|$ actually is the distribution of Y. Hence, by (2.35), with $x = z$, we get $G(x) = G^2(x)$, whatever x. This is possible only for degenerate distributions, and thus it contradicts that $G(x)$ is nondegenerate. The proof is complete. ∎

It is, of course, possible for $|U_n|$ to have an almost sure limit Y if $|U_n|$ converges completely to a degenerate limiting distribution $G^*(x)$. Simply transform the strong law of large numbers to a product: let R_j, $j \ge 1$, be independent and identically distributed random variables with finite expectation E. Set $X_j = e^{R_j}$, and $b_n = n$. Then $U_n = |U_n|$ converges a.s. to e^E as well as completely to $G^*(x)$ degenerated at e^E.

Remark 2.3 Observe the somewhat surprising property hidden in Theorem 2.8. We transformed Theorem 2.8 at (2.35) to show that $|U_n|$ is asymptotically independent of an arbitrary random variable Y whenever $|U_n|$ converges completely to a nondegenerate distribution function $G(x)$.

Theorem 2.8 can be utilized to transform the complete convergence of $|U_n|$ to a nondegenerate distribution $G(x)$ to a complete convergence with random sample size $M(n)$. That is, let $M(n)$ be an integer valued random variable with the following property:

(C1) there is a monotonic sequence $d(n) > 0$ of numbers such that $d(n) \to +\infty$ with n and $M(n)/d(n) \to T$ in probability, where T is a random variable with $P(T > 0) = 1$.

We want to show $|U_{M(n)}|$ converges completely to $G(x)$. For this purpose we make as assumption on the X_j:

(C2) to every $\delta > 0$, there is a number $v = v(\delta)$ such that

$$P\Big(\max\{||U_n| - |U_j|| : |n - j| \le vd(n)\} \ge a \Big| h_{k-1} \le T < h_k\Big) \le \delta,$$

where $h_{k-1} < h_k$ are fixed numbers, chosen arbitrarily, so we may assume that $h_k - h_{k-1} < v$.

If $P(X_j = 0) = 0$ and the variance $V(\log |X_j|) = \sigma_j^2 < +\infty$ and satisfies one of the classical conditions for the normalized sum of $\log |X_j|$ to be asymptotically normal, then (C2) is a consequence of Kolmogorov's inequality. Other cases (such as convergence of the normalized sum of $\log |X_j|$ to stable distributions) are also known to satisfy (C2). Our emphasis in the theorem that follows is that the asymptotic lognormality of $|U_n|$ remains asymptotically lognormal for $|U_{M(n)}|$, with $M(n)$ satisfying (C1), since (C2) is known to be valid in this case (just the conditional version of Kolmogorov's inequality). Note that we do not assume any stochastic relation between $M(n)$ and $U(n)$.

Theorem 2.11 *Assume that $P(X_j = 0) = 0$ for all j. Let the integer valued random variable $M(n)$ satisfy (C1). If with some constants a_n and $b_n > 0$ such that $b_n \to +\infty$ with n,*

$$V_n = e^{-a_n} \left(\prod_{j=1}^{n} |X_j| \right)^{1/b_n} \tag{2.36}$$

converges completely to a nondegenerate distribution function $G(x)$, *then, under (C2),*

$$V_{M(n)} = e^{-a_{M(n)}} \left(\prod_{j=1}^{M(n)} |X_j| \right)^{1/b_{M(n)}}$$

also converges completely to $G(x)$. *If* $G(x)$ *is lognormal, and the complete convergence of (2.36) is under a classical limit theorem for the normalized sum of* $\log |X_j|$, *then (C2) holds, so* $V_{M(n)}$ *remains asymptotically lognormal.*

Remark 2.4 If both a_n and $b_n > 0$ are monotonic in n and T is a degenerate random variable at a constant $c > 0$, then the random normalizations $a_{M(n)}$ and $b_{M(n)}$ can be modified to nonrandom constants a_n^* and $b_n^* > 0$ without effecting the limit $G(x)$. However, if T is nondegenerate, then the random sized product $(M(n))$ with nonrandom normalization a_n^* and $b_n^* > 0$ will have a limiting distribution different from $G(x)$. As a matter of fact, the distribution of T will enter in the limit. For another effect of the normalizing constants, see Section 6.5.1.

When the X_j are independent and identically distributed and $P(T = 1) = 1$, then Theorem 2.11, with $\log V_n$ and with finite variances of $\log |X_j|$, is due to Anscombe [7] (recall thet the limit $G(x)$ is lognormal, (C2) is just Kolmogorov's inequality, and the normalizing constants $a_{M(n)} = M(n)E(\log |X_1|)$ and $b_{M(n)} = \sqrt{M(n)V(\log |X_1|)}$ are kept random). The general case covers results of Rényi [129] and Mogyoródi [113]. The proof below utilizes steps from the proofs of Rényi and Mogyoródi.

PROOF OF THEOREM 2.11. Choose an arbitrary $\epsilon > 0$. Fix two numbers $0 < \alpha < \beta$ such that $P(\alpha \leq T \leq \beta) \geq 1 - \epsilon$. Divide the positive real line by points $0 \leq h_1 < h_2 < \cdots$, and let $K(\alpha, \beta)$ be the set of those h_j which belong to the interval $[\alpha, \beta]$. Set $e_{k,n} = h_k d(n)$. Let

$$A_{k,n} = \{e_{k,n} \leq M(n) < e_{k+1,n}\} \quad \text{and} \quad B_k = \{h_k \leq T < h_{k+1}\}.$$

By the total probability rule,

$$P(V_{M(n)} \le x) = \sum_{k=1}^{+\infty} P(V_{M(n)} \le x | A_{k,n}) \, P(A_{k,n}) \tag{2.37}$$

$$= \sum_{k=1}^{+\infty} P(V_{M(n)} \le x | B_k) \, P(B_k). \tag{2.38}$$

For a fixed k, rewrite the general term of (2.37) as

$$P\Big(V_{M(n)} \le x | A_{k,n}\Big) P\Big(A_{k,n}\Big) = P\Big(\{V_{M(n)} \le x\} \cap A_{k,n}\Big)$$
$$= P\Big(\{V_{M(n)} \le x\} \cap A_{k,n} \cap B_k\Big) + P\Big(\{V_{M(n)} \le x\} \cap A_{k,n} \cap B_k^c\Big)$$
$$= P\Big(\{V_{M(n)} \le x\} \cap B_k\Big) - P\Big(\{V_{M(n)} \le x\} \cap A_{k,n}^c \cap B_k\Big)$$
$$+ P\Big(\{V_{M(n)} \le x\} \cap A_{k,n} \cap B_k^c\Big).$$

Note that the last two terms converge to zero as $n \to +\infty$ and k remains fixed, entailing that (for fixed k), as $n \to +\infty$,

$$P\Big(V_{M(n)} \le x | A_{k,n}\Big) P\Big(A_{k,n}\Big) - P\Big(V_{M(n)} \le x | B_k\Big) P\Big(B_k\Big) \to 0. \tag{2.39}$$

We now apply condition (C2) (or Kolmogorov's inequality when $V(\log |X_j|)$ are finite and the normalized terms are asymptotically negligible). This allows us to change $V_{M(n)}$ to $V_{e_{k,n}}$ in (2.39), where the subscript $e_{k,n}$ means the integer closest to it (we could choose just any integer close to it). We refer to this new version of (2.39) as (2.39a). Finally, we reduce the infinite sums in (2.37) and (2.38) to finite ones by summing over those k for which $h_k \in K(\alpha, \beta)$. With a fixed division $\{h_k\}$ (it was initially arbitrary, but (C2) may force us to make the division finer; summation over members of $K(\alpha, \beta)$ keeps the number of terms in the sum finite nonetheless). In these finite sums, we can apply Theorem 2.8 to the terms

$$P\Big(V_{e_{k,n}} \le x | B_k\Big) \to G(x). \tag{2.40}$$

Hence, by the choice of $K(\alpha,\beta)$, and by (2.37)-(2.38), (2.39a) and (2.40), as $n \to +\infty$, the lim sup of

$$\left| P\Big(V_{M(n)} \le x\Big) - G(x) \sum_{h_k \in K(\alpha,\beta)} P(B_k) \right| \le \epsilon,$$

that is, $V_{M(n)}$ converges completely to $G(x)$. The proof is complete. ■

2.5 Triangular arrays and infinite divisibility

This section is devoted to convergence of products of random variables arising from triangular arrays. The notation and assumption of this section are the following. Let X_{nj} be a double indexed sequence of random variables, where $1 \le j \le k_n$, k_n finite, $n \ge 1$, and $k_n \to +\infty$ with n. For fixed n, the random variables X_{nj}, $1 \le j \le k_n$, are independent. We call the collection $\{X_{nj}\}$ a triangular array of random variables, or simply a triangular array. Throughout this section we assume that the variables X_{nj} are *product-negligible*: for every $\epsilon > 0$,

$$\max_{1 \le j \le k_n} P\left[\Big(|X_{nj}| - 1\Big)^2 \ge \epsilon\right] \to 0 \qquad \text{as } n \to +\infty. \tag{2.41}$$

Definition. The distribution function $F(x)$ of a random variable X is called product-infinitely divisible if there exists a sequence of positive integers n_i, $n_i \to +\infty$ with i, such that for every i, $X \stackrel{d}{=} X_{n_i 1} X_{n_i 2} \cdots X_{n_i n_i}$, where the $X_{n_i j}$, $1 \le j \le n_i$, are independent identically distributed.

We denote the collection of all product-infinitely divisible distribution functions by $\mathcal{M}$.

Elements in $\mathcal{M}$ can be further differentiated in terms of the parity of the cardinality of their decompositions. We say that a random variable X has an even decomposition if there exist i.i.d. random variables $X_{2t,j}$, $1 \le j \le 2t$, such that $X \stackrel{d}{=} X_{2t,1} X_{2t,2} \cdots X_{2t,2t}$.

We denote by $\mathcal{M}'$ all distribution functions in $\mathcal{M}$ that permit even decompositions with arbitrary large number of factors. That is, X has distribution function in $\mathcal{M}'$, if, and only if, $X \stackrel{d}{=} X_1X_2\cdots X_{2t}$, where the X_is are i.i.d., for an infinite number of t.

The aim of this section is to present a theory of complete convergence of the distribution function of $X_{n1}X_{n2}\cdots X_{nk_n}$, under assumption (2.41). We are going to show that the collection of all such limiting distributions coincides with $\mathcal{M}$. For related work without assumption (2.41) see Abramov [3].

If one deals with a single sequence of independent random variables, $X_1, X_2, \cdots$, then property (2.41) can be obtained by exponential normalization. That is, for an arbitrary sequence of real numbers b_n, $b_n \to +\infty$ with n, we define $X_{nj} = |X_j|^{1/b_n} sign(X_j)$, and convergence of the distribution function of

$$|X_1X_2\cdots X_n|^{\frac{1}{b_n}} sign(X_1X_2\cdots X_n) = X_{n1}X_{n2}\cdots X_{nn}$$

(see Theorem 1.28) can be viewed as a special case of the problem considered in this section.

As we have already noticed in Section 2.2, there is a connection between the convergence of $X_{n1}X_{n2}\cdots X_{nk_n}$ and the convergence of $\sum_{j=1}^{k_n} \log|\tilde{X}_{nj}|$. To exploit this to its full extent we introduce the following definitions and results.

Let $\{X_{nj}\}$ be a triangular array. The X_{nj} are said to be *sum-negligible*, or to satisfy the *uniform asymptotic negligibility*, if for every $\epsilon > 0$,

$$\max_{1\le j\le k_n} P\left[\left(|X_{nj}| \ge \epsilon\right] \to 0 \qquad \text{as } n \to +\infty. \tag{2.42}$$

Definition. The distribution function $G(x)$ of a random variable X is said to be sum-infinitely divisible (or simply infinitely divisible) if for every $n \ge 2$, $X \stackrel{d}{=} X_{n1} + X_{n2} + \cdots + X_{nn}$, where X_{nj}, $1 \le j \le n$ are independent identically distributed.

We denote the collection of all infinitely divisible distribution functions by M.

Let $\{X_{nj}\}$ be a triangular array, and $G_n^*(x) = P(\sum_{j=1}^{k_n} X_{nj} \le x)$. It is a classical result in the theory of infinite triangular arrays (see

[58], Chapter 5) that, under assumption (2.42), the family of possible limiting distributions for $G_n^*(x)$ coincides with M.

An alternative characterization of M can be obtained by introducing Poisson-type distribution functions.

A random variable Y is *Poisson-type* if $Y \stackrel{d}{=} A + BX$, where A and $B \neq 0$ are constants, and X is a Poisson random variable. The distribution function of Y is called a Poisson-type distribution.

It is well known (see [58], Theorem 4, p. 204) that a distribution function $G(x)$ is in M if, and only if, $G(x)$ is the complete limit of a finite convolution of Poisson-type distributions.

We start our investigation on the asymptotic properties of the distribution function of $X_{n1}X_{n2}\cdots X_{nk_n}$ by introducing the following definition.

Definition. Two triangular arrays $\{X_{nj}\}$ and $\{X'_{nj}\}$ are said to be equivalent if for every n there is a bijection H from X'_{nj} to X_{nj}, $1 \leq j \leq k_n$, with the following properties:

(a) $H(X'_{nj}) = -X'_{nj} = X_{nj}$ or $H(X'_{nj}) = X_{nt}$ for some $t, 1 \leq t \leq k_n$,

(b) for every n, $H(X'_{nj}) = -X'_{nj} = X_{nj}$ for an even number of j.

From the above definition we have the following property, which we state as a lemma.

Lemma 2.3 *Let $\{X'_{nj}\}$ and $\{X_{nj}\}$ be two equivalent triangular arrays. Then for every n,*

$$\prod_{j=1}^{k_n} X'_{nj} \stackrel{d}{=} \prod_{j=1}^{k_n} X_{nj}.$$

PROOF. Clearly for every pair of random variables X and Y,

$$XY \stackrel{d}{=} (-X)(-Y).$$

This, the independence of the X_{nj}, $1 \leq j \leq k_n$, k_n finite, and the property that the distribution of $\prod_{j=1}^{k_n} X_{nj}$ is preserved by any rearrangement of the factors, immediately imply the validity of the lemma. ∎

Next we discuss the limiting distribution of $sign(X_{n1}X_{n2}\cdots X_{nk_n})$. As we have done in Chapter 1 in dealing with $sign(X_1X_2\cdots X_n)$, we first discuss the case in which the X_{nj}s are continuous at zero.

Theorem 2.12 *Let $\{X_{nj}\}$ be a triangular array with $P(X_{nj}=0)=0$ for $n\geq 1$, $1\leq j\leq k_n$. Let*

$$\lim_{n\to+\infty} P\Big(sign(X_{n1}\cdots X_{nk_n})<0\Big)=P. \tag{2.43}$$

(i) P exists and is 1/2 if, and only if, either (a)

$$\liminf_{n\to+\infty}\sum_{j=1}^{k_n}\min\{P(X_{nj}<0),\,P(X_{nj}>0)\}=+\infty\,,$$

or (b) there exists a sequence X_{nj_n} such that

$$\lim_{n\to+\infty} P(X_{nj}<0)=\frac{1}{2},$$

or (c) there exists a partition of the nonnegative integers N, say $N=N_1\cup N_2$, such that

$$\liminf_{\substack{n\to+\infty\\ n\in N_1}}\sum_{j=1}^{k_n}\min\{P(X_{nj}<0),\,P(X_{nj}>0)\}=+\infty\,,$$

and a sequence X_{nj_n}, with $n\in N_2$, such that

$$\lim_{\substack{n\to+\infty\\ n\in N_2}} P(X_{nj_n}<0)=\frac{1}{2}.$$

(ii) P exists and is not equal to 1/2 if, and only if, there is a triangular array $\{X'_{nj}\}$ equivalent to $\{X_{nj}\}$, such that

$$\lim_{n\to+\infty} P\Big(sign(X'_{n1}\cdots X'_{nk_n})<0\Big)=P' \tag{2.44}$$

exists, $P'\neq 1/2$, and for all $\epsilon>0$, there exist a t^*, $t^*=t^*(\epsilon)$, and n_o, $n_o=n_o(\epsilon)$, such that for all $n\geq n_o$, there exist t_n, $1\leq t_n<k_n$, $k_n-t_n\to+\infty$, such that

$$\Big|P\Big(sign(X'_{n1}\cdots X'_{nt_n})<0\Big)-P\Big|\leq\epsilon, \tag{2.45}$$

$$\sum_{j=t_n+1}^{k_n} P(X'_{nj} < 0) < \epsilon, \tag{2.46}$$

and

$$P(X'_{nj} < 0) < \epsilon \quad \textit{for all } t^* < j \le k_n. \tag{2.47}$$

PROOF. Let $\nu_{nk_n} = sign(X_{n1}X_{n2}\cdots X_{nk_n})$. By proceeding as in the beginning of the proof of Theorem 1.21 we have that

$$\Big|P(\nu_{nk_n} = -) - P(\nu_{nk_n} = +)\Big| = \Big|\prod_{j=1}^{k_n}(1 - 2P(X_{nj} < 0))\Big|$$

$$= \prod_{j=1}^{k_n}\Big(1 - 2\min\{P(X_{nj} < 0),\, P(X_{nj} > 0)\}\Big),$$

which gives that

$$\lim_{n\to+\infty} P(\nu_{nk_n} = -) = \frac{1}{2}$$

if, and only if, for all $\epsilon > 0$ and all $T > 0$ there exists a N, $N = N(\epsilon, T)$, such that for all $n \ge N$ either there exists a X_{nj_n} such that

$$\Big|P(X_{nj_n} < 0) - \frac{1}{2}\Big| \le \epsilon,$$

or

$$\sum_{j=1}^{k_n} \min\{P(X_{nj} < 0),\, P(X_{nj} > 0)\} \ge T.$$

Since the above conditions are satisfied if, and only if, either (a) or (b) or (c) of part (i) of the theorem is satisfied, the validity of the first part of the theorem now follows.

Next we turn to the proof of part (ii). Assume that P in (2.43) exists and is not equal to 1/2. By again utilizing the calculation in the beginning of the proof of Theorem 1.21 we have

$$P(\nu_{nk_n} = -) - P(\nu_{nk_n} = +) = -\prod_{j=1}^{k_n}\Big(1 - 2P(X_{nj} < 0)\Big),$$

which implies the existence of n_1 such that for every $n \geq n_1$, $P(X_{nj} < 0) > 1/2$ for an even number of j if $P < 1/2$, and $P(X_{nj} < 0) > 1/2$ for an odd number of j if $P > 1/2$. A triangular array equivalent to X_{nj} which satisfies the theorem is constructed as follows. If $n < n_1$, let $Y_{nj} = X_{nj}$. If $n \geq n_1$ and $P < 1/2$, let $Y_{nj} = -X_{nj}$, if $P(X_{nj} < 0) > 1/2$, and $Y_{nj} = X_{nj}$ otherwise; if $n \geq n_1$ and $P > 1/2$, let $Y_{nj} = -X_{nj}$ if $P(X_{nj} < 0) > 1/2$ and $P(X_{nl} < 0) > 1/2$ for some $1 \leq l < j$, $Y_{nj} = X_{nj}$ otherwise. We now relabel the Y_{nj}, say X'_{nj}, in such a way that if $j < i$, $P(X'_{nj} < 0) \geq P(X'_{ni} < 0)$. Clearly X'_{nj} is equivalent to X_{nj}, and (2.44) holds. Since $P' \neq 1/2$, part (i) of the theorem implies

$$\limsup_{n\to+\infty} \sum_{j=1}^{k_n} \min\Big\{P(X_{nj} < 0),\, P(X_{nj} > 0)\Big\} < +\infty, \tag{2.48}$$

because, should the value in (2.48) be infinity, we could get $P' = 1/2$ on a subsequence, contradicting the existence of the limit P' and that this limit is not 1/2. Hence, (2.48) and the existence of the limit in (2.44) immediately imply the validity of (2.45), (2.46), and (2.47). The converse trivially holds, and therefore the proof is complete. ∎

Since by Lemma 2.3 equivalent triangular arrays do not effect the distribution function of $X_{n1}X_{n2}\cdots X_{nk_n}$, it is often convenient to assume that X_{nj}, $n \geq 1$, $1 \leq j \leq k_n$, satisfy the properties of the triangular array $\{X'_{nj}\}$ constructed at the end of the proof of Theorem 2.12. Whenever this is the case, we say that $\{X_{nj}\}$ is in standard form.

Next we extend the previous theorem to the general case.

Theorem 2.13 *Let $\{X_{nj}\}$ be a triangular array.*

$$(i) \quad \lim_{n\to+\infty} P\big(sign(X_{n1}X_{n2}\cdots X_{nk_n}) = 0\big) = 1$$

if, and only if,

$$\liminf_{n\to+\infty} \sum_{j=1}^{k_n} P(X_{nj} = 0) = +\infty. \tag{2.49}$$

(ii) Assume

$$\limsup_{n\to+\infty}\sum_{j=1}^{k_n} P(X_{nj}=0) < +\infty. \tag{2.50}$$

Let $\eta = -, 0, +$. *Then the following holds.*

(c) $$\lim_{n\to+\infty} P\big(sign(X_{n1}X_{n2}\cdots X_{nk_n}=\eta\big) = P(\nu=\eta),$$

exists and is symmetric if, and only if, either

$$\liminf_{n\to+\infty}\sum_{j=1}^{k_n}\min\{P(X_{nj}<0),\, P(X_{nj}>0)\} = +\infty\,,$$

or there exists a sequence $\{X_{nj_n}\}$ *such that*

$$\lim_{n\to+\infty}\Big(P(X_{nj}<0) - P(X_{nj}>0)\Big) = 0,$$

or there exists a partition of the positive integers N, *say* $N = N_1 \cup N_2$, *such that*

$$\liminf_{\substack{n\to+\infty\\ n\in N_1}}\sum_{j=1}^{k_n}\min\{P(X_{nj}<0),\, P(X_{nj}>0)\} = +\infty\,,$$

and a sequence $\{X_{nj_n}\}$, $n \in N_2$, *such that*

$$\lim_{\substack{n\to+\infty\\ n\in N_2}}\Big(P(X_{nj}<0) - P(X_{nj}>0)\Big) = 0.$$

(d) $$\lim_{n\to+\infty} P\big(sign(X_{n1}X_{n2}\cdots X_{nk_n}=\eta\big) = P(\nu=\eta),$$

exists and it is not symmetric if, and only if, there is a triangular array $\{X'_{nj}\}$ *equivalent to* $\{X_{nj}\}$, *such that*

$$\lim_{n\to+\infty} P\big(sign(X'_{n1}X'_{n2}\cdots X'_{nk_n}=\eta\big) = P(\nu'=\eta)$$

exists, is not symmetric, and for all $\epsilon > 0$, *there exist a* t^*, $t^* = t^*(\epsilon)$, *and* n_o, $n_o = n_o(\epsilon)$, *such that for all* $n \geq n_o$, *there exist* t_n, $1 \leq t_n < k_n$, $k_n - t_n \to +\infty$, *such that*

$$\Big|P\Big(sign(X'_{n1}\cdots X'_{nt_n}) = \eta\Big) - P(\nu'=\eta)\Big| \leq \epsilon, \tag{2.51}$$

$$\sum_{j=t_n+1}^{k_n} P(X'_{nj} < 0) < \epsilon, \tag{2.52}$$

and

$$P(X'_{nj} < 0) < \epsilon \quad \textit{for all } t^* < j \leq k_n. \tag{2.53}$$

(iii) If

$$\liminf_{n \to +\infty} \sum_{j=1}^{k_n} P(X_{nj} = 0) < +\infty, \tag{2.54}$$

but

$$\limsup_{n \to +\infty} \sum_{j=1}^{k_n} P(X_{nj} = 0) = +\infty, \tag{2.55}$$

then

$$\lim_{n \to +\infty} P\Big(sign(X_{n1}X_{n2}\cdots X_{nk_n}) = \eta\Big)$$

does not exist.

PROOF. We follow the steps of the proof of Theorem 1.22 and consider $P(\nu_{nk_n} = 0)$.

$$P(\nu_{nk_n} = 0) \;=\; 1 - P(\nu_{nk_n} \neq 0) = 1 - \prod_{j=1}^{k_n} \big(1 - P(X_{nj} = 0)\big),$$

which gives that $\lim_{n \to +\infty} P(\nu_{nk_n} = 0) = 1$ if, and only if, (2.49) is satisfied.

Next we turn to the proof of (ii), and consider $\tilde{X}_{nj}$. If $P(X_{nj} = 0) = 1$, $\tilde{X}_{nj} \overset{d}{=} 1$, and consequently for such $\tilde{X}_{nj}$, $P(sign(\tilde{X}_{nj}) = +) = 1$. The validity of part (ii) now follows by applying Theorem 2.12 to $\tilde{X}_{nj}$, and observing that the conditions Theorem 2.12 imposes on $\tilde{X}_{nj}$ are equivalent to the validity of the conditions expressed in parts (c) and (d) of (ii) in Theorem 2.12.

Clearly if (2.54) and (2.55) both hold, then by using part (i) and (ii), we can find two subsequences for which $P(\nu_{nk_n} = 0)$ has two different limit points, thus proving the validity of (iii). The proof is now complete. ∎

Next we discuss convergence of triangular arrays. Its theory was developed in a sequence of papers by Zolotarev [159], Bakštys [15] and Abramov [2] and [1], and their work is fully covered in the book by Bareikis and Šiaulys [22]. Our approach is different from the above works.

We start with the following statement.

Theorem 2.14 *Let* $\{X_{nj}\}$ *be a triangular array, and assume that* (2.41) *is satisfied. There exists a random variable* Y *such that*

$$X_{n1}X_{n2}\cdots X_{nk_n} \to Y \qquad \textit{completely}$$

if, and only if there exists a random variable V *such that*

$$Z_{n1}Z_{n2}\cdots Z_{nk_n} \to V \qquad \textit{completely,}$$

where $sign(Z_{nj}) \stackrel{d}{=} sign(X_{nj})$, $|Z_{nj}| \stackrel{d}{=} |X_{nj}|$, *and* $sign(Z_{nj})$ *and* $|Z_{nj}|$ *are independent. Whenever the above limits exist,* $Y \stackrel{d}{=} V$.

The above theorem is an easy consequence of the following theorem.

Theorem 2.15 *Let* $\{X_{nj}\}$ *be a triangular array that satisfies property* (2.41).

(i) If

$$\liminf_{n\to+\infty} \sum_{j=1}^{k_n} P(X_{nj}=0) = +\infty, \tag{2.56}$$

then

$$\lim_{n\to+\infty} P\Big(X_{n1}X_{n2}\cdots X_{nk_n}=0\Big) = 1. \tag{2.57}$$

(ii) Assume

$$\limsup_{n\to+\infty} \sum_{j=1}^{k_n} P(X_{nj}=0) < +\infty, \tag{2.58}$$

and $sign(X_{n1}X_{n2}\cdots X_{nk_n})$ *converges completely to a random variable* ν. *Then*

$X_{n1}X_{n2}\cdots X_{nk_n}$ *converges completely to a random variable* Y

if, and only if, there exists a random variable T *such that*

$$|X_{n1}X_{n2}\cdots X_{nk_n}| \to T \quad \textit{completely.}$$

Whenever the above limits exist, Y *is always of the form* $Y = \nu T$, ν *and* T *independent. Moreover, in the case* $Y \overset{d}{\neq} 0$,

$$\left|\tilde{X}_{n1}\tilde{X}_{n2}\cdots\tilde{X}_{nk_n}\right| \to \tilde{T} \qquad \textit{completely,} \quad P(\tilde{T}=0)=0.$$

Proof. Let $\{X_{nj}\}$ be a triangular array satisfying property (2.41). The validity of part (i) follows from part (i) of Theorem 2.13. So only part (ii) needs to be proved. We start with the sufficiency of the conditions. Let us assume that (2.58) holds, and that

$$sign(X_{n1}X_{n2}\cdots X_{nk_n}) \to \nu \quad \text{completely.}$$

We first consider the case $|X_{n1}X_{n2}\cdots X_{nk_n}|$ converges completely to a random variable T. If $P(T=0)=1$, then $X_{n1}X_{n2}\cdots X_{nk_n}$ also converges to 0 and we are done. So let us assume that $P(T=0)<1$. Let $W_{nk_n} = |\tilde{X}_{n1}\tilde{X}_{n2}\cdots\tilde{X}_{nk_n}|$. Since we are assuming that $sign(X_{n1}X_{n2}\cdots X_{nk_n})$ converges, we also have that W_{nk_n} converges completely to a random variable, which we denote by U. Hence the Mellin transform of W_{nk_n} converges to the Mellin transform of U. That is, for all real t,

$$\begin{aligned}\mathcal{M}_{W_{nk_n}}(t) &= \prod_{j=1}^{k_n}\mathcal{M}_{\tilde{X}_{nj}}(t) = \prod_{j=1}^{k_n} E\left[|\tilde{X}_{nj}|^{it}\right]\\ &= \prod_{j=1}^{k_n} E\left[e^{it\log|\tilde{X}_{nj}|}\right] \to \mathcal{M}_U(t) \quad \text{as } n\to+\infty.\end{aligned}$$

Since $E[e^{it\log|\tilde{X}_{nj}|}]$ can be viewed as the characteristic function of $\log|\tilde{X}_{nj}|$, the validity of the above limit for all real t implies the complete convergence of $\sum_{j=1}^{k_n}\log|\tilde{X}_{nj}|$ to a random variable, say

U_1, which in distribution must be equal to $\log|U|$. Hence $U > 0$, and $U \stackrel{d}{=} \tilde{T}$.

Let $\tilde{\nu}_{nk_n} = sign(\tilde{X}_{n1}\tilde{X}_{n2}\cdots\tilde{X}_{nk_n})$, and put

$$\lim_{n\to+\infty} P\Big(\tilde{\nu}_{nk_n} = +\Big) = q, \quad \lim_{n\to+\infty} P\Big(\tilde{\nu}_{nk_n} = -\Big) = 1 - q = p.$$

We first consider the case $q \neq p$, and, for convenience, we further assume that $\{X_{nj}\}$ is in standard form. For arbitrary n and $0 \le t \le k_n$, let

$$A_{n0} = \Big\{\omega : \tilde{X}_{nj} > 0 \text{ for } 1 \le j \le k_n\Big\},$$

and for $1 \le t \le k_n$,

$$A_{nt} = \Big\{\omega : \tilde{X_{nt}} < 0 \text{ and } \tilde{X_{nj}} > 0 \text{ if } t < j \le j_n\Big\}.$$

Then

$$P(\tilde{\nu}_{nk_n} = +) = \sum_{t=0}^{k_n} P\Big(\{\tilde{\nu}_{nk_n} = +\} \cap A_{nt}\Big).$$

Let $q_n(t) = P(\{\tilde{\nu}_{nk_n} = +\} \cap A_{nt})$, and consider $q_n(t)$ as a double indexed sequence on the positive integers. Since $q_n(t) \le 1$, we can find a subsequence $q_{n_1(i)}(t)$ such that $q_{n_1(i)}(0) \to q_\infty(0)$, as $i \to +\infty$, for some number $q_\infty(0)$. Now we consider the sequence $q_{n_1(i)}(t)$, and find a subsequence $q_{n_2(i)}(t)$ such that $q_{n_2(i)}(1) \to q_\infty(1)$, as $i \to +\infty$, for some number $q_\infty(1)$. Proceeding inductively, if $q_{n_k(i)}(t)$ is defined, we find a sequence $q_{n_{k+1}(i)}(t)$ with the following properties: (a) $q_{n_{k+1}(i)}(t)$ is a subsequence of $q_{n_k(i)}(t)$, and (b) $q_{n_{k+1}(i)}(t) \to q_\infty(t)$, as $i \to +\infty$, for $0 \le t \le k+1$. To obtain a sequence $q_{\hat{n}_i}(t)$ converging for every t, we take the diagonal sequence $\hat{n}_i = n_{i(i)}$. Then

$$\lim_{i\to+\infty} q_{\hat{n}_i}(t) = q_\infty(t) \text{ for } t \ge 0.$$

We claim

$$\begin{aligned} q = \lim_{n\to+\infty} P(\tilde{\nu}_{nk_n} = +) &= \lim_{i\to+\infty} P(\tilde{\nu}_{\hat{n}_i k_{\hat{n}_i}} = +) \\ &= \lim_{i\to+\infty} \sum_{t=0}^{k_{\hat{n}_i}} P\Big(\{\tilde{\nu}_{\hat{n}_i k_{\hat{n}_i}} = +\} \cap A_{\hat{n}_i t}\Big) \\ &= \sum_{t=0}^{+\infty} q_\infty(t) = q_\infty. \end{aligned}$$

Let $\epsilon > 0$ be given. Then there is a N_o, $N_o = N_o(\epsilon)$, such that for $n \geq N_o(\epsilon)$,

$$\Big|P\Big(\tilde{\nu}_{nk_n} = +\Big) - q\Big| < \frac{\epsilon}{2}.$$

Let $N \geq N_o$, and consider $\sum_{t=0}^{+\infty} q_\infty(t)$. Let $\hat{n}_l$ be such that $k_{\hat{n}_l} \geq N$ and

$$\Big|q_\infty(t) - q_{\hat{n}_l}(t)\Big| < \frac{\epsilon}{2(N+1)}, \qquad \text{for } 0 \leq t \leq N.$$

Then

$$\sum_{t=0}^{N} q_\infty(t) < \sum_{t=0}^{N} q_{\hat{n}_l}(t) + \frac{\epsilon}{2} \leq \sum_{t=0}^{k_{\hat{n}_l}} q_{\hat{n}_l}(t) + \frac{\epsilon}{2} \leq q + \epsilon.$$

Hence for all sufficiently large N,

$$\sum_{t=0}^{N} q_\infty(t) < q + \epsilon,$$

which implies $q_\infty \leq q + \epsilon$.

Next let $L > 0$ be arbitrary. Then there exists $N_1 > N_o(\epsilon)$, where $N_o(\epsilon)$ is as before, such that if $\hat{n}_l > N_1$ and $k_{\hat{n}_l} \geq N_1$

$$|q_{\hat{n}_l}(t) - q_\infty(t)| \leq \frac{\epsilon}{2^{t+2}}, \quad 0 \leq t \leq L.$$

Hence if $\hat{n}_l$ satisfies the above property,

$$\begin{aligned} q &\leq P\Big(\tilde{\nu}_{\hat{n}_l k_{\hat{n}_l}} = +\Big) + \frac{\epsilon}{2} = \sum_{t=0}^{k_{\hat{n}_l}} q_{\hat{n}_l}(t) + \frac{\epsilon}{2} \\ &\leq \sum_{t=0}^{L} q_\infty(t) + \sum_{t=L+1}^{k_{\hat{n}_l}} q_{\hat{n}_l}(t) + \sum_{t=0}^{L} \frac{\epsilon}{2^{t+2}} + \frac{\epsilon}{2}, \end{aligned}$$

which gives

$$q \leq q_\infty + \sum_{t=L+1}^{k_{\hat{n}_l}} q_{\hat{n}_l}(t) + \epsilon. \tag{2.59}$$

Note that $\lim_{l\to+\infty} \sum_{t=T+1}^{k_{\hat{n}_l}} q_{\hat{n}_l}(t)$ exists, is finite, non increasing and, by (2.48), eventually decreasing in L. Hence by letting $l \to +\infty$ and then $L \to +\infty$, (2.59) gives $q \leq q_\infty + \epsilon$.

Hence for arbitrary $\epsilon > 0$,

$$q \leq q_\infty + \epsilon \leq q + 2\epsilon,$$

which gives $q = q_\infty$.

Clearly if we define $p_n(t) = P\big(\{\tilde{\nu}_{nk_n} = -\} \cap A_{nt}\big)$, our previous results give

$$\lim_{i\to+\infty} p_{\hat{n}_i}(t) = p_\infty(t), \quad t \geq 0,$$

for some numbers $p_\infty(t)$, and

$$\sum_{t=0}^{+\infty} p_\infty(t) = p_\infty = p. \tag{2.60}$$

It is important to observe that from any given subsequence n' of n one can construct a diagonal subsequence $\hat{n}'_i$ which has the same properties as $\hat{n}_i$ (simply repeat the procedure leading to $\hat{n}_i$ by replacing n with n'). That is, by using the same notation as before,

$$\lim_{i\to+\infty} q_{\hat{n}'_i}(t) = q'_\infty(t) \text{ for } t \geq 0, \text{ and } \sum_{t=0}^{+\infty} q'_\infty(t) = q'_\infty = q.$$

It is of course possible that for specific t, $q'_\infty(t) \neq q_\infty(t)$. However the sum of the $q'_\infty(t)$ will still be q (simply repeat the same proof with $\hat{n}'_i$ replacing $\hat{n}_i$).

Since we are assuming that $|X_{n1}X_{n2}\cdots X_{nk_n}|e^{-a_n}$ converges completely to T, then $|\tilde{X}_{n1}\tilde{X}_{n2}\cdots\tilde{X}_{nk_n}|\,e^{-a_n}$ also converges completely to a random variable. We denote this latter limit by W. Let $W_{nk_n} = |\tilde{X}_{n1}\tilde{X}_{n2}\cdots\tilde{X}_{nk_n}|\,e^{-a_n}$. We claim that W_{nk_n} has the same limiting distribution as V_{nk_n}, where V_{nk_n} denotes a random variable with distribution function

$$P\Big(V_{nk_n} \le x\Big) = P\Big(W_{nk_n} \le x \Big| \tilde{X}_{nj} > 0,\, j \ge 2\Big).$$

Note that since we are assuming that $\{X_{nj}\}$ is in standard form, then $\{\tilde{X}_{nj}\}$ is also in standard form, and therefore $P(\tilde{X}_{nj} > 0) > 0$ for $j \ge 2$. Hence the distribution function of V_{nk_n} is well defined for all n.

To prove our claim we consider an arbitrary continuity point a of the distribution function of W, and show that

$$\lim_{n\to+\infty} E\left[I_{\{W_{nk_n}\le a\}}\right] = \lim_{n\to+\infty} E\left[I_{\{V_{nk_n}\le a\}}\right], \tag{2.61}$$

where $I_{\{\cdots\}}$ denotes the indicator of the event in $\{\cdots\}$. To this end we take an arbitrary convergent subsequence of the right hand side of (2.61), say

$$\lim_{i\to+\infty} E[I_{\{V_{n'_i,k_{n'_i}}\le a\}}] = l_a,$$

for some number l_a, and from n'_i construct a diagonal sequence $\hat{n}'_i$, with the same properties as $\hat{n}_i$ (see the comment after (2.60)). Now

$$\begin{aligned}\lim_{n\to+\infty} E\left[I_{\{W_{nk_n}\le a\}}\right] &= \lim_{i\to+\infty} E\left[I_{\{W_{\hat{n}'_i k_{\hat{n}'_i}}\le a\}}\right]\\ = \quad \lim_{i\to+\infty}\Bigg(E&\left[I_{\{W_{\hat{n}'_i k_{\hat{n}'_i}}\le a\}} I_{\{\tilde{\nu}_{\hat{n}'_i k_{\hat{n}'_i}}=+\}}\right]\\ &+E\left[I_{\{W_{\hat{n}'_i k_{\hat{n}'_i}}\le a\}} I_{\{\tilde{\nu}_{\hat{n}'_i k_{\hat{n}'_i}}=-\}}\right]\Bigg),\end{aligned} \tag{2.62}$$

and

$$\begin{aligned}&\lim_{i\to+\infty} E\left[I_{\{W_{\hat{n}'_i k_{\hat{n}'_i}}\le a\}} I_{\{\tilde{\nu}_{\hat{n}'_i k_{\hat{n}'_i}}=+\}}\right]\\ = \quad &\lim_{i\to+\infty}\sum_{t=0}^{k_{\hat{n}'_i}} E\left[I_{\{W_{\hat{n}'_i k_{\hat{n}'_i}}\le a\}} I_{\{\tilde{\nu}_{\hat{n}'_i k_{\hat{n}'_i}}=+\}} I_{A_{\hat{n}'_i t}}\right].\end{aligned} \tag{2.63}$$

From our assumptions on n_i' and $\hat{n}_i'$ we obtain

$$\begin{aligned} & \lim_{i\to+\infty} E\left[I_{\{W_{\hat{n}_i' k_{\hat{n}_i'}}\le a\}} I_{\{\tilde{\nu}_{\hat{n}_i' k_{\hat{n}_i'}}=+\}} I_{A_{\hat{n}_i' t}}\right] \qquad (2.64) \\ = & \lim_{i\to+\infty} E\left[I_{\{W_{\hat{n}_i' k_{\hat{n}_i'}}\le a\}} \Big| I_{\{\tilde{\nu}_{\hat{n}_i' k_{\hat{n}_i'}}=+\}} I_{A_{\hat{n}_i' t}}\right] q_{\hat{n}_i'}(t) \\ = & l_a q'_\infty(t), \end{aligned}$$

for $t > 0$, where in the derivation of the last equation we used the independence of the X_{nj}, and property (2.41). It is easy to see that $l_a q'_\infty(0)$ also describes the value of (2.64) in the case $t = 0$. This,

$$E\left[I_{\{W_{\hat{n}_i' k_{\hat{n}_i'}}\le a\}} I_{\{\tilde{\nu}_{\hat{n}_i' k_{\hat{n}_i'}}=+\}} I_{A_{\hat{n}_i' t}}\right] \le q_{\hat{n}_i'}(t),$$

and the convergence of $\sum_{t=0}^{+\infty} q_{\hat{n}_i'}(t)$ justify interchanging sum and limit in (2.63) obtaining

$$\begin{aligned} & \lim_{i\to+\infty} \sum_{t=0}^{k_{\hat{n}_i'}} E\left[I_{\{W_{\hat{n}_i' k_{\hat{n}_i'}}\le a\}} I_{\{\tilde{\nu}_{\hat{n}_i' k_{\hat{n}_i'}}=+\}} I_{A_{\hat{n}_i' t}}\right] \\ = & l_a \sum_{t=0}^{+\infty} q_{\hat{n}_i'}(t) = l_a q. \end{aligned}$$

If we repeat the same type of calculations to compute the value of (2.62) as i goes to infinity we obtain

$$\lim_{i\to+\infty} E\left[I_{\{W_{\hat{n}_i' k_{\hat{n}_i'}}\le a\}} I_{\{\tilde{\nu}_{\hat{n}_i' k_{\hat{n}_i'}}=-\}}\right] = l_a p,$$

thus proving (2.61).

Now that we have proved that $|\tilde{X}_{n1}\tilde{X}_{n2}\cdots\tilde{X}_{nk_n}|\, e^{-a_n}$ and V_{nk_n} have the same limiting distribution, from the previous calculations we immediately obtain that for continuity point a of the distribution function of W,

$$E\left[I_{\{|\tilde{X}_{n1}\tilde{X}_{n2}\cdots\tilde{X}_{nk_n}|e^{-a_n}\le a\}} I_{\{\tilde{\nu}_{nk_n}=+\}}\right] = qE\left[I_{\{W\le a\}}\right],$$

thus proving the complete convergence of $\tilde{X}_{n1}\tilde{X}_{n2}\cdots\tilde{X}_{nk_n}e^{-a_n}$ to a random variable of the form described by the theorem.

Next we consider the case $p = q = 1/2$. For any given t elements $\tilde{X}_{nj_i}$, $i = 1, 2, \cdots, t$, of X_{nj}, $1 \le j \le k_n$, put

$$B_{nt} = \left\{\tilde{X}_{nj_1}\tilde{X}_{nj_2}\cdots\tilde{X}_{nj_t} < 0\right\}. \tag{2.65}$$

Then, with W_{nk_n} as before,

$$\begin{aligned} &\left|E\left[I_{\{W_{nk_n}\le a\}}I_{\{\nu_{nk_n}=+\}}\right] - E\left[I_{\{W_{nk_n}\le a\}}I_{\{\nu_{nk_n}=-\}}\right]\right| = \quad (2.66)\\ = \;& \left|E\left[I_{\{W_{nk_n}\le a\}}I_{\{\nu_{nk_n}=+\}}I_{B_{nt}}\right] + E\left[I_{\{W_{nk_n}\le a\}}I_{\{\nu_{nk_n}=+\}}I_{B^c_{nt}}\right]\right.\\ &\left.-E\left[I_{\{W_{nk_n}\le a\}}I_{\{\nu_{nk_n}=-\}}I_{B_{nt}}\right] - E\left[I_{\{W_{nk_n}\le a\}}I_{\{\nu_{nk_n}=-\}}I_{B^c_{nt}}\right]\right|. \end{aligned}$$

Let $W'_{nk_n} = |\tilde{X}_{nt+1}\tilde{X}_{nt+2}\cdots\tilde{X}_{nk_n}|$. Then

$$\begin{aligned} &\lim_{n\to+\infty} E\left[I_{\{W_{nk_n}\le a\}}I_{\{\nu_{nk_n}=+\}}I_{B_{nt}}\right]\\ = \;& \lim_{n\to+\infty} E\left[I_{\{W_{nk_n}\le a\}}I_{\{sign(X_{nj_{t+1}}\cdots X_{nk_n})=-\}}I_{B_{nt}}\right]\\ = \;& \lim_{n\to+\infty} E\left[I_{\{W'_{nk_n}\le a\}}I_{\{sign(X_{nj_{t+1}}\cdots X_{nk_n})=-\}}I_{B_{nt}}\right] \quad (2.67)\\ = \;& \lim_{n\to+\infty} E\left[I_{\{W'_{nk_n}\le a\}}I_{\{sign(X_{nj_{t+1}}\cdots X_{nk_n})=-\}}\right]E\left[I_{B_{nt}}\right] \quad (2.68) \end{aligned}$$

where in (2.67) we used (2.41), and in (2.68) the independence of X_{nj}, $1 \le j \le k_n$. On the other hand, by using the same notation as above,

$$\begin{aligned} &\lim_{n\to+\infty} E\left[I_{\{W_{nk_n}\le a\}}I_{\{\nu_{nk_n}=-\}}I_{B^c_{nt}}\right]\\ = \;& \lim_{n\to+\infty} E\left[I_{\{W_{nk_n}\le a\}}I_{\{sign(X_{nj_{t+1}}\cdots X_{nk_n})=-\}}I_{B^c_{nt}}\right]\\ = \;& \lim_{n\to+\infty} E\left[I_{\{W'_{nk_n}\le a\}}I_{\{sign(X_{nj_{t+1}}\cdots X_{nk_n})=-\}}I_{B^c_{nt}}\right]\\ = \;& \lim_{n\to+\infty} E\left[I_{\{W'_{nk_n}\le a\}}I_{\{sign(X_{nj_{t+1}}\cdots X_{nk_n})=-\}}\right]E\left[I_{B^c_{nt}}\right]. \end{aligned}$$

Hence

$$\begin{aligned}
& \left|E\left[I_{\{W_{nk_n}\le a\}}I_{\{\nu_{nk_n}=+\}}I_{B_{nt}}\right]-E\left[I_{\{W_{nk_n}\le a\}}I_{\{\nu_{nk_n}=-\}}I_{B^c_{nt}}\right]\right| \\
= & \left|E\left[I_{\{W'_{nk_n}\le a\}}I_{\{sign(X_{nj_{t+1}}\cdots X_{nk_n})=-\}}\right]E\left[I_{B_{nt}}\right]\right. \\
& \left.-E\left[I_{\{W'_{nk_n}\le a\}}I_{\{sign(X_{nj_{t+1}}\cdots X_{nk_n})=-\}}\right]E\left[I_{B^c_{nt}}\right]\right|+\delta_n \\
\le & \left|E\left[I_{B_{nt}}\right]-E\left[I_{B^c_{nt}}\right]\right|+\frac{\delta_n}{2}
\end{aligned} \tag{2.69}$$

with δ_n going to zero as $n \to +\infty$. Similarly, the difference between the two other terms inside the absolute value following (2.66) is also bounded by the quantity in (2.69). Therefore

$$\begin{aligned}
& \left|E\left[I_{\{W_{nk_n}\le a\}}I_{\{\nu_{nk_n}=+\}}\right]-E\left[I_{\{W_{nk_n}\le a\}}I_{\{\nu_{nk_n}=-\}}\right]\right| \\
\le & 2\left|E\left[I_{B_{nt}}\right]-E\left[I_{B^c_{nt}}\right]\right|+\delta_n,
\end{aligned} \tag{2.70}$$

with $\delta_n \to 0$ as n goes to infinity.

Let $\{n\}$ be a sequence that satisfies one of the following properties: (a) there exists $\tilde{X}_{nj_i}$ such that $P(\tilde{X}_{nj_n} < 0) \to 1/2$ as $n \to +\infty$, or, in the case (a) does not apply, (b) for every $M > 0$ there exists a finite t, $t = t(M)$, such that for some choice of t elements of X_{nj}, say $\tilde{X}_{nj_i}$, $i = 1, 2, \cdots, t$,

$$\liminf_{n\to+\infty}\sum_{i=1}^{t}\min\left\{P(\tilde{X}_{nj_i}<0), P(\tilde{X}_{nj_i}>0)\right\} > M.$$

Note that this last assumption implies that for every $\epsilon > 0$ we can find t elements of X_{nj}, say $\tilde{X}_{nj_i}$, $i = 1, 2, \cdots, t$, such that

$$\limsup_{n\to+\infty}\left|P(\tilde{X}_{nj_1}\cdots\tilde{X}_{nj_t}<0)-P(\tilde{X}_{nj_1}\cdots\tilde{X}_{nj_t}>0)\right|<\epsilon. \tag{2.71}$$

Let $\epsilon > 0$ be given. Put $B_{nt} = B_{n1} = \{\tilde{X}_{nj_n} < 0\}$ if (a) applies, whereas B_{nt} is as in(2.65) if (b) applies, where the $\tilde{X}_{nj_i}$ are chosen to satisfy (2.71). Then for all continuity points a of T, (2.70) and the choice of the B_{nt} entail

$$\begin{aligned}
& \limsup_{n\to+\infty}\left|E\left[I_{\{W_{nk_n}\le a\}}I_{\{\nu_{nk_n}=+\}}\right]-E\left[I_{\{W_{nk_n}\le a\}}I_{\{\nu_{nk_n}=-\}}\right]\right| \\
\le & \limsup_{n\to+\infty}2\left|E\left[I_{B_{nt}}\right]-E\left[I_{B^c_{nt}}\right]\right|+\delta_n \\
\le & 2\epsilon,
\end{aligned}$$

thus showing the convergence of $X_{n1}X_{n2}\cdots X_{nk_n}$ to the desired limit along any sequence $\{n\}$ satisfying (a) or (b). So let us now assume that there exists a sequence $\{n\}$ that does not satisfy property (a) and (b). This implies that for any finite t,

$$\begin{aligned}&\lim_{n\to+\infty} P(sign(X_{nt+1}X_{nt+2}\cdots X_{nk_n}) = -)\\ = &\lim_{n\to+\infty} P(sign(X_{nt+1}X_{nt+2}\cdots X_{nk_n}) = +).\end{aligned}$$

Let $p_{nj} = P(X_{nj} < 0)$, $q_{nj} = P(X_{nj} > 0)$, and denote by $F_{nj}(x)$ the distribution function of X_{nj}. Let $\{Z_{nj}\}$ be a triangular array such that

$$\mathcal{M}_{Z_{nj}}(t) = q_{nj}\mathcal{M}_{|X_{nj}|}(t) + \gamma p_{nj}\mathcal{M}_{|X_{nj}|}(t).$$

That is, for every n and j, $sign(Z_{nj}) \overset{d}{=} sign(X_{nj})$, $|Z_{nj}| \overset{d}{=} |X_{nj}|$, and $sign(Z_{nj})$ and $|Z_{nj}|$ are independent. Given our initial assumptions it is clear that $Z_n = Z_{n1}Z_{n2}\cdots Z_{nk_n}$ converges completely to a random variable, say Y, of the desired form. We claim that $X_n = X_{n1}X_{n2}\cdots X_{nk_n}$ also converges completely to Y. We use Mellin transforms, which allows us to use the same technique as the one developed for sums, except that x is to be replaced by $\log x$ (see [58], Section 5.3). As can be seen a crucial step in showing that the Mellin transforms of Z_n and of the product $X_{n1}X_{n2}\cdots X_{nk_n}$ have the same limit is to show

$$\lim_{n\to+\infty}\max_{1\le j\le k_n}\left|\mathcal{M}_{X_{nj}^{\eta}}(t) - \mathcal{M}_{Z_{nj}^{\eta}}(t)\right| = 0, \tag{2.72}$$

where $\eta = -$ or $+$. Let first $\eta = +$. Then

$$\begin{aligned}&\left|\mathcal{M}_{X_{nj}^{\eta}}(t) - \mathcal{M}_{Z_{nj}^{\eta}}(t)\right|\\ \le\ &\left|p_{nj}\int_0^{+\infty} cos(t\log x)\,dF_{nj}(x) - q_{nj}\int_{-\infty}^{0} cos(t\log x)\,dF_{nj}(x)\right|\\ &+\left|p_{nj}\int_0^{+\infty} sin(t\log x)\,dF_{nj}(x) - q_{nj}\int_{-\infty}^{0} sin(t\log x)\,dF_{nj}(x)\right|\\ =\ &\left|p_{nj}q_{nj} + \epsilon_{nj}^{(1)} - q_{nj}p_{nj} + \epsilon_{nj}^{(2)}\right| + \left|p_{nj}q_{nj} + \epsilon_{nj}^{(3)} - q_{nj}p_{nj} + \epsilon_{nj}^{(4)}\right|\\ \le\ &|\epsilon_{nj}^{(1)}| + |\epsilon_{nj}^{(2)}| + |\epsilon_{nj}^{(3)}| + |\epsilon_{nj}^{(4)}|.\end{aligned}$$

Since by (2.41),

$$\lim_{n\to+\infty}\max_{1\le j\le k_n}\{|\epsilon_{nj}^{(i)}|\}=0,\quad i=1,2,3,4,$$

we immediately have that (2.72) holds. Similar calculations give that (2.72) continues to hold if X_{nj}^+ and Z_{nj}^+ are replaced by X_{nj}^- and Z_{nj}^-, respectively. Now the proof for establishing that $X_n = X_{n1}X_{n2}\cdots X_{nk_n}$ converges to Y along any sequence $\{n\}$ that does not satisfy properties (a) and (b) can be completed in the same manner in which the limits for sums are established, so we omit further details. Since the two cases for $\{n\}$ we considered lead to the same limit, $X_{n1}X_{n2}\cdots X_{nk_n}$ converges completely to a random variable Y of the claimed form. This completes the proof of the case $p=q=1/2$.

The validity of the necessity of the conditions of part (ii) of Theorem 2.15 is trivial, and the proof of the theorem is now complete. ∎

It is important to notice that in the setting of Theorem 2.15 the limiting distribution of $X_{n1}X_{n2}\cdots X_{nk_n}$ is symmetric if, and only if, the limiting distribution function of $sign(X_{n1}X_{n2}\cdots X_{nk_n})$ is symmetric. Clearly this is not true in the case of the limiting distribution of $X_1X_2\cdots X_n$, where the X_i come from a single sequence. For example, let $X_1 \stackrel{d}{=} X$, where X is as in Example 1.16, and let $X_i \stackrel{d}{=} 1$, $i\ge 2$. Then $X_1X_2\cdots X_n$ converges completely to $Y \stackrel{d}{=} X_1$, $P(Y<0)=1/2$, but Y is not symmetric.

Theorem 2.15 has several interesting implications. We start with the following result which immediately follows from Theorem 2.15.

Theorems 2.15 and 2.14 can be used to describe convergence to special distributions by independently considering the convergence of $sign(X_{n1}X_{n2}\cdots X_{nk_n})$, see Theorem 1.26, and the convergence of the sum $\sum_{j=1}^{k_n}\log|\tilde{X}_{nj}|$, for which a large number of results are already available in the literature (see [58], Chapter 5). For a related work, with emphasis on convergence to the logarithmically normal distribution, see Bakštys [16] and [17]. We do not investigate in detail this problem, but we illustrate the importance of Theorems 2.15 and 2.14 with the following example.

Example 2.5 Let $Y_j = -e,-1,1,e$ with $P(Y_j<0)=p_1$, $P(|Y_j| = e) = p_2$, $0<p_2<1$, $sign(Y_j)$ independent of $|Y_j|$. Let $Z_{nij} =$

$|Y_j|^{1/\sqrt{n_i p_2(1-p_2)}} e^{-p_2/\sqrt{n_i p_2(1-p_2)}} sign(Y_j)$, $1 \leq j \leq n_i$, $n_i \to +\infty$ with i, n_i even if $p_1 = 1$. Then (2.41) is satisfied, and we immediately have that the distribution function of $Z_{n_i1} Z_{n_i2} \cdots Z_{n_in_i}$ converges completely to a logarithmically normal distribution. The limit is symmetric if, and only if, $0 < p_1 < 1$.

On the other hand, if we let $Z_{nj} = Y_j$ with the further assumption that $P(|Z_{n_j}| = e) = \frac{a}{n}$, for some fixed positive number a, then the distribution function of $Z_{n_i1} Z_{n_i2} \cdots Z_{n_in_i}$ converges completely to a logarithmically Poisson distribution.

Theorem 2.14 extends the above results to any triangular array $\{X_{n_ij}\}$ that satisfies (2.41), and such that $sign(X_{n_ij}) \stackrel{d}{=} sign(Z_{n_ij})$ and $|X_{n_ij}| \stackrel{d}{=} |Z_{n_ij}|$.

Another consequence of Theorem 2.15 is that the collection of all possible limiting distribution functions of $X_{n1} X_{n2} \cdots X_{nk_n}$ consists of all those distribution functions whose Mellin transforms is of the form

$$\mathcal{M}_X(t) = q \mathcal{M}_{|X|}(t) + \gamma p \mathcal{M}_{|X|}(t), \tag{2.73}$$

where q and p are nonnegative real numbers, $p + q \leq 1$, and $W = \log|\tilde{X}| \in M$.

Next we explore the connection between distribution functions which satisfy (2.73) and distribution functions which are product-infinitely divisible. Assume $\mathcal{M}_X(t)$ satisfies (2.73). Since $W = \log|\tilde{X}|$ is infinitely divisible, then for every $n \geq 2$ there exist n independent identically distributed random variables, say $W_j = W_{nj}$, $1 \leq j \leq n$, such that $W \stackrel{d}{=} W_1 + W_2 + \cdots + W_n$. Let $Y_j = Y_{nj} = \epsilon_j e^{W_j}$, $1 \leq j \leq n$, $\epsilon_j = 0, 1$, $P(\epsilon_j = 0) = r_n$, ϵ_j independent of ϵ_i, $i \neq j$, and independent of W_i, $1 \leq i \leq n$, with r_n chosen such that $(1 - r_n)^n = P(X \neq 0)$. Then $|X| \stackrel{d}{=} Y_1 Y_2 \cdots Y_n$. Let $0 \leq x \leq 1$. Clearly, $(x - (1 - x))^n = 1$. For arbitrary $0 \leq p \leq 1$, let

$$f_p(x) = \sum_{\substack{k=0 \\ k \text{ odd}}}^{n} \binom{n}{k} x^k (1 - x)^{n-k}.$$

If n is odd, $f_p(0) = 0$ and $f_p(1) = 1$, and the continuity of $f_p(x)$ implies that the equation $f_p(x) = p$ has a solution in $[0, 1]$. If n is even, it is convenient to view $f_p(x) = P(X_1X_2\cdots X_n = -1)$, where the X_is are independent identically distributed random variables, $X_1 = -1, +1$, $P(X_1 = -1) = x$. By considering the difference

$$P(X_1X_2\cdots X_n = -1) - P(X_1X_2\cdots X_n = +1) = \delta$$

(see the proof of Theorem 2.15) one immediately gets that $\delta \le 0$, with equality if, and only if, X_i is symmetric. Hence, if n is odd, $f_p(0) = 0$, $f_p(x) \le x$ for $0 \le x \le 1$, with equality holding if, and only if, $x = 1/2$. This and the continuity of $f_p(x)$ imply that $f_p(x) = p$ has a solution in $[0, 1]$ if, and only if, $0 \le p \le 1/2$.

Let $V, V_1, V_2, \cdots, V_n$ be independent identically distributed non-negative random variables, and let $0 \le x \le 1$. Then

$$\begin{aligned}
&\prod_{j=1}^{n} \Big((1-x)\,\mathcal{M}_{V_j}(t) + \gamma x\,\mathcal{M}_{V_j}(t)\Big) \\
&= \Big((1-x)\,\mathcal{M}_V(t) + \gamma x\,\mathcal{M}_V(t)\Big)^n \\
&= \Big((1-x) + \gamma x\Big)^n \prod_{j=1}^{n} \mathcal{M}_{V_j}(t),
\end{aligned}$$

where in the last expression we implicitly assume the property $\gamma^2 = \gamma\gamma = 1$. Hence if $F(x)$ satisfies (2.73), $F(x) \in \mathcal{M}$. Moreover if $F(0-) \le 1/2$, then $F(x)$ is also in $\mathcal{M}'$.

If $F(x)$ is product-infinitely divisible, then by definition there exists a triangular array $\{X_{n_ij}\}$, $1 \le i \le n_i$, such that $F(x)$ is the limit distribution function of $X_{n_i1}X_{ni2}\cdots X_{n_in_i}$. Clearly the X_{n_ii} can be chosen to satisfy (2.41). This and a result in the next chapter (Theorem 3.9) prove the validity of the following result.

Theorem 2.16 *(i) A distribution function $F(x)$ is product-infinitely divisible if, and only if, there is a triangular array $\{X_{nj}\}$ of random variables that satisfies* (2.41) *and the distribution function of $X_{n1}X_{n2}\cdots X_{nk_n}$ converges completely to $F(x)$. (ii) $\mathcal{M}' \not\subset \mathcal{M}$.*

The above results imply a further characterization of $\mathcal{M}$. That is, a distribution function $F(x)$ is product-infinitely divisible if, and only if, $F(x)$ is the complete limit of the distribution function of a finite product of logarithmically Poisson-type random variables (see definition after Theorem 2.1 and Example 2.4).

2.6 Independent identically distributed random variables

In this section we shall characterize a subset of $\mathcal{L}$ obtained by requiring that the factors in $U(n, b_n)$ (see(2.11)) are independent and identically distributed (i.i.d.). This collection is related to a corresponding collection of distribution in L.

Definition. A distribution function F(x) is said to be stable if there are independent identically distributed random variables $Y_1, Y_2, \cdots$ and constants A_n and B_n such that $F(x)$ is the limiting distribution function of the normalized sum

$$\frac{1}{B_n}\Big(Y_1 + Y_2 + \cdots + Y_n\Big) - A_n.$$

Hence, by definition, stable distributions are of class L, and consequently they are infinitely divisible. An alternate characterization of stable distributions can be given in terms of characteristic functions. That is, a distribution function F(x) is stable if, and only if, its characteristic function $\phi(t)$ satisfies the following property: for every pair of real numbers b_1 and b_2 there exist real numbers a and $b > 0$ such that

$$\phi(b_1 t)\phi(b_2 t) = e^{-iat}\phi(bt) \qquad \text{for all real } t. \tag{2.74}$$

We now extend the above definition to elements of class $\mathcal{L}$.

Definition. We say that the distribution function $G(x)$ is product-stable if $G(x)$ is the limiting distribution of

$$\left\{\prod_{j=1}^{n} \Big|X_j\Big|^{\frac{1}{b_n}} sign(X_j)\right\} e^{-a_n}$$

where the factors are independent identically distributed random variables, a_n and b_n are sequences of real numbers, and $b_n \to +\infty$ as $n \to +\infty$.

Recall that a random variable X has distribution function in $\mathcal{L}$ if, and only if, $X \stackrel{d}{=} \nu W$, $\nu = -, 0, +$, $W \geq 0$, W and ν are independent, and $\log \tilde{W}$ is of class L. This and Theorem 1.19 imply the following equivalent definition of product-stable distribution.

Definition. We say that a distribution function $G(x)$ is product-stable if either $G(x)$ is degenerate at zero, or $G(x)$ is logarithmically stable, continuous at zero, and either $G(x)$ is symmetric, or $G(0) = 1$, or $1 - G(0-) = 1$.

From the above definition we immediately have that for a product-stable distribution function $G(x)$ Theorem 2.3 holds with the further restriction that either $p + q = 0$, or $p = q = 1/2$, or $pq = 0$ and $p + q = 1$.

The characterization of product-stable distributions can also be given in terms of their Mellin transforms on the line of (2.74). We restrict our discussion to the nondegenerate case. Let $\mathcal{M}_X(t)$ be the Mellin transform of a nondegenerate product-stable distribution function. Then

$$\mathcal{M}_X(t) = q\mathcal{M}_{|X|}(t) + \gamma p\mathcal{M}_{|X|}(t), \tag{2.75}$$

where either $q = p = 1/2$, or $q + p = 1$ and $qp = 0$. We claim that for arbitrary real numbers b_1, b_2, there exist real numbers a, $b > 0$ such that

$$\mathcal{M}_X(b_1 t)\,\mathcal{M}_X(b_2 t) = \mathcal{M}_X(bt)e^{-iat}.$$

Let b_1 and b_2 be arbitrary real numbers. We start by assuming that $q = p = 1/2$. Then by using (2.75) one obtains

$$\begin{aligned}
\mathcal{M}_X(b_1 t)\,\mathcal{M}_X(b_2 t) &= \Big(\frac{1}{2}\mathcal{M}_{|X|}(b_1 t) + \gamma\frac{1}{2}\mathcal{M}_{|X|}(b_1 t)\Big) \\
&\quad \cdot\Big(\frac{1}{2}\mathcal{M}_{|X|}(b_2 t) + \gamma\frac{1}{2}\mathcal{M}_{|X|}(b_2 t)\Big) \\
&= \frac{1}{2}\mathcal{M}_{|X|}(b_1 t)\,\mathcal{M}_{|X|}(b_2 t) \\
&\quad +\gamma\frac{1}{2}\mathcal{M}_{|X|}(b_1 t)\,\mathcal{M}_{|X|}(b_2 t) \\
&= \frac{1}{2}\mathcal{M}_{|X|}(bt)e^{-iat} + \gamma\frac{1}{2}\mathcal{M}_{|X|}(bt)e^{-iat} \\
&= \Big(\frac{1}{2}\mathcal{M}_{|X|}(bt) + \gamma\frac{1}{2}\mathcal{M}_{|X|}(bt)\Big)e^{-iat}, \quad (2.76)
\end{aligned}$$

for some real numbers a and $b > 0$, which follows from (2.74) after observing that since $P(X = 0) = 0$, $\mathcal{M}_{|X|}(t) = E(|X|^{it}) = E(e^{it \log |X|})$ can be viewed as the characteristic function of a stable distribution function. In terms of distribution functions, (2.76) says that if $F(x)$ is a product-stable distribution function, and X, Y and V are independent random variables with common distribution function $F(x)$, then $|X|^{b_1} |Y|^{b_2} sign(XY)$ has the same distribution function as $|V|^b e^a sign(V)$. Clearly the above result continues to hold in the case $q + p = 1$ and $pq = 0$.

Conversely, let $F(x)$ be a distribution function in $\mathcal{L}^*$ continuous at zero, and assume that its Mellin transfom $\mathcal{M}_X(t)$ satisfies (2.76). That is,

$$\begin{aligned}\mathcal{M}_X(b_1 t)\, \mathcal{M}_X(b_2 t) &= \Big(q\mathcal{M}_{|X|}(b_1 t) + \gamma\, p\mathcal{M}_{|X|}(b_1 t)\Big) \\ &\quad \cdot\Big(q\mathcal{M}_{|X|}(b_2 t) + \gamma\, p\mathcal{M}_{|X|}(b_2 t)\Big) \\ &= (q^2 + p^2)\, \mathcal{M}_{|X|}(b_1 t)\, \mathcal{M}_{|X|}(b_2 t) \\ &\quad +\gamma\, 2pq\, \mathcal{M}_{|X|}(b_1 t)\, \mathcal{M}_{|X|}(b_2 t) \\ &= q\mathcal{M}_{|X|}(bt)e^{-iat} + \gamma\, p\mathcal{M}_{|X|}(bt)e^{-iat}.\end{aligned}$$

By letting $t = 0$, the last equation gives that

$$q^2 + p^2 = q \quad \text{and} \quad 2pq = p,$$

which can hold if, and only if, either $q = p = 1/2$, or $q + p = 1$ and $qp = 0$. This further implies that for all real numbers b_1 and b_2,

$$E(e^{ib_1 t \log |X|})E(e^{ib_2 t \log |X|}) = E(e^{ibt \log |X|})e^{-iat},$$

for some real numbers a and $b > 0$. That is, $\log |X|$ is a stable distribution. Hence $F(x)$ is product-stable. We have proved the following result.

Theorem 2.17 *A distribution function $F(x)$ continuous at zero is product-stable if, and only if, $F(x)$ is of class $\mathcal{L}^*$, and for arbitrary real numbers b_1, b_2, its Mellin transform $\mathcal{M}_X(t)$ satisfies*

$$\mathcal{M}_X(b_1 t)\, \mathcal{M}_X(b_2 t) = \mathcal{M}_X(bt)e^{-iat}$$

for some real numbers $a, b > 0$.

The relationship between product-stable and stable distribution functions can be further exploited to characterize the form of all possible limiting distribution functions of normalized products of i.i.d. random variables. The following result immediately follows from our previous results and classical results on stable distribution functions (see [58], p. 247).

Theorem 2.18 *Let $X_1, X_2, \cdots$ be independent identically distributed random variables with common distribution function $F(x)$ continuous at 0, $F(0) < 1$. Then there exist sequences of real numbers $\{a_n\}$ and $\{b_n\}$, with $b_n \to +\infty$ with n, and a random variable Y such that*

$$|X_1 X_2 \cdots X_n|^{\frac{1}{b_n}} e^{-a_n} sign(X_1 X_2 \cdots X_n) \to Y \qquad completely,$$

if, and only if, for all real t,

$$\lim_{n\to+\infty} (\mathcal{M}_{|T_n|}(t))^n = \phi(t),$$

where $|T_n| = |X|^{\frac{1}{b_n}} e^{-\frac{a_n}{n}}$, and $\phi(t)$ has the form

$$\phi(t) = e^{ist - c|t|^\alpha \left[1 + i\beta \frac{t}{|t|} u(t,\alpha)\right]}$$

where $s, c \geq 0$, $0 < \alpha \leq 2$, and $-1 \leq \beta \leq 1$ are constants and

$$u(t,\alpha) = \begin{cases} tan(\frac{\pi\alpha}{2}) & if\ \alpha \neq 1 \\ \frac{2}{\pi}\log|t| & if\ \alpha = 1. \end{cases}$$

The quantities involved in the definition of $\phi(t)$ are defined by continuity at $t = 0$, yielding $\phi(0) = 1$.

2.7 Order statistics as products

Let $X_1, X_2, \cdots, X_n$ be independent and identically distributed random variables with (continuous) distribution function $F(x)$. Rearrange the X_j into increasing order, and denote the new sequence by $X_{1:n} \leq X_{2:n} \leq \cdots \leq X_{n:n}$. By the assumed continuity of $F(x)$, equations between $X_{r:n}$ and $X_{r+1:n}$ for any r occur only with probability zero. The variable $X_{r:n}$ is called the r-th order statistic of the

X_j, $1 \le j \le n$. We call the terms $X_{j:n}$ and $X_{n-j:n}$ extremes if j is bounded in n, moderately extremes if j goes to infinity with n but $j/n \to 0$, central if $0 < \liminf j/n \le \limsup j/n < 1$ as $n \to +\infty$, and a p-th ($(1-p)$-th) quantile if $\lim j/n \to p$ with $0 < p < 1$.

Transform the variables X_j to $U_j = F(X_j)$. The U_j are uniformly distributed on $(0, 1)$. Since monotonic transformations preserve inequalities, $U_{r:n} = F(X_{r:n})$, $1 \le r \le n$. A further transformation $V_j = 1 - U_j = 1 - F(X_j)$ preserves the uniform distribution, but reverses ordering: $V_{r:n} = 1 - F(X_{n-r:n})$. One more transformation is useful in distributional studies of the order statistics: $Y_j = -\log F(X_j)$, or, which is the same, $Y_j = -\log U_j$, becomes unit exponential, and so does $Y_j^* = -\log(1-U_j) = -\log(1-F(X_j))$. Here, the star transformation preserves the ordering of the variables: $Y_{r:n}^* = -\log(1 - U_{r:n}) = -\log(1 - F(X_{r:n})$. Therefore, for most asymptotic results, or even for exact distributions, it suffices to deal with either uniform or exponential variables from which, via the inverse of $F(x)$, general results can be obtained.

For the general theory of order statistics, and for a variety of applications, we refer to N. Balakrishnan and C. R. Rao [18], and H. A. David and H. N. Nagaraja [36]. Here, in the present section, we concentrate on the relation of order statistics to products of independent random variables.

Let us start with basic formulas of distribution. By the binomial distribution,

$$P\Big(X_{r:n} \le x\Big) = \sum_{k=r}^{n} \binom{n}{k} F^k(x)(1 - F(x))^{n-k},$$

which can also be written as

$$P\Big(X_{r:n} \le x\Big) = \frac{n!}{(r-1)!(n-r)!} \int_0^{F(x)} t^{r-1}(1-t)^{n-r}\, dt. \quad (2.77)$$

The preceding formulas do not assume even continuity for $F(x)$. On the other hand, if $F(x)$ is differentiable with density function $F'(x) = f(x)$, then (2.77) implies that $F_{r:n}(x) = P(X_{r:n} \le x)$ also has a density $f_{r:n}(x)$ and

$$f_{r:n}(x) = \frac{n!}{(r-1)!(n-r)!} F^{r-1}(x)(1 - F(x))^{n-r} f(x). \quad (2.78)$$

When evaluating the joint density of $X_{r:n}$ and $X_{j:n}$, $r < j$, for $F(x)$ with density $f(x)$, we note that in small intervals $(x, x + \delta x)$ and $(y, y + \delta y)$ the probability of having more than one random variables is of smaller magnitude than $(\delta x)(\delta y)$ as both δx and δy go to zero (which we denote $o((\delta x)(\delta y)))$. Therefore, the probability that $x < X_{r:n} \leq x + \delta x$ and $y < X_{j:n} \leq y + \delta y$, for $x < y$, equals, upto $o((\delta x)(\delta y))$, the probability that one of the X_k falls into the interval $(x, x + \delta x)$, another into $(y, y + \delta y)$, $r - 1$ falling below x, $j - r - 1$ of them to be between $x + \delta x$ and y, and the rest, $n - j$ of them, exceed $y + \delta y$. Hence, the joint density $f_{r,,j:n}(x, y)$ of $X_{r:n}$ and $X_{j:n}$, $r < j$, $x < y$, is given by

$$\begin{aligned} f_{r,,j:n}(x, y) &= \frac{n!}{(r-1)!(j-r-1)!(n-j)!} \\ &\cdot F^{r-1}(x)\Big[F(y) - F(x)\Big]^{j-r-1}\Big[1 - F(y)\Big]^{n-j} f(x)f(y). \end{aligned} \tag{2.79}$$

The joint density of several order statistics, assuming that the density of $F(x)$ exists, can be computed by similar reasoning. Upon combining (2.78) and (2.79), one gets the conditional density of $X_{j:n}$, given $X_{r:n} = x$, which we utilize for computing the distribution of ratios of consecutive order statistics for a uniform population. We, in fact, present the following result.

Theorem 2.19 *Let $U_1, U_2, \cdots, U_n$ be independent random variables, uniformly distributed on $(0, 1)$. Then the random variables*

$$R_j = \Big(\frac{U_{j:n}}{U_{j+1:n}}\Big)^j, \quad 1 \leq j \leq n-1, \qquad \text{and } R_n = U_{n:n}^n \tag{2.80}$$

are independent random variables each uniformly distributed on $(0, 1)$.

Corollary 2.1 *For the U_j of Theorem 2.19*

$$U_{j:n} = R_j^{1/j} R_{j+1}^{1/(j+1)} \cdots R_n^{1/n}, \quad 1 \leq j \leq n \tag{2.81}$$

where R_j, $1 \leq j \leq n$, are independent and identically distributed uniform random variables. Furthermore, for arbitrary continuous population distribution $F(x)$,

$$F(X_{j:n}) = R_j^{1/j} R_{j+1}^{1/(j+1)} \cdots R_n^{1/n}, \quad 1 \leq j \leq n \tag{2.82}$$

where the R_j are as in (2.81).

Theorem 2.19 has been discovered several times, the first one appears to be Sukhatme [146] but it was presented by Sukhatme just as a curiosity and it was not picked up again until the 1950s. The works of Malmquist [109], Rényi [127], and Epstein and Sobel [45] added significance to Theorem 2.19. See Gather et al. [70] for a further analysis of the historical side of Theorem 2.19 as well as for a further review of characterization theorems using order statistics.

PROOF OF THEOREM 2.19. First note that

$$P(U_{n:n} \leq x) = P(U_j \leq x,\ 1 \leq j \leq n) = x^n,$$

so R_n of (2.80) is indeed uniformly distributed on $(0, 1)$. Next, we apply (2.78) and (2.79) with $r < n$ and $j = r + 1$, and compute the conditional density of $X_{r+1:n}$ at y, given that $X_{r:n} = x$ for random samples X_j from a population with distribution $F(x)$ that has a density $f(x)$. Upon denoting this conditional density by $f_{r+1|r:n}(y|x)$, we get

$$f_{r+1|r:n}(y|x) = (n-r)\left[\frac{1-F(y)}{1-F(x)}\right]^{n-r-1} \frac{f(y)}{1-F(x)} \quad \text{for } y > x. \tag{2.83}$$

Let us transform the variables U_j to $Y_j^* = -\log(1 - U_j)$. Then the Y_j^* are unit exponential, that is, their distribution $1 - F(x) = f(x) = e^{-x}$. Hence, by (2.83),

$$f_{r+1|r:n}(y|x) = (n-r)e^{-(n-r)(y-x)}, \quad y > x,$$

which, by the transformation $d^*_{r:n} = (n-r)(Y^*_{r+1:n} - Y^*_{r:n})$ yields that $d^*_{r:n}$ is unit exponential as well. Without regard to transformations but just with a reference to exponentiality of distribution, we can utilize the preceding result for the order statistics of $Y_j = -\log U_j$, which are exponential but reverses ordering, and then transform the result back to the sequence $U_{r:n}$, entailing that the ratios in (2.80) are indeed uniform variables on $(0, 1)$.

The independence of the R_j of (2.80) now easily follows. Consider all order statistics of a sample of size n from a general population $F(x)$ with density $f(x)$. Let $x_1 < x_2 < \cdots < x_n$ be fixed distinct points in the support of $F(x)$ (that is, points of increase of $F(x)$). Build small intervals (a_j, b_j) of length δx_j around x_j, $1 \leq j \leq n$, choosing the length of each δx_j so that the intervals (a_j, b_j) be disjoint. Then the probability that exactly one of the X_k falls into each of the intervals (a_j, b_j), $1 \le j \le n$, divided by the product of the δx_j, and passing to zero by each δx_j provides the joint density

$$f_{(n)}(x_1, x_2, \cdot\cdot, x_n) = n! f(x_1) f(x_2) \cdot \cdot f(x_n), \quad x_1 < x_2 < \cdot\cdot < x_n, \tag{2.84}$$

where the factor $n!$ counts for the permutation of the variables X_k among the intervals (a_j, b_j) in our preceding construction. Obviously, the density above for any other $(x_1, x_2, \cdots, x_n)$ is zero. Now, if we specialize to $f(x) = e^{-x}$, $x > 0$, the right hand side of (2.84) becomes

$$n!\, e^{-x_1 - x_2 - \cdots - x_n}, \quad x_1 < x_2 < \cdots < x_n,$$

which, for the difference $d_{j:n} = (n-j)(X_{j+1:n} - X_{j:n})$, results in independence. This is equivalent to the independence of the ratio R_j of (2.80) for uniformly distributed variables by utilizing our logarithmic transformation as before. This completes the proof. ∎

In applying the representations (2.81) and (2.82) we should ignore the way in which the variables R_j are constructed. What remains important is the fact that they are independent and identically distributed uniform random variables on $(0, 1)$. This makes a significant difference in the situations when j is bounded, for example, $j = 1$, representing the minimum as a product, or when j is asymptotically np, $0 < p < 1$, leading to p-th quantile. As a matter of fact, in the case of the minimum, the product in (2.81) becomes an infinite product as n goes to infinity, while for the quantile, the product in (2.81) is a triangular array, due to the exponent $1/t$, where t starts at j which is around np. We add here that, for $j = n$, the case of the maximum, (2.81) is not a product but just a single term. However, if we recalculate the formula of Theorem 2.19 with the variables $1 - U_j$ which are also independent uniform variables, then the roles of minimum and maximum reverse, and the maximum will ultimately be an infinite product. One more fact is to be understood about the sequence R_j, $1 \le j \le n$. Each R_j depends on both n and j, however their distributions do not, that is, they are all uniform on $(0, 1)$. Hence, for distributional purposes, we can treat the sequence R_j as if they form an infinite sequence of independent and identically distributed random variables. However, in any conclusion of complete convergence one cannot concurrently conclude almost sure convergence. In fact, in most cases, almost sure convergence is not true.

The following theorem contains typical limit distributions for uniform populations.

Theorem 2.20 *Let $U_1, U_2, \cdots, U_n$ be independent random variables uniformly distributed on $(0,1)$. Then, as $n \to= \infty$,*

(i)

$$\lim P\Big(U_{1:n} \leq \frac{x}{n}\Big) = 1 - e^{-x}, \quad x > 0\,;$$

(ii)

$$\lim P\Big(U_{[np]:n} \leq p + x\sqrt{p(1-p)/n}\Big) = N(x)$$

where $0 < p < 1$, $[\,]$ is integer part, and $N(x)$ is the standard normal distribution function;

(iii)

$$\lim P\Big(U_{n:n} \leq 1 + \frac{x}{n}\Big) = e^{x}, \quad x < 0.$$

PROOF. In (i) the limit in the form of $e^{-a_n} U_{1:n}$ converging completely follows from Simonelli [139]'s theorem (Theorem 1.26), while in (ii) we face triangular arrays in the form of (1.42), and thus the existence of the limit follows from Theorem 1.28 ($b_n = n$). However, we reverse the approach here, and use classical (and simple) proofs, and thus the results will serve as examples for possible limits in the quoted theorems. Now,

$$P\Big(U_{1:n} \leq \frac{x}{n}\Big) = 1 - P\Big(U_{1:n} > \frac{x}{n}\Big) = 1 - \big(1 - \frac{x}{n}\big)^n \to 1 - e^{-x},$$

as $n \to +\infty$, which proves (i). For (ii) we go back to the general formula (2.77), or its equivalent formula just preceding (2.77), and observe that, since $F(x) = x$ for $0 < x < 1$, the probability on the left hand side in (ii) is the sum of the terms of a binomial distribution with parameters n and $p^* = p + x\sqrt{p(1-p)/n}$. More precisely, if V

is a random variable with a binomial distribution with parameters n and p^*, then, by the formula preceding (2.77),

$$\begin{aligned} P\Big(U_{[np]:n} \le p^*\Big) &= P\Big(V \ge [np]\Big) \\ &= P\Big(V \ge np^* - x\sqrt{np^*(1-p^*} + o(1)\Big) \end{aligned}$$

which converges to $1-N(-x) = N(x)$, as stated in part (ii). Finally, for (iii), note the earlier observation that with U_j, the variables $1-U_j$ are also uniformly distributed and for this latter sequence the minimum is one minus the original maximum. Hence, as $n \to +\infty$

$$\lim P\Big(1 - U_{n:n} \le \frac{x}{n}\Big) = \lim P\Big(U_{1:n} \le \frac{x}{n}\Big) = 1 - (1 - e^{-x}) = e^{-x},$$

for $x > 0$, which is the claimed form in (iii). The theorem is established. ∎

It is instructive to directly use Theorem 2.4 in the representation (2.82) after taking logarithm, obtaining the lognormal limiting distribution in the form

$$\lim P\Big(\log U_{[np]:n} \le \log p + x\sqrt{(1-p)/(np)}\Big) = N(x)\ (n \to +\infty),$$

which is the same as the limiting distribution of

$$P\Big(U_{[np]:n} \le pe^{g(x,n,p)}\Big) = P\Big(U_{[np]:n} \le p + p(e^{g(x,n,p)} - 1)\Big) \quad (2.85)$$

where $g(x,n,p) = x\sqrt{(1-p)/np}$, that is, its limit is $N(x)$. However, since $g(x,n,p)$ tends to zero as $n \to +\infty$,

$$e^{g(x,n,p)} - 1 = g(x,n,p) + O(\frac{1}{n})$$

and thus (2.85), with its limit $N(x)$, becomes part (ii) of the theorem above.

In both arguments for part (ii) of Theorem 2.20, it is not really used that in the index of $U_{j:n}$, j is exactly $[np]$; both the statement in (ii) of Theorem 2.20 and its proofs remain unchanged if we only assume that $j = np + o(\sqrt{n})$.

The second proof of Theorem 2.20 (ii), based on Theorem 2.4, entails that if we replace $U_{j:n}$ by $F(X_{j:n})$, the expression in (2.85) remains asymptotically normal (even with j chosen as stated in the preceding paragraph). This is due to Corollary 2.1, that is, to the equivalence of (2.81) and (2.82). Next, we turn to the inverse of $F(x)$ and use the finite Taylor expansion of $F^{-1}(y)$ in the form

$$F^{-1}\big(pe^{g(x,n,p)}\big) = Q + \frac{p\big(e^{g(x,n,p)} - 1\big)}{f(Q\, r_n)},$$

where $r_n \to 1$ as $n \to +\infty$, and Q is defined as $Q = F^{-1}(p)$, an expansion that is justified by the conditions of the next theorem. We thus proved

Theorem 2.21 *Let $X_1, X_2, \cdots, X_n$ be independent random variables with common distribution function $F(x)$ whose density $f(x) = F'(x)$ exists and is continuous and positive in a neighborhood of p, $0 < p < 1$. Set $Q = F(p)$, and let $j = j_n = np + o(\sqrt{n})$. Then, as $n \to +\infty$,*

$$\lim P\Big(X_{j:n} - Q \le \frac{x}{f(Q)}\sqrt{p(1-p)/n}\Big) = N(x).$$

Chapter 3

Characterizations of probability distribution

This chapter is devoted to characterization problems. The main emphasis is on characterizing distributions,or families of distribution, by properties associated with products of independent random variables. However, we include some classical characterization results partly for historical reason and partly as tools for making our new characterization results complete. For several extensions and for further references we recommend the now classical books by Linnik and Rao [90], Galambos and Kotz [61], Azlarov and Volodin [13] and Kakosyan, Klebanov and Melamed [91]. Several recent conference proceedings and special issues of periodicals also deal with this subject matter, which we do not include here. These are specified in the subsections of the present chapter.

3.1 General description of characterization

Let $\mathcal{F}$ be the family of all distribution functions of random variables. Let A be a property that some members F of $\mathcal{F}$ posses while others do not. Let $\mathcal{F}(A)$ be the subset of $\mathcal{F}$ such that if $F \in \mathcal{F}(A)$ then F has the property A and if $F \notin \mathcal{F}(A)$ then F does not satisfy A. This identification of the subset $\mathcal{F}(A) \subset \mathcal{F}$ is called a characterization result or theorem by (property) A. A property A may lead to a too large a family $\mathcal{F}(A)$, while for others $\mathcal{F}(A)$ may even be empty (although in the definition of $\mathcal{F}(A)$ we implicitly expressed that 'some members of $\mathcal{F}$' possess A, some time it is meaningful to consider the case when $\mathcal{F}(A)$ is empty). Although in this chapter we consider characterizations mainly when $\mathcal{F}(A)$ is not empty and neither is $\mathcal{F}(A)$ a large subset of $\mathcal{F}$, let us look at examples to show that we should not restrict the size of $\mathcal{F}(A)$ in order to get meaningful and nontrivial results.

Example 3.1 Let $X_1, X_2, \cdots$ be i.i.d. random variables with distribution function $F(x)$. Let A be the property that

$$P\Big(\lim_{n\to+\infty} \frac{X_1 + \cdots + X_n}{n} = X\Big) = 1 \tag{3.1}$$

when X is a finite random variable.

Then it is the most fundamental result of probability theory (Kolmogorov's strong law of large numbers) that $F \in \mathcal{F}(A)$ if, and only if,

$$E(X) = \int_\Omega X\,dP = \int_{-\infty}^{+\infty} x\,dF(x) \quad \text{is finite.}$$

Example 3.2 Let X be a random variable with distribution function $F(x)$. Let A^* be the property that $F(x) = 1$ or 0 for all x.

Clearly, $F(x) \in \mathcal{F}(A^*)$ if, and only if, $F(x)$ is degenerate at some point c; that is, $P(X = c) = 1$ with some constant c.

While this example appears to be more of a definition than a theorem, a return to Example 3.1 gives a real characterization theorem behind $\mathcal{F}(A^*)$.

Example 3.3 Let X be a random variable appearing in (3.1), and A^{**} denotes the property that $F(x)$ is the distribution function of such an X.

By the same law of Kolmogorov as in Example 3.1, we have that $\mathcal{F}(A^{**}) = \mathcal{F}(A^*)$ of Example 3.2.

In order to see that even empty $\mathcal{F}(A)$ can cover a significant theorem, we look at an extension of the strong law of large numbers.

Example 3.4 Let $X_1, X_2, \cdots$ be i.i.d. random variables with distribution function $F(x)$. Let B be the property that there is a sequence

$b_n > 0$ of numbers such that $b_n \to +\infty$ with n and either b_n/n or $n/b_n \to 0$ as $n \to +\infty$, and

$$P\Big(\lim_{n\to+\infty} \frac{X_1 + \cdots + X_n}{b_n} = Y\Big) = 1 \tag{3.2}$$

where Y is a finite random variable.

It is known (Chow and Robbins [30]) that $\mathcal{F}(B)$ is empty. In other words, in order to have an almost sure finite limit for $(X_1 + \cdots + X_n)/b_n$, b_n must essentially be n.

Characterization problems started in statistics, and the emphasis in statistics is to identify a property A for which $\mathcal{F}(A)$ is a parametric family of distribution, or at most a few (in some sense a finite number of) parametric families. The next example is of this nature.

Example 3.5 Let $X_1, X_2, \cdots$ be i.i.d. random variables with finite expectation E and finite variance V. Assume that there are sequences a_n and $b_n > 0$ of real numbers such that

$$\lim_{n\to+\infty} P\Big(\frac{X_1 + \cdots + X_n - a_n}{b_n} \le x\Big) = G(x) \tag{3.3}$$

exists and $G(x)$ is nondegenerate. Let C be the property that $G(x)$ appears in (3.3).

By the classical central limit theorem, $\mathcal{F}(C)$ is the family of normal distributions $N(m, \sigma)$. This fact leads to two new questions of characterization by looking at (3.3) as follows.

Let $0 < r < 1$ be a fixed number, and choose the integer $n_1 \sim rn$. Split the sum

$$\begin{aligned} S_n = X_1 + \cdots + X_n &= (X_1 + \cdots + X_{n_1}) + (X_{n_1+1} + \cdots + X_n) \\ &= S_{n_1} + S^*_{n-n_1}, \quad \text{say.} \end{aligned}$$

Note S_{n_1} and $S^*_{n-n_1}$ are independent, and (3.3) applies to both of them. Consequently, after normalization (note that $a_n = nE$ and $b_n = \sqrt{nV}$) and passage to the limit we have

$$G(x) = G_1(x) * G_2(x), \tag{3.4}$$

where $*$ signifies convolution and each of $G(x)$, $G_1(x)$ and $G_2(x)$ belongs to the normal family $N(m, \sigma)$. Another form following from the preceding argument is

$$\text{normal distribution } = G_1(x) * G_2(x), \tag{3.5}$$

where both $G_1(x)$ and $G_2(x)$ are nondegenerate distribution functions. Now, without reference to the central limit theorem at (3.3), just looking at (3.4) and (3.5) one can ask:

which non-degenerate family $V(x, a, b)$ of distributions satisfies (3.4); that is, assuming only that each of $G(x)$, $G_1(x)$ and $G_2(x)$ belongs to the same parametric family $V(x, a, b)$ of distributions, can we identify $V(x, a, b)$ by (3.4)? The solution of (3.4) in $V(x, a, b)$ is not unique. We already know that the normal family $N(m, \sigma)$ is a solution, and the reader can easily check (use characteristic functions) that both the family of Poisson distributions and the Cauchy family $V(x, a, b) = V(\frac{x-a}{b})$, with $b > 0$,

$$v(x) = V'(x) = \frac{1}{\pi}\frac{1}{1+x^2},$$

satisfy (3.4). Therefore, one can seek all solutions of (3.4) or one can impose further conditions on $G(x)$ and asks which conditions lead to a unique (family) of solutions for (3.4).

The same discussion is valid for (3.5). The solutions for (3.4) and (3.5) are given in Section 3.3.

3.2 The Cauchy functional equation and some associated characterizations

A random quantity measured by the random variable X terminates by an accident, or another term for it is that X is of no-ageing if, for every nonnegative real numbers $x \geq 0$ and $y \geq 0$,

$$P\Big(X > x + y \big| X > x\Big) = P\Big(X > y\Big). \tag{3.6}$$

The equation (3.6) is also known as the lack of memory property. The preceding terms for X comes from interpreting (3.6) as saying that a person of age x wants to buy accident insurance for the next y

time units, and therefore this person is treated equally with a newly-born: that is, age does not matter because the future length (y) of life is terminated by an accident from the insurer's point of view.

Let the distribution function of X be $F(x)$. Then (3.6) can be written

$$\frac{1-F(x+y)}{1-F(x)} = 1 - F(y)$$

or, with $G(x) = 1 - F(x)$,

$$G(x+y) = G(x)G(y), \quad x \geq 0, y \geq 0. \tag{3.7}$$

Notice that (3.7) as it now stands is a monotonic function.

Theorem 3.1 *The only solution of (3.7) among monotonic functions is $G(x) = e^{-\lambda x}$, $x \geq 0$, λ is constant, assuming $G(x) \not\equiv 0$.*

PROOF. By iterating (3.7) and by induction we get, for any fixed $n \geq 2$ and $x_j \geq 0$, $1 \leq j \leq n$,

$$G(x_1 + x_2 + \cdots + x_n) = G(x_1)G(x_2)\cdots G(x_n).$$

Choose each $x_j = 1/n$, $1 \leq j \leq n$, we get $G(1) = G^n(\frac{1}{n})$, while the choice $x_j = 1/m$, $1 \leq j \leq n$, leads to $G(\frac{n}{m}) = G^n(\frac{1}{m}) = G^{n/m}(1)$. We thus have that, if $G(1) = 0$, then $G(x) = 0$ for all rational x. Because $G(x)$ is monotonic, $G(x) \equiv 0$. Next, we assume that $G(1) \neq 0$. Set $\lambda = -\log G(1)$. Then, for $x = n/m$,

$$G(x) = G^x(1) = e^{-\lambda x}, \quad x \geq 0 \text{ rational}, \tag{3.8}$$

and, by monotonicity again, $G(x) = e^{-\lambda x}$ for all $x \geq 0$. The proof is complete. ∎

Remark 3.1 Note that both parts of the theorem ($G(x) = 0$ or (3.8) applies) rely solely on (3.7) for x rationals. The monotonicity assumption is used only to conclude that the forms valid for rationals remain valid for all real x.

Corollary 3.1 *Life-length terminated by an accident is always exponentially distributed* ($F(x) = 1 - e^{-\lambda x}$, $x \geq 0$, $\lambda > 0$ *constant).*

The Corollary is immediate from Theorem 3.1 by the equivalence of (3.6) and (3.7) for monotonic functions and from $G(0) = 1$ for life-length distributions, i.e. $G(x) \not\equiv 0$.

In view of Remark 3.1, we can change 'monotonic' to 'continuous' in Theorem 3.1 in order to have the same conclusion. That is, without change in the proof, we have

Theorem 3.2 *The solution of (3.7) among continuous functions are (i)* $G(x) \equiv 0$ *or (ii)* $G(x) = e^{-\lambda x}$, $x \geq 0$, *and* λ *is a constant.*

Let us move away from the relation between (3.6) and (3.7), and let us study the functional equation

$$G(x + y) = G(x)G(y), \qquad x, y \in D, \tag{3.9}$$

where D is a set of real numbers such that $x, y \in D$ implies that $x + y \in D$. When D is the whole real line, and $G(x)$ is assumed to be real valued Lebesgue measurable function, (3.9) is called *Cauchy's functional equation.*

Theorem 3.3 *Let* $G(x)$ *be a Lebesgue measurable real valued function satisfying (3.9) with* $D =$ *real line or* $D = \{x : x \geq 0\}$. *Then either* $G(x) \equiv 0$ *or* $G(x) = e^{ax}$ *with some constant* a, $x \in D$.

PROOF. We follow an argument of J. Aczél [5] and M. Kac [89].

Obviously, $G(x) \equiv 0$ satisfies (3.9). Note that if $G(x) = 0$ for one x, then for any real y, $G(x + y) = 0$ by (3.9), yielding $G(x) \equiv 0$. Therefore, $G(x) > 0$ for all x if $G(x) \not\equiv 0$ (negative values of $G(x)$ are not possible at (3.9) with our D, since every $x = x/2 + x/2$, and $G(x) = G(x/2)G(x/2) > 0$). We write

$$g(x) = \log G(x), \text{ i.e., } G(x) = e^{g(x)}, \text{ and } G^*(x) = e^{ig(x)}, \tag{3.10}$$

where i is the complex imaginary unit.

Note that $G^*(x)$ also satisfies (3.9), and that $|G^*(x)| = 1$ for all real x. Since $G(x)$ is Lebesgue measurable, so is $G^*(x)$, and by its boundedness, $G^*(x)$ is Lebesgue integrable on any finite interval.

Since $G^*(x)$ cannot be identically zero ($|G^*(x)| = 1$) there is a value h such that

$$d = \int_0^h G^*(x)\,dx \neq 0.$$

Fix h, and let us integrate (3.9) in y. We get

$$\begin{aligned} d\,G^*(x) &= \int_0^h G^*(x+y)\,dy = \int_x^{x+h} G^*(t)\,dt \\ &= \int_0^{x+h} G^*(t)\,dt - \int_0^x G^*(t)\,dt. \end{aligned}$$

By the measurability of $G^*(t)$, the extreme right hand side is continuous, and thus so is the left hand side. That is, $G^*(x)$ is a continuous function. But then the extreme right hand side is differentiable, yielding that the left hand side, $G^*(x)$ is a differentiable function. Let us now go back to (3.9) with $G^*(x)$, and let us differentiate (3.9) in y, and then set $y = 0$. We have

$$(G^*)'(x) = (G^*)'(0)G^*(x). \tag{3.11}$$

If we take a fixed branch of the logarithm, say, the one for which $\log 1 = 0$, then $\log G^*(x)$ is differentiable and (3.11) yields $G^*(x) = ce^{iax}$, where c and a are real constants. In fact, after substituting back into (3.9), we obtain $c = 1$.

Next, from (3.10) we conclude that $g(x) = ax + 2\pi R(x)$, where $R(x)$ is some integer valued function. We apply (3.9) once more resulting in the equation (after simplifying by the contributions of e^{ax}),

$$e^{2\pi R(x+y)} = e^{2\pi R(x)} e^{2\pi R(y)}. \tag{3.12}$$

This is (3.9) again, and thus, by Remark 3.1, for all rational $x \geq 0$, $e^{2\pi R(x)} = e^{-\lambda x}$ with some constant λ, and integer valued $R(x)$. This is possible only if $\lambda = 0$ and $R(x) = 0$ for all $x \geq 0$. By choosing $x > 0$ and $y = -x$ in (3.12), $R(-x) = 0$ follows as well. The proof is completed. ∎

An easy iteration of (3.9) with integer x and y only, leads to the following theorem. The details are left to the reader.

Theorem 3.4 *The solutions of (3.9) with* D = *set of positive integers are (i)* $G(x) \equiv 0$ *or (ii)* $G(x) = q^x$ *with* $x \geq 1$ *integer, and* $q > 0$ *constant.*

When Theorem 3.4 is used for solving the problem at (3.6), part (ii) shows that the discrete no-ageing distribution is

$$P(X = x) = P(X > x - 1) - P(X > x) = q^{x-1} - q^x = (1 - q)q^{x-1},$$

the geometric distribution. The theorem that the geometric distribution is the only one on the positive integers that has the lack of memory property also is frequently used.

Example 3.6 Let $X_1, X_2, \cdots$ be i.i.d. random variables with common distribution function $F(x)$. Set $Z_n = \max(X_1, X_2, \cdots, X_n)$. Assume that there are sequences a_n and $b_n > 0$ of constants such that $(Z_n - a_n)/b_n$ converges completely to a nondegenerate distribution function $H(x)$. Then, for every $s > 0$, there are constants $A(s)$ and $B(s) > 0$ such that

$$H^s\big(A(s) + B(s)x\big) = H(x).$$

Furthermore, if $B(s) \equiv 1$, $s > 0$, then $H(x) = e^{-e^{-a-bx}}$ with some real constants a and $b > 0$.

We include only the major steps leading to the conclusion of the example. The reader can easily fill in the details or may consult our basic reference [58] , pp. 267-269. The aim here is to show the relation of the example to the Cauchy equation.

Since

$$\begin{aligned} P(Z_n \leq a_n + b_n x) &= P(X_j \leq a_n + b_n x,\ 1 \leq j \leq n) \\ &= F^n(a_n + b_n x) \rightarrow H(x) \end{aligned}$$

by assumption, if we choose $n = mn_1$, $m > 1$ fixed integer, one gets by elementary arguments

$$H^m(A_m + B_m x) = H(x) \quad \text{for all } x \text{ and } m \geq 2 \text{ integer.} \qquad (3.13)$$

This can be extended to a continuous version: for any $s > 0$,

$$H^s\big(A(s) + B(s)x\big) = H(x) \quad \text{for all } x \text{ and all } s > 0. \qquad (3.14)$$

Put

$$\alpha(H) = \inf\{x : H(x) > 0\}$$

and

$$\omega(H) = \sup\{x : H(x) < 1\}.$$

Then, for $\alpha(H) < x < \omega(H)$, both $g(x) = \log[-\log H(x)]$ and $g^{-1}(y) = \inf\{x : g(x) \geq y\}$ are well defined, and both are monotonic functions on their domains. The fundamental equation (3.14) takes the form

$$g\big(A(s) + B(s)x\big) + \log s = g(x)$$

and

$$\frac{g^{-1}(y - \log s) - A(s)}{B(s)} = g^{-1}(y). \tag{3.15}$$

Set $y = 0$ in (3.15), and subtract the resulting equation from (3.15). Then the change of variables $u = 1/s$, $y = \log v$, and $g_o^{-1} = g^{-1}(y) - g^{-1}(0)$, and the assumption $B(s) = 1$ yield

$$g_1^{-1}(vu) - g_1^{-1}(u) = g_1^{-1}(v),$$

where $g_1^{-1}(z) = g_o^{-1}(\log z)$. This is (3.7) via the final transformation $M(x) = e^{g_1^{-1}(x)}$. Since $g_1^{-1}(z)$, and thus $M(x)$, are monotonic, by Theorem 3.1, $H(x)$ is as stated.

Should we use (3.13) as the starting point, then the conclusion of Example 3.6 is to be interpreted that if, for a population distribution $H(x)$, the distribution of $Z_m - A_m$, with some constant sequence A_m for all $m \geq 2$, is $H(x)$ itself then $H(x)$ belongs to the unique family $e^{-e^{-a-bx}}$, $a, b > 0$ constants, of distribution functions.

The equation (3.13) is the characterizing equation for max −stable distributions. Not much modification of the preceding argument beyond (3.14) is needed to conclude that $B^*(t) = B(e^t)$, which is no longer identically one, satisfies the Cauchy equation (3.9) with $D =$ whole real line. This time Theorem 3.3 is needed to have a unique solution for $B^*(t)$, from which the intermediate steps identify $H(x)$. An interesting relation of products to extremes is pointed out in Simonelli [140]

Example 3.7 Let X and Y be independent random variables with common continuous distribution function $F(x)$. Assume that $F(0) = 0$, and that $F(x)$ is strictly increasing for all $x > 0$. Set $U = \min(X, Y)$ and $Z = \max(X, Y)$. Then the independence of U and $Z - U$ implies $F(x) = 1 - e^{-ax}$, $a > 0$, $x \geq 0$.

This example is a remarkable characterization of the exponential distribution. Our emphasis on this classical characterization theorem is once again to show its relation to the Cauchy equation. By the assumed independence

$$P\big(Z - U \leq x \big| U = y\big) = P(Z - U \leq x)$$

for almost all y with respect to Lebesgue measure. The fact that we can refer to Lebesgue measure in y is the consequence of the assumption that $F(x)$ is strictly increasing for $x > 0$. On the other hand, by the continuity of $F(x)$,

$$P\big(Z - U \leq x \big| U = y\big) = P\big(Z \leq x + y | U = y\big) = \frac{F(x + y) - F(y)}{1 - F(y)}$$

for almost all $y \geq 0$ and all $x \geq 0$. We thus have

$$P\big(Z - U \leq x\big) = \frac{F(x + y) - F(y)}{1 - F(y)}, \tag{3.16}$$

almost all $y \geq 0$ and all $x \geq 0$. Let $y \to 0$. The right hand side of (3.16) becomes $F(x)$ while the left hand side is not effected by any change of y. That is, the left hand side of (3.16) is $F(x)$, yielding

$$F(x) = \frac{F(x + y) - F(y)}{1 - F(y)} = 1 - \frac{1 - F(x + y)}{1 - F(y)}$$

or, equivalently, $(1 - F(x))(1 - F(y)) = 1 - F(x + y)$ for almost all (Lebesgue) $y \geq 0$ and all $x \geq 0$. Hence, Theorem 3.3 is applicable, which entails the claimed exponentiality.

3.3 Characterizations by decompositions

We return to equations (3.4) and (3.5), and determine their solutions under some restrictions on the functions involved. We shall refer to

(3.4) as Pólya's characterization problem (a reference to [118]), and (3.5) will be referred to as Cramér's decomposition problem (Cramér [34]).

We alternatively write random variables, characteristic functions and distribution functions with the following cast: X, X_j, ϕ, ϕ_j, $F(x)$ and $F_j(x)$ denote random variables and their respective characteristic functions and distribution functions. Whenever no further conclusions are drawn from it, we refer to convolution equation by saying that the underlying random variables are independent.

With the notation

$$m_k = E(X^k) = \int_{-\infty}^{+\infty} x^k \, dF(x)$$

and

$$\mu_k = E(|X|^k) = \int_{-\infty}^{+\infty} |x|^k \, dF(x),$$

whenever they are finite, we have (recall Section 1.1.3)

$$\phi^{(k)}(0) = i^k E(X^k)$$

if the right hand side is finite. If $\phi(t)$ is differentiable any number of times

$$\phi(t) = \sum_{k=0}^{+\infty} \frac{i^k t^k m_k}{k!} \tag{3.17}$$

for all t for which the Taylor expansion on the right hand side is convergent. If the series in (3.17) is convergent for all t, we call $\phi(t)$ analytic. We shall use double indeces for the moments $m_{j,k}$ and $\mu_{j,k}$ if ϕ (or X) has a subscript j.

As a preparation towards solving (3.4) and (3.5), we prove the following lemmas.

Lemma 3.1 *A characteristic function $\phi(t)$ is analytic if, and only if, the moment generating function*

$$M(t) = \int_{-\infty}^{+\infty} e^{t|x|} \, dF(x) < +\infty \qquad \textit{for all real } t.$$

PROOF. Since the absolute moments μ_k satisfy the property that $\mu_k^{1/k}$ is nondecreasing (a consequence of Hölder inequality-Theorem1.9), we have that the sum in (3.17) remains finite for an analytic function $\phi(t)$ if m_k is replaced by μ_k. Simply, first note that $m_{2k} = \mu_{2k}$, and then estimate

$$|m_{2k-1}| \le \mu_{2k-1} \le \mu_{2k}^{(2k-1)/(2k)} \le \mu_{2k} + 1,$$

where one is added for the case $\mu_{2k} < 1$. Hence

$$M(t) = \int_{-\infty}^{+\infty} \sum_{k=0}^{+\infty} \frac{|x|^k t^k}{k!} \, dF(x) = \sum_{k=0}^{+\infty} \frac{\mu_k t^k}{k!} \quad \text{is finite.}$$

Conversely, if $M(t)$ is finite then

$$E\left(\sum_{k=0}^{+\infty} \frac{|tX|^k}{k!}\right) = E\left(e^{|tX|}\right) < +\infty$$

and thus, by the dominated convergence theorem,

$$\phi(t) = E\left(e^{itX}\right) = E\left(\sum_{k=0}^{+\infty} \frac{(itX)^k}{k!}\right) = \sum_{k=0}^{+\infty} \frac{(it)^k}{k!} E(X^k)$$

for all t. The lemma is established. ∎

Lemma 3.2 *Let $\phi_o(t)$ be an analytic characteristic function. Assume that $\phi_o(t) = \phi_1(t)\phi_2(t)$ for all real t and $\phi_j(t)$, $j = 1, 2$, are characteristic functions. Then both $\phi_1(t)$ and $\phi_2(t)$ are analytic.*

PROOF. By Lemma 3.1

$$\int_{-\infty}^{+\infty} e^{t|x|} \, dF_o(x) < +\infty \qquad \text{for all } t. \tag{3.18}$$

On the other hand, the assumed convolution equation of the lemma implies (first prove the equation below for step functions in place of $e^{t|x|}$)

$$\int_{-\infty}^{+\infty} e^{t|x|} \, dF_o(x) = \int_{-\infty}^{+\infty} e^{t|x|} \, dF_1(x) \cdot \int_{-\infty}^{+\infty} e^{t|x|} \, dF_2(x), \tag{3.19}$$

for all t. Since the above integrals do not vanish for any t, by (3.18), the integrals on the right hand side must be finite. One more appeal to Lemma 3.1 entails that $\phi_j(t)$, $j = 1, 2$, are analytic. ∎

We are now in the position of proving Cramér's decomposition theorem.

Theorem 3.5 *The functional equation* $e^{-t^2/2} = \phi_1(t)\phi_2(t)$, *where* $\phi_1(t)$ *and* $\phi_2(t)$ *are nondegenerate characteristic functions, has a unique family of solution: both* $\phi_1(t)$ *and* $\phi_2(t)$ *are characteristic functions of a normal distribution. Equivalently, if* $X = X_1 + X_2$ *is normal, where* X_1 *and* X_2 *are independent nondegenerate random variables, then both* X_1 *and* X_2 *are normal.*

PROOF. Because $e^{-t^2/2}$ is an analytic characteristic function, so are $\phi_1(t)$ and $\phi_2(t)$ by Lemma 3.2. This further entails that all moments associated with $\phi_1(t)$ and $\phi_2(t)$ are finite, and we may assume that $E(X_j) = 0$, $j = 1, 2$. That is,

$$\phi_1(0) = \phi_2(0) = 1 \text{ an } \phi_1'(t) = \phi_2'(0) = 0. \tag{3.20}$$

Next, we turn to complex variables $z = u + iv$, and we show that both $\phi_1(z)$ and $\phi_2(z)$ are finite for all z. Now, the estimate

$$\left|\phi_j(z)\right| \leq \int_{-\infty}^{+\infty} |e^{izx}| \, dF_j(x) = \int_{-\infty}^{+\infty} e^{-vx} \, dF_j(x) = \phi_j(iv) \tag{3.21}$$

shows that $\phi_j(iv)$ must be estimated for all real v in order to conclude that $\phi_j(z)$ is finite for all complex $z = u + iv$. This task will be achieved by deducing from the basic convolution equation of the theorem that

$$\begin{aligned} \phi_1(iv)\phi_2(iv) &= \int_{-\infty}^{+\infty} e^{-vx} \, dF_1(x) \cdot \int_{-\infty}^{+\infty} e^{-vx} \, dF_2(x) \\ &\leq \int_{-\infty}^{+\infty} e^{-vx} \, dN(x) \end{aligned} \tag{3.22}$$

where $N(x)$ is the standard normal distribution function, and we utilized the last equation in (3.21). By arguing first on finite intervals and by utilizing that the convolution of $F_1(x)$ and $F_2(x)$ is equal to $N(x)$, one easily obtains by writing up Riemann-Stieltjes sums that, for finite $a < b$ and $A < B$,

$$\begin{aligned} \int_a^b e^{-vy} \left[\int_A^B e^{-v(x-y)}\, dF_1(x-y) \right] dF_2(y) \; &\leq \; \int_A^B e^{-vx}\, dN(x) \\ &\leq \; \int_{-\infty}^{+\infty} e^{-vx}\, dN(x). \end{aligned}$$

Since the last right hand side does not depend on the constants $a < b$ and $A < B$, and upon observing that the innermost integral on the left hand side depends on $x - y$ only, passage to the limit with $A \to -\infty$, $B \to +\infty$ and then $b \to +\infty$ and $a \to -\infty$, (3.22) obtains (with an argument just as at (3.19), actually equation can be shown to hold in (3.22)).

The combination of (3.21) and (3.22) now tells us that $\phi_j(z)$, $j = 1, 2$, are finite. Since $E(X_j)$ is finite, by the dominated convergence theorem we have that $\phi_j(z)$ is differentiable, so $\phi_1(z)$ and $\phi_2(z)$ are regular and (3.20) holds. Finally, since $e^{-t^2/2} - \phi_1(t)\phi_2(t) = 0$ on the whole real line, and all the three functions are regular on the complex plane, the equation

$$e^{-z^2/2} = \phi_1(z)\phi_2(z), \quad z \text{ complex} \tag{3.23}$$

holds. The regularity of $\phi_j(z)$, together with (3.20), imply that $\psi_j(z) = \big(\log \phi_j(z)\big)/z^2$ is a regular function ($\log s$ here means the main branch of the logarithm, i.e., $\log 1 = 0$). If we show that $\psi_j(z)$, $j = 1, 2$, are bounded on the whole complex plane, then Liouville's theorem entails $\psi_j(z) = c_j$, a constant, i.e., $\phi_j(z) = e^{c_j z^2}$. But for real $t = z$, $\phi_j(t)$ is a characteristic function, and thus $|\phi_j(t)| \leq 1$ and $\phi_j(t)\phi_j(-t)$ is a real valued function, we must have $c_j = -v$, $v > 0$ real ($v = 0$ is not possible, because $\phi_j(t)$ is nondegenerate). This will complete the proof. Therefore, the only step remaining is to show that $\psi_j(z)$ is bounded. Because $\psi_j(z)$ is regular, Schwarz's integral formula

$$\psi_j(z) = Im\big[\psi_j(0)\big] + \frac{1}{2\pi} \int_0^{2\pi} Re\big[\psi_j(Ce^{iy})\big] \frac{Ce^{iy} + z}{Ce^{iy} - z}\, dy,$$

valid for all $C > 0$, reduces the investigation of the boundedness of

$\psi_j(z)$ to that of $Re[\psi_j(z)]$. But by (3.21) and (3.23)

$$\begin{aligned} |\phi_2(z)| &\leq \phi_2(iu) = \frac{e^{u^2/2}}{\phi_1(iu)} \leq 2e^{u^2/2+d|u|} \\ &\leq 2e^{|z^2|+d|z|} \end{aligned}$$

because

$$\phi_1(iu) = \int_{-\infty}^{+\infty} e^{-ux}\, dF_1(x) \geq \int_{-d}^{d} e^{-|u||x|}\, dF_1(x) \geq \frac{1}{2}\, e^{-d|u|},$$

where d is chosen such that $F_1(d) - F_1(-d) \geq 1/2$. Hence

$$\big|Re\big[\log \phi_2(z)\big]\big| = \big|\log \phi_2(z)\big| \leq |z^2| + |z| + \log 2$$

so $\psi_2(z)$ is bounded. But then so is $\psi_1(z)$ by (3.23). The theorem is established. (For the fundamental theorems of complex analysis, we refer to [111]). ∎

We now turn to characterizations by (3.4). We have indicated that (3.4) has several solutions, so further assumption is needed for making (3.20) a characterization theorem. The following result is essentially due to Pólya [118], but its proof that follows is completely different from the original one.

Theorem 3.6 *Let a and b be real constants such that $a^2 + b^2 = 1$. Let X be a nondegenerate random variable, and let X_1 and X_2 be independent copies of X. Then X and $aX_1 + bX_2$ are identically distributed if, and only if, each of X, X_1 and X_2 is a normal random variable.*

In the proof we follow several steps of Linnik [102] and Kakosyan, Klebanov and Melamed [91], although we do not develop the full theory of operators of the latter authors. In order to make the proof clearer to follow, we separate one step into a Lemma.

Lemma 3.3 *Let $\psi_1(t) = e^{-t^2/2}$ and set $\psi_c(t) = \psi_1(t/c)$, $c > 0$. Assume that $F(x)$ is a symmetric distribution function and let its* *characteristic function be $\psi(t)$. If there is a sequence t_k of real numbers such that $t_k \to 0$ as $k \to +\infty$ and $\psi(t_k) = \psi_c(t_k)$ with some $c > 0$ for all $k \geq 1$, then $\psi(t) = \psi_c(t)$ for all t. Consequently, $F(x)$ is normal.*

PROOF. Let Y be a random variable whose distribution function is $F(x)$, and let Z_c be a normal random variable with characteristic function $\psi_c(t)$. We shall show that the assumptions of the lemma imply that all moments of Y are finite and equal those of Z_c. This, via Taylor expansions for characteristic functions, yields the conclusion of the lemma.

Since the distributions of both Y and Z_c are symmetric,

$$\begin{aligned} 1 - \psi_1(t_k) &= \int_{-\infty}^{+\infty} (1 - \cos t_k x)\, dF(x) = 1 - \psi_c(t_k) \\ &= 1 - e^{-t_k^2/(2c^2)} \end{aligned} \tag{3.24}$$

which is bounded by a constant times t_k^2. Hence

$$\int_{-u}^{u} \frac{x^2(1 - \cos t_k x)}{t_k^2 x^2}\, dF(x) \quad \text{is bounded in } u$$

and

$$\frac{1 - \cos t_k x}{t_k^2 x^2} \to \frac{1}{2} \quad \text{as } t_k \to 0.$$

These implies that

$$\int_{-\infty}^{+\infty} x^2\, dF(x) < +\infty,$$

and, consequently, $\psi(t)$ is twice differentiable and by (3.24), $\psi''(0) = \psi_c''(0)$. In addition, by symmetry, $\psi'(0) = \psi_c'(0) = 0$. We now prove by induction that, for every even integer $2s$,

$$\int_{-\infty}^{+\infty} x^{2s}\, dF(x) < +\infty \quad \text{and} \quad \psi^{(2s)}(0) = \psi_c^{(2s)}(0). \tag{3.25}$$

When (3.25) is established then we have that odd derivatives $\psi^{(2s-1)}(0)$ also exist and $\psi^{(2s-1)}(0) = \psi_c^{(2s-1)}(0) = 0$ by symmetry. But then the Taylor expansion of $\psi(t)$ converges for all t and equals the Taylor expansion of $\psi_c(t)$. By the uniqueness of characteristic functions $F(x)$ is normal.

Now, for proving (3.25), we assume that (3.25) is valid for all $k \leq s-1$. Because the underlying distributions are symmetric, the characteristic functions $\psi(t)$ and $\psi_c(t)$ are real valued continuous functions, and by assumptions $\psi(t_k) = \psi_c(t_k)$, and $\psi^{(2s-2)}(t)$ exists. Hence, by arguing with finite Taylor expansions we have from Rolle's theorem that the sequence t_k generates a sequence t_k^* such that $\psi^{(2s-2)}(t_k^*) = \psi_c^{(2s-2)}(t_k^*)$ and $t_k^* \to 0$ as $k \to +\infty$. Besides, we already know $\psi^{(2s-2)}(0) = \psi_c^{(2s-2)}(0)$. Therefore, we have

$$\begin{aligned} O(t_k^*) &= \psi_c^{(2s-2)}(0) - \psi_c^{(2s-2)}(t_k^*) = \psi^{(2s-2)}(0) - \psi^{(2s-2)}(t_k^*) \\ &= (-1)^{s-1} \int_{-\infty}^{+\infty} x^{2s-2}(1 - cos\, t_k^* x)\, dF(x), \end{aligned}$$

which, just the way we deduced (3.25) from (3.24), implies (3.25) with $2s$. The lemma is established. ∎

Remark 3.2 Note that, in the preceding lemma, it is not essential that $\psi_c(t)$ be generated from the standard normal distribution. If, instead, we assume that $\psi_1(t)$ is a symmetric characteristic function which is uniquely determined by its associated moments via the derivatives of $\psi_1(t)$, then $\psi(t_k) = \psi_c(t_k)$, $t_k \to 0$, implies that $\psi(t) = \psi_c(t)$ for all t.

Proof of Theorem 3.6. Let $\phi(t)$ be the characteristic function of X. If X is normal, $\phi(t) = \phi(at)\phi(bt)$ for all t, assuming $a^2 + b^2 = 1$. We turn to the converse. We assume that $\phi(t)$ is nondegenerate and that

$$\phi(t) = \phi(at)\,\phi(bt), \quad a^2 + b^2 = 1, \quad \text{all real } t, \tag{3.26}$$

holds, and deduce that $\phi(t)$ is normal. Set $\psi(t) = \phi(t)\phi(-t) - |\phi(t)|^2$. By (3.26)

$$\psi(t) = \psi(at)\psi(bt) \quad a^2 + b^2 = 1, \quad \text{all real } t, \tag{3.27}$$

and $\psi(t)$ is a symmetric nondegenerate characteristic function. let $\psi_c(t)$ be as in Lemma 3.3. The following possibilities exist:

(i) $\psi(t) = \psi_c(t)$ for all t with some $c > 0$;

(ii) for a fixed T, there exists a $0 < t_o \leq T$ such that

$$\psi(t_o) = \psi_{c_o}(t_o) \quad \text{with some } c_o > 0,$$

but part (i) fails, and $u = \inf\{t > 0 : \psi(t) = \psi_{c_o}(t)\} > 0$.

The logical third possibility of $\psi(t) \neq \psi_c(t)$ for any $0 < t \leq T$ is not actually possible if we show the first part of (ii). We start with part (ii). Since $\psi(t)$ is a nodegenerate characteristic function, obviously there is a t_o such that $0 < t_o \leq T$ and $\psi(t_o) < 1$, whatever the fixed value $T > 0$ be. But then we can determine a value $t_1 > 0$ such that $\psi_1(t_1) = e^{-t_1^2/2} = \psi(t_o)$. In other words, $\psi(t_o) = \psi_1(t_1) = \psi_1(t_o t_1/t_o) = \psi_{c_o}(t_o)$ with $c_o = t_o/t_1 > 0$. If $u = \inf\{t > 0 : \psi(t) = \psi_{c_o}(t)\} = 0$, then we have a sequence $t_k \to 0$ such that $\psi(t_k) = \psi_{c_o}(t_k)$, which, by Lemma 3.3, would conclude $\psi(t) = \psi_{c_o}(t)$ for all t, contradicting the assumption that part (i) fails.

We thus have $u > 0$, $\psi(u) = \psi_{c_o}(u)$, for all $0 < t < u$, $\psi(t) \neq \psi_{c_o}(t)$. This means that the difference $\delta(t) = \psi(t) - \psi_{c_o}(t)$ is either strictly positive or strictly negative on $(0, u)$. Choose $t = u$ in (3.27). Write up (3.27) at $t = u$ as well with $\psi_{c_o}(t)$ (for which (3.27) is evidently true), and subtract the two resulting equations; we get

$$\begin{aligned} 0 &= \psi(u) - \psi_c(u) = \psi(au)\psi(bu) - \psi_c(au)\psi_c(bu) \\ &= \left[\psi(au) - \psi_c(au)\right]\psi(bu) + \psi_c(au)\left[\psi(bu) - \psi_c(bu)\right]. \end{aligned}$$

However, because both $|a|u$ and $|b|u$ are inside the interval $(0, u)$, and the characteristic functions $\psi(t)$ and $\psi_c(t)$ are those of symmetric distributions, i.e., real valued even functions, the extreme right hand side above cannot be zero. In other words, the case of the second part of part (ii) is not possible either. That is, (i) applies, and thus $\psi(t)$ is normal. But $\psi(t) = \phi(t)\phi(-t)$ with nondegenerate $\phi(t)$, so, Theorem 3.5 applies entailing that $\phi(t)$ is normal. This completes the proof. ∎

A decomposition theorem like Theorem 3.5 is valid for the Poisson distribution as well (Raikov [122])

Theorem 3.7 *Let X_1 and X_2 be independent and identically distributed random variables with $P(X_j \geq 0) = 1$ and $P(X_j = 0) < 1$, $j = 1, 2$. If $X = X_1 + X_2$ is a Poisson random variable, then so are X_1 and X_2.*

The idea of the proof is very similar to that of Theorem 3.5, except that it is much simpler by the nature of the distribution involved.

Proof of Theorem 3.7. X is a convolution of X_1 and X_2, hence both X_1 and X_2 take only the nonnegative integers almost surely. Put $p_k = P(X_1 = k)$, $r_k = P(X_2 = k)$, $H(z) = \sum_{k=0}^{+\infty} r_k z^k$ and $G(z) = \sum_{k=0}^{+\infty} p_k z^k$. Then

$$G(z)H(z) = e^{\lambda(z-1)} \quad \text{and} \quad \frac{\lambda^n e^{-\lambda}}{n!} = \sum_{k=0}^{n} p_k r_{n-k}, \ \lambda > 0. \quad (3.28)$$

Since $p_o r_o \neq 0$ by assumption,

$$|H(z)| \leq e^{\lambda(|z|-1)}/p_o$$

and

$$|G(z)| \leq e^{\lambda(|z|-1)}/r_o.$$

Hence, both $H(z)$ and $G(z)$ are finite for all complex z, and they are regular functions (since they have infinite power series representation). By (3.28), neither $H(z)$ nor $G(z)$ vanishes, one can consider $\log|H(z)|$ and $\log|G(z)|$. Since $\log p_o + \log r_o = -\lambda$, we have

$$\log|H(z)| \leq \log H(|z|) \leq -\log p_o + \lambda(|z| - 1) = \lambda|z| + \log r_o,$$

and $\log|H(z)| > 0$ for $|z| > 1$. Hence, $[\log|H(z)| - \log r_o]/|z|$ is bounded, and thus, by Liouville's theorem, it is a constant: $H(z) = r_o e^{cz}$. Since $H(1) = 1$, $c > 0$, and thus $H(z) = e^{c(z-1)}$. One more appeal to (3.28) yields $G(z) = e^{(\lambda-c)(z-1)}$, $\lambda - c > 0$, so the theorem is valid. ∎

Just to see what a difference a slight change in the conditions of Theorem 3.6 makes in solving Pólya's problem we prove the following result.

Theorem 3.8 *Let X be a random variable with a nondegenerate distribution function $F(x)$, and assume that $F(x)$ is symmetric about zero. Let X_1 and X_2 be independent copies of X. Then, for all constants $a > 0$ and $b > 0$ with $a + b = 1$, X and $aX_1 + bX_2$ are identically distributed if, and only if, $F(x)$ is a Cauchy distribution function.*

PROOF. Let $\phi(t)$ be the characteristic function of X. Then $\phi(t)$ is real valued and an even function, due to the assumed symmetry of $F(x)$. Furthermore, the basic assumption of the theorem translates to,

$$\phi(t) = \phi(at)\phi(bt), \quad a > 0, b > 0, a + b = 1, t \geq 0. \tag{3.29}$$

Upon renaming $at = x$, $bt = y$, $t = x + y$, (3.29) becomes equivalent to

$$\phi(x + y) = \phi(x)\phi(y), \quad \text{all } x \geq 0, y \geq 0. \tag{3.30}$$

This is Cauchy's functional equation (3.7), and since $\phi(t)$ is a continuous function, by Theorem 3.2, $\phi(t) = e^{-\lambda t}$, $t \geq 0$. Because $|\phi(t)| \leq 1$, $\lambda > 0$. By symmetry, $\phi(t) = e^{-\lambda|t|}$, t real, which is the characteristic function of a Cauchy distribution. By the uniqueness theorem for characteristic functions, $F(x)$ is as claimed. ∎

There is a general solution to Pólya's problem when one assumes at (3.4) that $G(x)$ is of the form $V((x-a)/b)$, that is, the parameters are location and scale parameters. In this form, (3.4) is just the definition of (sum) stable distributions. In such a form, we refer to [58], pp. 242-247. In the theorems presented here the nature of the parameters have been replaced by other conditions which lead to unique solutions (normal and Cauchy, respectively). That is, we separated the normal and Cauchy distributions from the sum-stable distributions by our choice of conditions in Theorems 3.6 and 3.8.

The Poisson distribution is not a sum-stable distribution. Its sole parameter is neither a location nor a scale parameter. Yet, the Poisson distribution can be characterized by (3.4) among distributions on the nonnegative integers with some further restrictions. We do not discuss such characterizations.

Closer to products and their associated limit theorems are the logarithmically normal and logarithmically Poisson distributions. In their decomposition theorems the signs of the underlying random variables play important role.

We start with a simple lemma.

Lemma 3.4 *Let Z be a random variable such that $sign(Z)$ and $|Z|$ are independent. Let U be another random variable, and assume that $Z \stackrel{d}{=} U$. Then $sign(U)$ and $|U|$ are independent, furthermore, $sign(U) \stackrel{d}{=} sign(Z)$. In particular, if Z is symmetric, then $|Z|$ and $sign(Z)$ are always independent.*

PROOF. By the assumed identical distribution of U and Z, we immediately have that $|U| \stackrel{d}{=} |Z|$ and $sign(U) \stackrel{d}{=} sign(Z)$. That is, only the independence of $sign(U)$ and $|U|$ needs proof. However, if we write down the Mellin transforms of Z and U, for U using the definition of its Mellin transform while for Z exploiting the independence assumption for its sign and absolute value, we get

$$\mathcal{M}_U(t) = \mathcal{M}_{U^+}(t) + \gamma\mathcal{M}_{U^-}(t) = (1 - p + \gamma p)\mathcal{M}_{|Z|}(t),$$

where $p = P(sign(U) = -) = P(sign(Z) = -)$, and $\mathcal{M}_{|Z|}(t) = \mathcal{M}_{|U|}(t)$. From the uniqueness theorem for Mellin transforms, we get that $|U|$ and $sign(U)$ are independent. The particular case of symmetric distributions follows by observing if $|Z|$ is any random variable and X is a random variable taking the values of $-1, 0$ and $+1$ with $P(X = -1) = P(X = +1) = P(Z < 0)$, and if X is independent of Z, then $X\,|Z|$ is symmetric with the same absolute value as Z. So, if Z is symmetric, then its distribution is the same as that of $X\,|Z|$, that is, $sign(Z)$ and X exhibit the same property. The lemma is established. ∎

We now prove a general characterization theorem by a decomposition into a product. Recall the definition of class $\mathcal{L}^*$ of distributions from Section 2.2. See also G. Bareikis and J. Šiaulys [22], pp. 70-73.

Theorem 3.9 *Let W be a nonsymmetric random variable in $\mathcal{L}^*$, and let X and Y be independent. We further assume that either*

$P(|Y| = a) > 0$, for some $a \geq 0$, or that there exists a region symmetric about zero in which Y has a nonzero density function. Then $W \stackrel{d}{=} XY$ if, and only if, $|W| \stackrel{d}{=} |XY|$, both X and Y are in $\mathcal{L}^*$, and $sign(XY) \stackrel{d}{=} sign(W)$.

PROOF. Let $\nu = sign(W)$, $P(\nu = -) = p = 1 - q \neq 1/2$, $\nu_1 = sign(X)$, and $\nu_2 = sign(Y)$. We start by assuming

$$W = \nu\,|W| \stackrel{d}{=} XY = \nu_1\,\nu_2\,|X|\,|Y|,$$

which implies

$$\nu \stackrel{d}{=} \nu_1\,\nu_2 \qquad \text{and} \qquad |W| \stackrel{d}{=} |X||Y|,$$

and by Lemma 3.4, $\nu_1\nu_2$ is independent of $|XY|$. In the case one factor in the product XY, say X, is in $\mathcal{L}^*$, there must exist independent random variables ν_2' and $|Y'|$, with the same distribution as ν_2 and $|Y|$, respectively, such that

$$\nu_2'|Y'| \stackrel{d}{=} Y.$$

By Lemma 3.4, $sign(Y)$ must be independent of $|Y|$. Next we assume that $sign(X)$ and $sign(Y)$ are not independent of $|XY|$, and let $|W| = |X'|\,|Y'|$, where $|X'|$ and $|Y'|$ are independent and with the same distribution as $|X|$ and $|Y|$, respectively. We first consider the case $P(|Y| = a) > 0$, for some $a \geq 0$. Then the conditional distribution function of W given $|Y'| = a$ is identical to the conditional distribution function of XY given $|Y| = a$, and hence

$$\mathcal{M}_{W||Y'|=a}(t) = \mathcal{M}_{XY||Y|=a}(t).$$

This gives

$$\begin{aligned}&(q\,a^{it}\,\mathcal{M}_{|X'|}(t) + \gamma\,p\,a^{it}\,\mathcal{M}_{|X'|}(t))\\ &= (q'\,a^{it} + \gamma\,p'a^{it})\,(\mathcal{M}_{X^+}(t) + \gamma\,\mathcal{M}_{X^-}(t)),\end{aligned}$$

where p' and q' are uniquely determined by the distribution function of Y, and such that $p' \geq 0$, $q' \geq 0$, $p' + q' = 1$. The left hand side and the first factor in the right hand side of the above equation are the

Mellin transforms of random variables with the signs independent of the absolute values, and this and our previous result imply that X must also have the same property, which contradicts our assumption on X. We now turn to the case of assuming that there exists a region symmetric about zero in which Y has a nonzero density function, say $f(y)$. Hence, for any y in this region, we can assume that the conditional distribution functions of W given $|Y'| = |y|$ and of XY given $|Y| = |y|$ are well defined and identical. We have

$$\begin{aligned}\mathcal{M}_{W||Y'|=|y|}(t) &= (q\,|y|^{it}\,\mathcal{M}_{|X'|}(t) + \gamma p\,|y|^{it}\,\mathcal{M}_{|X'|}(t)) \\ &= \mathcal{M}_{XY||Y|=|y|}(t) \\ &= \left(\frac{f(-|y|)}{f(-|y|)+f(|y|)}\,|y|^{it} + \gamma\,\frac{f(|y|)}{f(-|y|)+f(|y|)}\,|y|^{it}\right) \\ &\quad \cdot(\mathcal{M}_{X^+}(t) + \gamma\,\mathcal{M}_{X^-}(t)).\end{aligned}$$

As in the previous case, all the quantities inside parantheses in the above equation can be viewed as Mellin transforms of random variables, with two of them having the property that the sign is independent of the absolute value. Hence the remaining random variable, which is X, must also have this property, again contradicting our initial assumption on X. Hence if $sign(X)$ and $sign(Y)$ are not independent of $|XY|$, $sign(W)$ is not independent of $|W|$. This completes the proof of the necessity part of the theorem. The validity of the sufficiency part is immediate, and the proof of the theorem is complete. ∎

Let X be a logarithmically-Poisson random variable in $\mathcal{L}^*$. Direct computation gives

$$\mathcal{M}_X(t) = q_1\, e^{-\lambda_1(1-e^{it})} + \gamma p_1\, e^{-\lambda_1(1-e^{it})},$$

where λ_1 is the parameter of the Poisson distribution of $\ln|X|$, and $p_1 = 1 - q_1 = P(sign(X) = -)$. If Y is another logarithmically-Poisson random variable in $\mathcal{L}^*$ independent of X, and $p_2 = 1 - q_2 = P(sign(Y) = -)$, λ_2 the parameter of the distribution function of $\ln|Y|$, then

$$\mathcal{M}_Y(t) = q_2\, e^{-\lambda_2(1-e^{it})} + \gamma p_2\, e^{-\lambda_2(1-e^{it})},$$

and

$$
\begin{aligned}
\mathcal{M}_{XY}(t) &= \mathcal{M}_X(t)\,\mathcal{M}_Y(t) \\
&= (q_1\, e^{-\lambda_1(1-e^{it})} + \gamma\, p_1\, e^{-\lambda_1(1-e^{it})}) \\
&\quad \cdot (q_2\, e^{-\lambda_2(1-e^{it})} + \gamma\, p_2\, e^{-\lambda_2(1-e^{it})}) \\
&= (q_1 q_2 + p_1 p_2)\, e^{-(\lambda_1+\lambda_2)(1-e^{it})} \\
&\quad +\gamma\,(p_1 q_2 + q_1 p_2)\, e^{-(\lambda_1+\lambda_2)(1-e^{it})}.
\end{aligned}
$$

Hence $W = XY$ is also logarithmically-Poisson in $\mathcal{L}^*$. This, Theorems 3.7 and 3.9 imply the validity of the following result.

Theorem 3.10 *Let X and Y be independent random variables.*
(i) $W = XY$ is a symmetric logarithmically-Poisson random variable if, and only if, X or Y is symmetric, and either the distribution function of each of X and Y is logarithmically-Poisson, or one of X and Y is logarithmically-Poisson, the other is logarithmically degenerate at zero.
(ii) $W = XY$ is a non symmetric logarithmically-Poisson random variable in $\mathcal{L}^*$ if, and only if, neither X nor Y is symmetric, and either the distribution function of each of X and Y is logarithmically-Poisson, and both X and Y are in $\mathcal{L}^*$, or one of the distribution functions of X or Y is logarithmically-Poisson and in $\mathcal{L}^*$, and the other is logarithmically degenerate at zero.

It is worth noting that if X is positive and logarithmically-Poisson, then it has the property that its sign is independent of $|X|$, and indeed this is possible in Theorem 3.10 as X in (i) and (ii), in which case Y has the same independent sign as W.

Among the continuous distribution functions, a similar decomposition theorem can be obtained for the logarithmically-normal distribution.

Let X and Y be independent logarithmically-normal random variables in $\mathcal{L}^*$. Then $\ln|X| \sim N(\mu_1, \sigma_1^2)$, and $\ln|Y| \sim N(\mu_2, \sigma_2^2)$. Therefore if $P(X<0) = p_1 = 1 - q_1$ and $P(Y<0) = p_2 = 1 - q_2$,

$$
\begin{aligned}
\mathcal{M}_X(t) &= q_1\, e^{t\mu_1} e^{\frac{-t^2\sigma_1^2}{2}} + \gamma\, p_1\, e^{t\mu_1} e^{\frac{-t^2\sigma_1^2}{2}} \\
\mathcal{M}_Y(t) &= q_2\, e^{t\mu_2} e^{\frac{-t^2\sigma_2^2}{2}} + \gamma\, p_2\, e^{t\mu_2} e^{\frac{-t^2\sigma_2^2}{2}},
\end{aligned}
$$

and

$$\begin{aligned}\mathcal{M}_{XY}(t) &= \mathcal{M}_X(t)\mathcal{M}_Y(t)\\ &= (p_1p_2+q_1q_2)\,e^{t(\mu_1+\mu_2)}e^{\frac{-t^2(\sigma_1^2+\sigma_2^2)}{2}}\\ &\quad +\gamma\,(q_1p_2+q_2p_1)\,e^{t(\mu_1+\mu_2)}e^{\frac{-t^2(\sigma_1^2+\sigma_2^2)}{2}}.\end{aligned}$$

That is, XY is also logarithmically-normal and is in $\mathcal{L}^*$. Clearly the above result continues to hold in the case Y (or X) is a logarithmically-degenerate random variable. This, together with Theorems 3.5 and 3.9 entail the following result.

Theorem 3.11 *Let X and Y be independent random variables.*
(i) $W = XY$ is a symmetric logarithmically-normal random variable if, and only if, X or Y is symmetric, and either both X and Y are logarithmically-normal, or one of X and Y is logarithmically-normal and the other is logarithmically-degenerate.
(ii) Assume either $P(|Y| = a) > 0$, for some $a \geq 0$, or that there exists a region symmetric about zero in which Y has a nonzero density function. Then $W = XY$ is a non symmetric logarithmically-normal random variable in $\mathcal{L}^*$ if, and only if, both X and Y are in $\mathcal{L}^*$, neither X nor Y is symmetric, and either both X and Y are logarithmically-normal, or one of X and Y is logarithmically-normal and the other is logarithmically-degenerate.

3.4 Further characterizations of logarithmically normal and Cauchy distributions

With the notation of Section 3.1, we denote by $\mathcal{F}$ all distribution functions continuous at zero. Let A be the property that a random variable X with distribution from $\mathcal{F}$ satisfies that $|X|$ and $1/|X|$ are identically distributed. Then $\mathcal{F}(A)$ contains the distribution of X if, and only if, $\log|X|$ has a symmetric distribution. In other words, A characterizes log-symmetry. Furthermore, if the distribution of X belongs to $\mathcal{F}(A)$ and if Y is an independent copy of X, then XY and X/Y are identically distributed, assuming that $sign(X)$ and $|X|$ are independent. Hence, distributional properties of XY and X/Ycannot lead to characterization of distribution. However, by assuming independence of XY and X/Y, we can get the following characterization theorem.

Theorem 3.12 *Let X and Y be independent and identically distributed nondegenerate random variables with distribution function $F(x)$ continuous at zero. Then XY and X/Y are independent if, and only if, $F(x)$, or the distribution function of $-X$, is traditional lognormal.*

PROOF. Let $U = \log|XY| = \log|X| + \log|Y|$ and $V = \log|X/Y| = \log|X| - \log|Y|$. If $\log|X|$ is normal then U and V are independent (see [58], p. 135). In fact, it is a classical result in characterization theory that if U and V are independent then $\log|X|$ is normal (see Lukacs and Laha [105], p. 75). We therefore have to show that necessarily $X \geq 0$ or $X \leq 0$ a.s.. Without loss of generality we may assume that $\log|X|$ is normalized to be standard normal. We compute $E[(XY)(X/Y)]$ in two different ways. First, $E[(XY)(X/Y)] = E[X^2] = e^2$. On the other hand, by independence,

$$\begin{aligned} E[(XY)(X/Y)] &= E(XY)\,E(X/Y) = E(X)\,E(Y)\,E(X)\,E(1/Y) \\ &= E^3(X)E(1/Y). \end{aligned}$$

We thus have from the preceding calculation,

$$\begin{aligned} E(X^2) &= e^2 = |E^3(X)|\,|E(1/Y)| \\ &\leq E^3(|X|)\,E(|1/Y|) = E^4(|X|), \end{aligned} \tag{3.31}$$

by the earlier cited log-symmetry of $|X|$. But $E^4(|X|) = (e^{1/2})^4 = e^2$, and thus equation holds in (3.31), entailing $|E(X)| = E(|X|)$, that is, either $P(X \geq 0) = 1$ or $P(X \leq 0) = 1$. The theorem is established. ∎

Next, we study characterizations in connection with the geometric mean. In general, for a constant $b > 0$, and for random variables X_1 and X_2, we write

$$(X_1X_2)^b = |X_1X_2|^b sign(X_1X_2) = (-1)^{\nu_1+\nu_2}|X_1X_2|^b \tag{3.32}$$

where ν_1 and ν_2 are random variables taking the values 0 and 1, and $sign(X_j) = (-1)^{\nu_j}$, $j = 1, 2$. For the sequel in this discussion we assume that X_1 and X_2 are independent and identically distributed random variables with common distribution function $F(x)$, continuous at zero. We look for solutions of the equation (see Galambos and Simonelli [66])

$$X_1 \stackrel{d}{=} (X_1 X_2)^b, \quad b > 0. \tag{3.33}$$

Clearly, with $Y_j = \log |X_j|$,

$$|X_1| \stackrel{d}{=} |X_1 X_2|^b, \quad \text{and} \quad Y_1 \stackrel{d}{=} b(Y_1 + Y_2). \tag{3.34}$$

If we further assume that the variance $0 < V(Y_1) < +\infty$, then by equating expectations and variances in (3.34) we get that $E(Y_j) = 0$ and $b = 1/\sqrt{2}$. In addition, if we put $p = P(\nu_j = 0)$, $j = 1, 2$, (3.32) and (3.33) yield, just as in Example 3.1, and regardless of the value of $b > 0$, that

$$\text{in (3.33)}, \ p = 1 \ \text{ or } \ p = \frac{1}{2} \ \text{ whatever } b > 0. \tag{3.35}$$

Indeed, just note that

$$P(\nu_1 + \nu_2 = 0 \text{ or } 2) = p^2 + (1-p)^2$$

which becomes $P(\nu_1 = 0) = p$ only if $p = 1$ or $p = 1/2$. Now, if we iterate (3.34), the classical central limit theorem entails that Y_1 is normal with parameters $E(Y_1) = 0$ and $0 < V(Y_1) < +\infty$ arbitrary. Consequently, X_j is logarithmically normal with rule for its sign at (3.35).

There are several logarithmically normal distributions, even with the simple rule for signs at (3.35), but not all such distributions are solution of (3.33). Indeed, if we iterate (3.33) itself, in light of (3.32), $F(x)$ must belong to $\mathcal{L}^*$, that is, $sign(X_1)$ and $|X_1|$ are independent. One point of view can be expressed by observing that if $\phi(t)$ is the characteristic function of Y_1, then by (3.34)

$$\phi(t) = \phi(\frac{t}{\sqrt{2}})\phi(\frac{t}{\sqrt{2}}),$$

that is, (3.4) applies (with specific constants and in the form of characteristic functions). On the other hand, if we translate (3.32) into an equation of Mellin transforms, formally it also looks like (3.4), but its iteration, via Section 2.2, does give the proper solution.

It is instructive to look at a logarithmically normal distribution of a random variable U with $P(U < 0) = P(U > 0) = 1/2$, but $|U|$ and $sign(U)$ are not independent. Such a U, as we have just concluded, is not a solution of (3.32). The example is as follows.

Let W be a lognormal random variable, $\ln W \in N(0,1)$, with density function $g(w)$, and let

$$f(x) = \begin{cases} \frac{g(-x)}{4} & x < -1 \\ \frac{3g(-x)}{4} & -1 \leq x < 0 \\ \frac{g(x)}{4} & 0 \leq x < 1 \\ \frac{3g(x)}{4} & x \geq 1. \end{cases}$$

Since $P(0 < W \leq 1) = 1/2$, $f(x)$ is a density function. Let U have density function $f(x)$. Clearly, $P(U \leq 0) = 1/2$. Moreover

$$P(|U| \leq a) = \frac{1}{4}\int_0^a g(x)\,dx + \frac{3}{4}\int_{-a}^0 g(-x)\,dx = P(W \leq a),$$

if $0 < a \leq 1$, and

$$\begin{aligned} P(|U| \leq a) &= P(W \leq 1) + \frac{3}{4}\int_1^a g(x)\,dx + \frac{1}{4}\int_{-a}^1 g(-x)\,dx \\ &= P(W \leq a), \end{aligned}$$

if $a > 1$. Hence U is logarithmically normal, but clearly $|U|$ and $sign(U)$ are not independent.

In order to complete the solution of (3.33) with finite variance for Y at (3.34), we note that if $V(Y_j) = 0$, then $Y_j = \log|X_j|$ is constant with probability one, and the constant must be zero by (3.34). This yields $X_j = 1$ or -1, and the rule of signs at (3.35) applies. Of course, $b > 0$ is arbitrary.

We summarize what we have proved in connection with (3.33) in the following theorem.

Theorem 3.13 *X_1 and X_2 are independent and identically distributed random variables with distribution function $F(x)$, assumed continuous at zero. Let the variance $V = V(\log|X_j|) < +\infty$. Then (3.33), with notation at (3.32) is valid if, and only if, one of the following conditions holds* *(i)* $V > 0$, $b = 1/\sqrt{2}$, $F(x)$ *is lognormal, and* $\log|X_j|$ *has zero expectation and variance* V*;*

(ii) $V > 0$, $b = 1/\sqrt{2}$, $X_1 = (-1)^{\nu_1} e^{Y_1}$, Y_1 *is normal* $N(0, V)$*, and* X_1 *is symmetric (recall Lemma 3.4);*

(iii) $V = 0$, $b > 0$ *arbitrary, and* X *takes the values 1 and -1 with the rule for signs as at (3.35);*

(iv) $V = 0$, $b = 1/2$, *and* X_1 *takes the arbitrary contants c and* $-c$ *with the rule for signs as at (3.35).*

We see from Theorem 3.13 that if we specify $b = 1/2$ in (3.33) (the case of the geometric mean) then either X_1 is logarithmically degenerate (parts (iii) and (iv) above), or $Y_1 = \log|X_1|$ does not have a finite variance. Actually, even if we assume that $E = E(\log|X_1|)$ is finite, repeated application of (3.34), together with the strong law of large numbers, leads to $Y_1 = E$ a.s., and thus, once more part (iv) of Theorem 3.13 applies to X_1.

Let us now assume that $E(\log|X_1|)$ is not finite. Then, from (3.34), Y_1 is the limit of the arithmetical mean of the Y_j, a property characterizing the Cauchy family of distributions. This fact, together with the rule (3.35) for the signs yields

Theorem 3.14 *Let* $b = 1/2$ *in (3.32) and (3.33). Then* X_1 *is a solution of (3.33) if, and only if, either* $X_1 \geq 0$ *or* X_1 *is symmetric and one of the conditions*

(i) $\log|X_j|$ *is degenerated*

(ii) X_1 *is logarithmically Cauchy*

holds.

3.5 A characterization of the normal distribution

In this section we turn to characterizations obtained by random normalization. In terms of products, this is about taking a product divided by another statistics of the same variables as appearing in the product, and investigate the distributional characteristics of such ratio. We start by proving the following limit theorem due to Galambos and Simonelli [65].

Theorem 3.15 *Let $X_1, X_2, \cdots$ be a sequence of independent random variables with $P(X_j = 0) = 0$, $j \geq 1$, and assume that*

$$\lim_{n\to+\infty} P\Big(sign(X_1X_2\cdots X_n) = -\Big) = p$$

exists. There exists a function $g(n)$, $g(n) \to +\infty$ with n, and a random variable U such that

$$\frac{g(n)X_1X_2\cdots X_n}{\sqrt{\sum X_{i_1}^2 X_{i_2}^2 \cdots X_{i_{n-1}}^2}} \longrightarrow U \qquad \textit{completely,} \tag{3.36}$$

where $\sum$ is summation over all integers $1 \leq i_1 < i_2 < \cdots < i_{n-1} \leq n$, if, and only if, there exists a random variable Y such that

$$\frac{\frac{1}{X_1^2} + \frac{1}{X_2^2} + \cdots + \frac{1}{X_n^2}}{g(n)^2} \longrightarrow Y \qquad \textit{completely.} \tag{3.37}$$

Whenever the above limits exist, U is always of the form $U \stackrel{d}{=} \nu W$, $W \geq 0$, ν and W are independent, and $P(W \leq a) = 1 - P(Y \leq \frac{1}{a^2})$, for all continuity points $a > 0$ of Y.

Proof. Let

$$T_n = \frac{g(n)X_1X_2\cdots X_n}{\sqrt{\sum X_{i_1}^2 X_{i_2}^2 \cdots X_{i_{n-1}}^2}} = \frac{g(n)sign(X_1X_2\cdots X_n)}{\sqrt{\frac{1}{X_1^2} + \frac{1}{X_2^2} + \cdots + \frac{1}{X_n^2}}},$$

and for arbitrary $1 \leq i < j$ let

$$V_{i,j} = \frac{\frac{1}{X_i^2} + \frac{1}{X_{i+1}^2} + \cdots + \frac{1}{X_j^2}}{g(n)^2},$$

and

$$\nu_{i,j} = sign(X_iX_{i+1}\cdots X_j).$$

Since for arbitrary $a > 0$,

$$I_{\{|T_n|\leq a\}} = I_{\{V_{1,n}\geq \frac{1}{a^2}\}},$$

where $I_{\{\dots\}}$ is the indicator of the expression in the dotted space, (3.36) trivially implies (3.37). So we only need to show that (3.37) implies (3.36). Let us assume that $V_{1,n}$ converges completely to a random variable Y. The continuity of $F_{X_i}(x)$ at $x = 0$, $i \geq 1$, implies that for arbitrary k, $V_{1,k}$ converges to zero in probability, as $n \to +\infty$, which gives

$$\lim_{n\to+\infty} \left| E\left(I_{\{V_{1,n}\geq a\}} I_{\{\nu_{1,n}=\eta\}}\right) - E\left(I_{\{V_{k,n}\geq a\}} I_{\{\nu_{1,n}=\eta\}}\right)\right| = 0, \quad (3.38)$$

$\eta = -, +$, for all continuity points a of Y. We start by considering the case

$$\lim_{n\to+\infty} P(sign(X_1X_2\cdots X_n) = -) = \frac{1}{2}. \quad (3.39)$$

For all continuity points $a > 0$ of Y,

$$\begin{aligned}
&\lim_{n\to+\infty} \left|E\left(I_{\{|T_n|\leq a\}} I_{\{\nu_{1,n}=-\}}\right) - E\left(I_{\{|T_n|\leq a\}} I_{\{\nu_{1,n}=+\}}\right)\right| \\
= &\lim_{n\to+\infty} \left|E\left(I_{\{V_{1,n}\geq \frac{1}{a^2}\}} I_{\{\nu_{1,n}=-\}}\right) - E\left(I_{\{V_{1,n}\geq \frac{1}{a^2}\}} I_{\{\nu_{1,n}=+\}}\right)\right| \\
= &\lim_{n\to+\infty} \left|E\left(I_{\{V_{k,n}\geq \frac{1}{a^2}\}} I_{\{\nu_{1,n}=-\}}\right) - E\left(I_{\{V_{k,n}\geq \frac{1}{a^2}\}} I_{\{\nu_{1,n}=+\}}\right)\right| \\
= &\lim_{n\to+\infty} \Big|E(I_{\{\nu_{1,k-1}=-\}}) E\left(I_{\{V_{k,n}\geq \frac{1}{a^2}\}} I_{\{\nu_{k,n}=+\}}\right) + \\
&\qquad E(I_{\{\nu_{1,k-1}=+\}}) E\left(I_{\{V_{k,n}\geq \frac{1}{a^2}\}} I_{\{\nu_{k,n}=-\}}\right) - \\
&\qquad E(I_{\{\nu_{1,k-1}=-\}}) E\left(I_{\{V_{k,n}\geq \frac{1}{a^2}\}} I_{\{\nu_{k,n}=-\}}\right) - \\
&\qquad E(I_{\{\nu_{1,k-1}=+\}}) E\left(I_{\{V_{k,n}\geq \frac{1}{a^2}\}} I_{\{\nu_{k,n}=+\}}\right)\Big| \\
\leq &\; 2\left|E(I_{\{\nu_{1,k-1}=-\}}) - E(I_{\{\nu_{1,k-1}=+\}})\right|,
\end{aligned}$$

where in the second equation we have exploited (3.38) and in the third equation the independence of the X_i's. By (3.39) the right hand side of the above inequality can be made arbitrarily small by choosing k sufficiently large, thus proving the validity of (3.36).

Next let us assume that

$$\lim_{n\to+\infty} P(sign(X_1X_2\cdots X_n) = -) = p, \qquad p \neq \frac{1}{2}, \quad (3.40)$$

or, equivalently (see Theorem 1.21),

$$\sum_{i=1}^{+\infty} P(X_i \leq 0) < +\infty. \tag{3.41}$$

In this case for all continuity points $a > 0$ of Y,

$$\begin{aligned}
&\lim_{n\to+\infty} \left|E\left(I_{\{|T_n|\leq a\}} I_{\{\nu_{1,n}=-\}}\right) - pE\left(I_{\{|T_n|\leq a\}}\right)\right| \\
=& \lim_{n\to+\infty} \left|E\left(I_{\{V_{1,n}\geq \frac{1}{a^2}\}} I_{\{\nu_{1,n}=-\}}\right) - pE\left(I_{\{V_{1,n}\geq \frac{1}{a^2}\}}\right)\right| \\
=& \lim_{n\to+\infty} \left|E\left(I_{\{V_{k,n}\geq \frac{1}{a^2}\}} I_{\{\nu_{1,n}=-\}}\right) - pE\left(I_{\{V_{k,n}\geq \frac{1}{a^2}\}}\right)\right| \\
=& \lim_{n\to+\infty} \Big|E(I_{\{\nu_{1,k-1}=-\}})E\left(I_{\{V_{k,n}\geq \frac{1}{a^2}\}} I_{\{\nu_{k,n}=+\}}\right) + \\
& E(I_{\{\nu_{1,k-1}=+\}})E\left(I_{\{V_{k,n}\geq \frac{1}{a^2}\}} I_{\{\nu_{k,n}=-\}}\right) - \\
& pE\left(I_{\{V_{k,n}\geq \frac{1}{a^2}\}}\right)\Big| \\
\leq& \lim_{n\to+\infty} \left(|E(I_{\{\nu_{1,k-1}=-\}}) - p| + 2E(I_{\{\nu_{k,n}=-\}})\right) \\
\leq& |E(I_{\{\nu_{1,k-1}=-\}}) - p| + 2\sum_{i=k}^{+\infty} P(X_i < 0),
\end{aligned}$$

where in the second and third equation we have again used (3.38) and the independence of the X_i's, respectively. The right hand side of the last inequality can be made arbitrarily small by choosing k to be sufficiently large, see (3.40) and (3.41), which implies the validity of (3.36). The above discussion also proves the claimed form of U and the relationship between W and Y. The proof of the theorem is complete. ∎

We now apply Theorem 3.15 to solve characterization problems. The first theorem is due to Galambos and Simonelli [65]. In the special case of symmetry the result was proved by Quine [121].

Theorem 3.16 *Let $X_1, X_2, \cdots$ be a sequence of independent identically distributed random variables with common distribution function*

$F(x)$, $0 < P(X_1 < 0) < 1$. Then

$$\frac{g(n)X_1X_2\cdots X_n}{\sqrt{\sum X_{i_1}^2 X_{i_2}^2 \cdots X_{i_{n-1}}^2}} \longrightarrow U \qquad \textit{completely,} \tag{3.42}$$

where U is a normal random variable with mean 0 and variance σ^2, iff

(i) $T(x) = F(x) - F(-x)$ is regularly varying at $x = 0$ with exponent one;

and

$$(ii) \lim_{n\to+\infty} \frac{g(n)^2}{\inf\left\{x : F(\frac{1}{\sqrt{x}}) - F(\frac{-1}{\sqrt{x}}) = \frac{1}{n\sigma\sqrt{2\pi}}\right\}} = 1.$$

PROOF. Let T_n and $V_{1,n}$ to be as in the proof of Theorem 3.15. We start by assuming T_n converges to U completely, U normal with mean 0 and variance σ^2. Then $V_{1,n}$ converges to $1/U^2$ completely, where $1/U^2$ has characteristic function

$$\phi_{\frac{1}{U^2}}(t) = e^{-\frac{|t|^{1/2}}{\sigma}(1-isign(t))} \tag{3.43}$$

(see Moran [114], p. 271, for $\sigma = 1$). The validity of conditions (i) and (ii) of Theorem 3.37 can be proved by using well known results for sums of independent random variables (our proof is along the line of Quine's [121]). $F_{1/X_1^2}(x)$ is in the domain of attraction of the stable law $F_{1/U^2}(x)$ with characteristic exponent $\alpha = 1/2$ (see (3.43)), iff $1-F_{1/X_1^2}(x)+F_{1/X_1^2}(-x)$ is regularly varying at infinity with exponent -1/2 (see [58], p. 260), or, equivalently, iff $F_{X_1}(x) - F_{X_1}(-x)$ is regularly varying at 0 with exponent 1. Thus condition (i) holds. This implies $F_{X_1}(x)$ is continuous at $x = 0$, and hence a suitable normalizing sequence b_n such that $(1/X_1^2 + 1/X_2^2 + \cdots + 1/X_n^2)/b_n$ converges to $1/U^2$ completely is given by

$$\begin{aligned} b_n &= \inf\left\{x : P(\frac{1}{X_1^2} \geq x) = \frac{c_1 + c_2}{n}\right\} \\ &= \inf\left\{x : F_{X_1}(\frac{1}{\sqrt{x}}) - F_{X_1}(\frac{-1}{\sqrt{x}}) = \frac{c_1 + c_2}{n}\right\}, \end{aligned}$$

where c_1 and c_2 are parameters determined by the Lévy-Khintchine spectral function of (3.43) (see [58], p. 244). From $P(1/X_1^2 > 0) = 1$ and (3.43) we obtain $c_2 = 0$, and $\sqrt{2\pi}c_1 = 1/\sigma$, which gives $c_1 = 1/(\sigma\sqrt{2\pi})$ (see [58], p. 250 and p. 261). Hence $g(n)$ must satisfy condition (ii). For the converse, let us assume the validity of conditions (i) and (ii). Then by our previous discussion we immediately have that $V_{1,n}$ converges to $1/U^2$ completely. Since $0 < P(X_1 \leq 0) < 1$, and by (i), $P(X_1 = 0) = 0$,

$$\lim_{n\to+\infty} P(sign(X_1X_2\cdots X_n) = -) = \frac{1}{2}.$$

From this and Theorem 3.15 the validity of Theorem 3.16 immediately follows. The proof is now complete. ∎

From our previous discussion it is easy to see that if X is a normal random variable with expected value zero, then Theorem 3.16 applies with $g(n) = \text{const.}\, n$. However, as the next example demonstrates, linear functions are not the only choice.

Let X be a random variable with symmetric distribution function $F_X(x)$ given by

$$F_X(x) = \begin{cases} \frac{1}{2} & x = 0 \\ 1 - \frac{1}{2}e^{-x(\log x)^2} & 0 < x < e^{-2} \\ 1 - e^{-10x} & x \geq e^{-2} \end{cases}$$

$(F(x) = 1 - F(-x)$ if $x < 0)$. Let $k > 0$ be arbitrary. Then

$$\begin{aligned}\lim_{x\to 0} \frac{F_X(kx) - F_X(-kx)}{F_X(x) - F_X(-x)} &= \lim_{x\to 0} \frac{1 - e^{-kx(\log kx)^2}}{1 - e^{-x(\log x)^2}} \\ &= \lim_{x\to 0} \frac{kx(\log kx)^2(1 + +o(1))}{x(\log x)^2(1 + o(1))} = k.\end{aligned}$$

Hence $F_X(x)$ satisfies condition (i) of Theorem 3.16. To find a function $g(n)$ that satisfies condition (ii), we must require for $g(n)^2$ to be asymptotic to

$$\inf\left\{x : F_X(\frac{1}{\sqrt{x}}) - F_X(-\frac{1}{\sqrt{x}}) = \frac{\text{const.}}{n}\right\},$$

which can be assumed to be equal to

$$\inf\left\{x:\ 1-e^{-\frac{1}{\sqrt{x}}(\log\frac{1}{\sqrt{x}})^2}=\frac{\text{const.}}{n}\right\}.$$

To determine the form of $g(n)$ we consider the quantity

$$1-e^{-\frac{1}{\sqrt{x}}(\log\frac{1}{\sqrt{x}})^2}=\frac{1}{\sqrt{x}}(\log\frac{1}{\sqrt{x}})^2\Big(1+o(1)\Big).$$

For this quantity to be asymptotic to const./n we must have x asymptotic to $n^2h(n)^2$, where $h(n)$ is chosen so that

$$\frac{(\log\frac{1}{nh(n)})^2}{h(n)}\longrightarrow 1, \qquad \text{as } n\to+\infty,$$

which can be obtained by letting $h(n)=(\log n)^2$, for example. Hence a suitable choice for $g(n)$ is $Cn(\log n)^2$, where C is a constant determined by σ^2, the variance of U.

Next we turn to the finite case, $n=2$, and consider the statistic

$$U=\frac{X_1X_2}{\sqrt{X_1^2+X_2^2}}. \tag{3.44}$$

Characterization problems for (3.44) were considered by Shepp [137], who proved that if X_1 and X_2 are independent, and X_i is normal $N(0,V_i)$, $i=1,2$, then U is normal $N(0,V_1V_2/(\sqrt{V_1}+\sqrt{V_2})^2)$ (see also Quine [120]).

Let X_1 and X_2 be independent identically distributed random variables. We consider the distributional equation

$$X_1\stackrel{d}{=}b\frac{X_1X_2}{\sqrt{X_1^2+X_2^2}}, \qquad b>0, \tag{3.45}$$

and ask whether for some constant $b>0$ (3.45) has any other solution besides the normal distribution. The following theorem is due to Galambos and Simonelli [65] and [66].

Theorem 3.17 *Let X_1,X_2 be independent random variables with common distribution function $F_{X_1}(x)$, $0<F_{X_1}(0)<1$. We assume*

$F_{X_1}(x) - F_{X_1}(-x)$ is regularly varying at $x = 0$ with exponent one. Then

$$X_1 \stackrel{d}{=} b\frac{X_1X_2}{\sqrt{X_1^2+X_2^2}}, \qquad b > 0, \tag{3.46}$$

has a solution iff $b = 2$ and X_1 is normal with mean 0 and arbitrary variance.

PROOF. Let X_1 and X_2 be as in the theorem. Clearly (3.46) implies

$$\frac{1}{|X_1|} \stackrel{d}{=} b^{-1}\left(\sqrt{\frac{1}{X_1^2}+\frac{1}{X_2^2}}\right),$$

which implies

$$W_1^2 \stackrel{d}{=} b^{-2}\left(W_1^2 + W_2^2\right), \tag{3.47}$$

where $W_i = 1/X_i^2$, $i = 1, 2$. By iterating (3.47) one obtains

$$\begin{aligned} W_1 &\stackrel{d}{=} b^{-4}\left(W_1 + W_2 + W_3 + W_4\right) \stackrel{d}{=} \cdots \\ &\stackrel{d}{=} \lim_{n\to+\infty} b^{-2n}\left(W_1 + W_2 + \cdots + W_{2n}\right) \end{aligned}$$

which implies W_1 is sum-stable. To determine W_1 we exploit the representation of the characteristic function of a stable distribution which is uniquely defined by the values of four numerical parameters, s, c, α, and γ (see [58], p. 247). As in the proof of Theorem 3.16, our initial assumption that $F_{X_1}(x) - F_{X_1}(-x)$ is regularly varying at $x = 0$ with exponent one immediately give that $\alpha = 1/2$. To determine the other parameter we exploit (3.47) which implies that the characteristic function of W_1, say $\phi(t)$, must satisfy

$$\phi(t) = \phi\left(\frac{t}{b}\right)^2.$$

The above equation gives that

$$\frac{1}{b} = \frac{1}{2} \quad \text{and} \quad s = 0$$

(see [58], Exercise 2 p. 251 and its solution, p. 439, for example). Since W_1 is sum-stable with $\alpha = 1/2$, there exist constants $c_1 \geq 0$ and $c_2 \geq 0$, $c_1 + c_2 > 0$, such that $\gamma = (c_2 - c_1)/(c_2 + c_1)$ (see [58], p. 250). $W_1 \geq 0$ gives $c_2 = 0$ (see [58], pp. 260 and 261), and hence $\gamma = -1$ and

$$\phi_{W_1}(t) = e^{-c|t|^{1/2}(1-i\frac{t}{|t|})}. \tag{3.48}$$

To complete our proof we show that any solution of (3.46) must be symmetric. In fact, if we iterate (3.46) we obtain

$$\begin{aligned} X_1 &\stackrel{d}{=} b\frac{\frac{bX_1X_2}{\sqrt{X_1^2+X_2^2}}\frac{bX_3X_4}{\sqrt{X_3^2+X_4^2}}}{\sqrt{\frac{b^2X_1^2X_2^2}{X_1^2+X_2^2}+\frac{b^2X_3^2X_4^2}{X_1^2+X_2^2}}} \\ &\stackrel{d}{=} b^2\frac{X_1X_2X_3X_4}{\sqrt{X_1^2X_2^2X_3^2+X_1^2X_2^2X_4^2+X_1^2X_3^2X_4^2+X_2^2X_3^2X_4^2}} \stackrel{d}{=} \cdots \\ &\stackrel{d}{=} \lim_{n\to+\infty} b^n\frac{X_1X_2\cdots X_{2^n}}{\sqrt{\sum X_{i_1}^2X_{i_2}^2\cdots X_{i_{2^n-1}}^2}}, \end{aligned}$$

where $X_1, X_2, \cdots$ are i.i.d., and $\sum$ is as in Theorem 3.15. Hence, if $b > 1$, Theorem 3.15 implies X_1 must be symmetric.

We have already mentioned that X_1 normal $N(0, V)$ is a solution of (3.46). If Y is another solution, then by (3.48) we must have $1/Y^2 \stackrel{d}{=} c'(1/Z^2)$, for some $c' > 0$, which gives, under the additional requirement that Y is symmetric, $Y \in N(0, V)$, V uniquely determined by c'. The proof is now complete. ∎

3.6 Bivariate distributions

Let (X, Y) be a random vector with distribution function $F(x, y) = P(X \leq x, Y \leq y)$. While $F(x, y)$ uniquely determines the distribution functions $F_X(x) = P(X \leq x)$ and $F_Y(y) = P(Y \leq y)$ of the components X and Y by simply letting $y \to +\infty$ and $x \to +\infty$ in $F(x, y)$, the converse is, of course, not true. There are whole families of bivariate distributions $F(x, y)$ with specified marginal distributions $F_X(x)$ and $F_Y(y)$. This fact appears in several textbooks dealing with multivariate distributions, or the reader may want to consult

Mardia [110] and Ord [115] for more detailed accounts. One can easily generate a family of bivariate distributions from the marginals by the simple form

$$\begin{aligned} F(x,y) &= P(X \le x, Y \le y) \\ &= 1 - P(X > x) - P(Y > y) + P(X > x, Y > y) \\ &= 1 - \Big(1 - F_X(x)\Big) - (1 - F_Y(y)\Big) + P(X > x, Y > y). \end{aligned}$$

One can now choose the survival function $G(x,y) = P(X > x, Y > y)$ in a variety of ways, always making sure that the resulting expression should give a distribution function for $F(x,y)$. For example, if both $F_X(x) = 1 - e^{-x}$, $x \ge 0$, and $F_Y(y) = 1 - e^{-y}$, $y \ge 0$, are unit exponential variables, the choices (in all cases $x \ge 0$, $y \ge 0$)

$$\begin{aligned} G_1(x,y) &= e^{-x-y} \\ G_2(x,y) &= e^{-\max(x,y)} \\ G_3(x,y) &= e^{-(x^2+y^2)^{1/2}} \end{aligned}$$

all have appeared in the literature, and each resulting $F_j(x,y)$, $1 \le j \le 3$, is known as a bivariate exponential distribution. The case $G_1(x,y)$ for $F(x,y)$ is, of course, the case of independence of X and Y.

Since marginals are inadequate for characterizing $F(x,y)$, one should look for some other properties of the vector (X,Y) that leads to characterizing $F(x,y)$. One successful direction turned out to be using the conditional distributions

$$F_{(Y)}(x|y) = P(X \le x | Y = y) \tag{3.49}$$
$$F_{(X)}(y|x) = P(Y \le y | X = x). \tag{3.50}$$

While the conditional distributions $F_{(Y)}(x|y)$ and $F_{(X)}(y|x)$ alone do not yet characterize $F(x,y)$, they provide a nice way of describing all $F(x,y)$ with given conditionals in (3.49) and (3.50), from which slight additional assumptions lead to characterization theorems. While scattered results had been available in the literature, applicable to specific $F(x,y)$, the first fundamental result appeared in Castillo and Galambos [27], [29], and [28], which lead to a variety of generalizations and, consequently, to a vast literature. We present in this section the method and the result of Castillo and Galambos. It is fitting to deal with the normal case because of the frequent appearance of the logarithmically normal distribution in this book. The transformation to the logarithmically normal distribution is quite straightforward, while extensions and generalizations to other bivariate distributions can be found in a nicely written book by Arnold, Castillo, and Sarabia [9], which contains a comprehensive list of references.

Our aim is to characterize the traditional bivariate normal density function

$$n(x,y) = \frac{e^{-\frac{1}{2(1-\rho^2)}\left[(\frac{x-m_1}{\sigma_1})^2+(\frac{y-m_2}{\sigma_2})^2-2\rho(\frac{x-m_1}{\sigma_1})(\frac{y-m_2}{\sigma_2})\right]}}{2\pi\sigma_1\sigma_2\sqrt{1-\rho^2}} \tag{3.51}$$

among all bivariate densities $f(x,y) = f_{X,Y}(x,y)$ whose conditional densities (the derivatives of (3.49) and (3.50))

$$f_{(Y)}(x|y) = \frac{1}{\sqrt{2\pi}\sigma_1(y)} e^{-\frac{(x-m_1(y))^2}{2\sigma_1^2(y)}} \tag{3.52}$$

and

$$f_{(X)}(y|x) = \frac{1}{\sqrt{2\pi}\sigma_2(x)} e^{-\frac{(y-m_2(x))^2}{2\sigma_2^2(x)}} \tag{3.53}$$

are normal. Let the marginal densities of X and Y be $f_X(x)$ and $f_Y(y)$, respectively. Then, by

$$f(x,y) = f_X(x) f_{(X)}(y|x) = f_Y(y) f_{(Y)}(x|y), \tag{3.54}$$

(3.52) and (3.53) lead to the equation

$$\frac{f_X(x)}{\sigma_2(x)} e^{-\frac{(y-m_2(x))^2}{2\sigma_2^2(x)}} = \frac{f_Y(y)}{\sigma_1(y)} e^{-\frac{(x-m_1(y))^2}{2\sigma_1^2(y)}}. \tag{3.55}$$

By definition, $\sigma_2(x) > 0$ and $\sigma_1(y) > 0$ for all x and y, while (3.55) implies that $f_X(x) > 0$ and $f_Y(y) > 0$. We take logarithms of both sides. We get

$$g(x) - \frac{1}{2}\Big(\frac{y - m_2(x)}{\sigma_2^2(x)}\Big)^2 = h(y) - \frac{1}{2}\Big(\frac{x - m_1(y)}{\sigma_1^2(y)}\Big)^2, \tag{3.56}$$

where $g(x) = \log[f_X(x)/\sigma_2(x)]$ and $h(y) = \log[f_Y(y)/\sigma_1(y)]$. Upon expanding the terms in (3.56), after rearrangement, we have

$$\sigma_1^2(y)[2g(x)\sigma_2^2(x) - m_2^2(x)] + 2m_2(x)y\sigma_1^2(y) - y^2\sigma_1(y) \quad (3.57)$$
$$= \sigma_2^2(x)[2h(y)\sigma_1^2(y) - m_1^2(y)] + 2m_1(y)x\sigma_2^2(x) - x^2\sigma_2^2(x).$$

Note that the functions $\sigma_1^2(y)$, $y\sigma_1^2(y)$ and $y^2\sigma_1^2(y)$ are linearly independent, and so are the functions $\sigma_2^2(x)$, $x\sigma_2^2(x)$ and $x^2\sigma_2^2(x)$. It then follows from (3.57) (see Aczél [4], p.161) that each of $2g(x)\sigma_2^2(x)-m_2^2(x)$, $2m_2(x)$ and -1 is a linear combination of the functions $\sigma_2^2(x)$, $x\sigma_2^2(x)$ and $x^2\sigma_2^2(x)$. Similarly, each of the functions $2h(y)\sigma_1^2(y) - m_1^2(y)$, $2m_1(y)$ and -1 is a linear combination of $\sigma_1^2(y)$, $y\sigma_1^2(y)$ and $y^2\sigma_1^2(y)$. Furthermore, these two systems of 'linear equations' are related through (3.57), restricting the coefficients appearing in them. One gets (the details are given in Aczél [4]), with appropriate constants A, B, C, D, E, F, G, H and J,

$$g(x) = \frac{1}{2}\Big[A + Bx + Cx^2 + \frac{(G + Hx + Jx^2)^2}{D + Ex + Fx^2}\Big], \quad (3.58)$$

$$h(y) = \frac{1}{2}\Big[A + 2Gy - Dy^2 - \frac{(B + 2Hy - Ey^2)^2}{4(C + 2Jy + Fy^2)}\Big], \quad (3.59)$$

$$m_1(y) = -\frac{B + 2Hy - Ey^2}{2(C + 2Jy - Fy^2)}, \quad (3.60)$$

$$m_2(x) = \frac{G + Hx + Jx^2}{D + Ex + Fx^2}, \quad (3.61)$$

$$\sigma_1^2(y) = -\frac{1}{C + 2Jy - Fy^2}, \quad (3.62)$$

$$\sigma_2^2(x) = \frac{1}{D + Ex + Fx^2}, \quad (3.63)$$

$$f_X(x) = \Big(D + Ex + Fx^2\Big)^{-\frac{1}{2}} e^{g(x)}, \quad (3.64)$$

$$f_Y(y) = \Big(-C - 2Jy + Fy^2\Big)^{-\frac{1}{2}} e^{h(y)}, \quad (3.65)$$

from which system (3.52)-(3.54) entail the general solution

$$f(x,y) = \frac{1}{2\pi} e^{\frac{1}{2}Q(x,y)} \tag{3.66}$$

with

$$\begin{aligned} Q(x,y) &= A + Bx + Cx^2 + 2Gy - Dy^2 + 2Hxy \\ &\quad +2Jx^2y - Exy^2 - Fx^2y^2. \end{aligned} \tag{3.67}$$

The coefficients in (3.58)-(3.67) are restricted by the nature of the functions involved in (3.58)-(3.67): variances are positive, densities are nonnegative; furthermore, $f_X(x)$, $f_Y(y)$, and $f(x,y)$ must be integrable. These facts impose the following rules on A, B, C, D, E, F, G, H and J:

either (a) $E = F = J = 0$, $C < 0$, $D > 0$, and $H^2 < -CD$;

or (b) $F > 0$, $E^2 < 4DF$, and $J^2 < -CF$ (with $D > 0$, $C < 0$).

Note that (a) leads to the classical bivariate normal distribution for $f(x,y)$ of (3.66), and then the marginal and conditional distributions $f_X(x)$, $f_Y(y)$, and $f_{(X)}(y|x)$, $f_{(Y)}(x|y)$, respectively, are normal, the conditional variances $\sigma_1(y)$ and $\sigma_2(x)$ of (3.62)-(3.63) are positive constants, and the regression lines $m_1(y)$ and $m_2(x)$ of (3.60)-(3.61) are linear.

On the other hand, in case (b), $f(x,y)$ of (3.66)-(3.67) is not traditional bivariate normal distribution. We take an example from Castillo and Galambos [29].

Example 3.8 Let $f(x,y)$ be the density function in (3.66) with

$$\frac{1}{2} Q(x,y) = A - x^2 - y^2 - 2x^2y - 2xy^2 - 2x^2y^2, \quad x, y \text{ all real.}$$

We then know that the conditionals $f_{(X)}(y|x)$ and $f_{(Y)}(x|y)$ are normal, as in the case for all functions in (3.66). However, the conditional variances $\sigma_1^2(y)$ and $\sigma_2^2(x)$ are not constant, and the regression lines $m_1(y)$ and $m_2(x)$ are not linear. In fact, since our choice of $Q(x,y)$ yields $B = G = H = 0$, $C = -D = J = -2$, and $E = F = 4$, we get by substitution into (3.61) and (3.63),

$$m_2(x) = -\frac{x^2}{1 + 2x + 2x^2} \quad \text{and} \quad \sigma_2^2(x) = \frac{1}{2 + 4x + 4x^2}.$$

The value of A is to be determined by integrating $f(x, y)$ on the whole plane and setting it equal to one.

There are other quite surprising densities $f(x, y)$ under the choice (b) for the parameters. One can make regression lines constant but the conditional variances are not constant, and one can generate densities $f(x, y)$ which are not unimodal. We do not analyze this further. We concentrate here on characterizations, and establish the following result of Castillo and Galambos [29] and [28] (we use the notation of (3.52)-3.53)).

Theorem 3.18 *The density function $f(x, y)$ is a classical bivariate normal density if, and only if, both $f_{(X)}(y|x)$ and $f_{(Y)}(x|y)$ are normal and any one of the following properties hold:*

(i) either $\sigma_2^2(x)$ or $\sigma_1^2(y)$ is constant;

(ii) either $\lim_{x\to+\infty} x^2\sigma_2^2(x) = +\infty$ or $\lim_{y\to+\infty} y^2\sigma_1^2(y) = +\infty$;

(iii) $\liminf_{x\to+\infty} \sigma_2^2(x) \neq 0$ or $\liminf_{y\to+\infty} \sigma_1^2(y) \neq 0$;

(iv) $m_1(y)$ or $m_2(x)$ is linear and not constant.

PROOF. The preparation leading to the statement of the theorem makes the proof almost immediate. First note that all properties in the theorem are valid for the classical bivariate density function $n(x, y)$ of (3.51). Hence, only the converse needs proof. So we assume that (3.52) and (3.53) are valid. Therefore, $f(x, y)$ has the form (3.66) and (3.67), and all expressions (3.58)-(3.65) are applicable. Furthermore, either Rule (a) or Rule (b) applies to all coefficients in (3.58)-(3.67). Now, if $\sigma_1^2(y)$ is a constant, then (3.62) entails that $J = F = 0$, which in turn implies that Rule (a) is to be followed. That is, $f(x, y)$ is bivariate normal. The assumption that $\sigma_2^2(x)$ is a constant leads similarly to the classical case.
(ii) If $y^2\sigma_1^2(y) \to +\infty$ with y, (3.62) yields that $F = 0$. Once again, Rule (a) applies to the constants, which implies classical normality. The case $x^2\sigma_2^2(x) \to +\infty$ with x is similar.
(iii) If $\liminf_{x\to+\infty} \sigma_2^2(x) \neq 0$, then, in view of (3.63), $E = F = 0$, which once again leads to Rule (a), and to classical normality.
(iv) If $m_1(y)$ is linear but not a constant, we have from (3.60) that

$E = J = F = 0$ which implies Rule (a) and classical bivariate normality for $f(x, y)$
The theorem is established. ■

We conclude with another example.

Example 3.9 Take $B = E = G = H = J = 0$ in (3.67), and apply Rule (b) for the rest of the coefficients. We have from (3.66) and (3.67),

$$f(x, y) = Ke^{-\frac{1}{2}\left[C^* x^2 + Dy^2 + Fx^2 y^2\right]} \tag{3.68}$$

where $C^* = -C > 0$, $D > 0$ and $F > 0$, and $K > 0$ is the normalizing constant.

By substitution into (3.64) and (3.65), with (3.58) and (3.59), we get the following forms for the marginals

$$f_X(x) = K_1 \big(D + Fx^2\big)^{-\frac{1}{2}} e^{-\frac{1}{2}C^* x^2}, \quad K_1 > 0 \text{ normalizing constant,}$$

and

$$f_Y(y) = K_2 \big(C^* + Fy^2\big)^{-\frac{1}{2}} e^{-\frac{1}{2}Dy^2}, \quad K_2 > 0 \text{ normalizing constant.}$$

These are not normal whenever $F > 0$ (and normal if $F = 0$ which takes us back to Rule (a)).

By simple computation one can demonstrate that (X, Y) with density (3.68) can be uncorrelated and yet not independent. For an actual data evaluation in which (3.68) is chosen as the underlying density see Arnold and Strauss [9].

Chapter 4

Applications to interacting particle systems

4.1 Introduction

Products of indicator random variables arise naturally in the study of the ergodic theory of interacting particle systems (IPS) and cellular automata. We stress the first of these topics, and use products to develop a unifying method to study the ergodic theory of several models of IPS. At the end of the chapter we give an example of the role of products in cellular automata, and we consider a simple version of what is known as the game of life of Conway [69].

The aim of this section is to introduce the main definitions and properties of IPS which will be needed in the rest of the chapter. We don't go into the details of the proofs, and we refer the interested reader to the book of Liggett [99], Chapter 1, where one can find the proofs of all the results presented below.

Let Z^d represent the d-dimensional integer lattice, and $\Lambda = \{0, 1\}^{Z^d}$. In what follows η, ζ will denote elements of Λ, and they are called configurations, and x, y, z will denote elements of Z^d, and they are called sites. The value of η at x is denoted by $\eta(x)$. By $|\eta|$ we denote the number of x's in Z^d such that $\eta(x) = 1$. Note that each $\eta \in \Lambda$ can be uniquely identified with a subset A of Z^d, i.e., A is the collection of all $x \in Z^d$ such that $\eta(x) = 1$. In fact when we want to stress the finiteness property of a configuration η, we use this subset representation of elements of Λ. IPS are a special type of continuous time Markov processes on Λ, which we denote by $\{\eta_t^\zeta\}_{t\in[0,+\infty)}$. η_t^ζ describes the configuration of the process at time t, and ζ describes its initial configuration, i.e., $\eta_0^\zeta = \zeta$. We can view such a system as a collection of particles that lives in Z^d, which evolves in time according to some specified but random dynamics. The dynamics of the system can be described by giving nonnegative uniformly bounded functions

$$c_T : \Lambda \times \Lambda \to [0, +\infty) \qquad T \subset Z^d,\ |T| < +\infty,$$

where $c_T(\eta, \zeta)$ describes the rate at which configuration η changes into configuration ζ, where ζ differs from η only at sites in T. This collection of functions determines the evolution of the system through the equation

$$\begin{aligned} &P\left[\eta_{t+h}(x) \neq \eta_t(x) \text{ for some } x \in T | \eta_t = \eta\right] \\ &\quad = \sum_{\zeta} c_T(\eta, \zeta)\, h + o(h) \quad \text{as } h \to 0. \end{aligned} \tag{4.1}$$

Let $D(\Lambda) = \{f_A\}$, where $f_A(\eta)$ is a continuous function on Λ, whose values only depend on the value of η at sites in A. For any η, ζ, and any finite $T \subset Z^d$, we define

$$\eta^{\zeta(T)}(x) = \begin{cases} \zeta(x) & \text{if } x \in T \\ \eta(x) & \text{if } x \notin T. \end{cases}$$

The functions $c_T(\eta, \zeta)$ can be utilized to define an operator Ψ on $D(\Lambda)$ by

$$\Psi f_A(\eta) = \sum_T \sum_\zeta c_T(\eta, \zeta)\left[f_A(\eta^{\zeta(T)}) - f_A(\eta)\right].$$

If the functions $c_T(\eta, \zeta)$ satisfy some regularity conditions, then Ψ is a Markov generator. From classical results from the theory of Markov processes, a Markov generator on $D(\Lambda)$ generates a Feller semigroup $\{S(t),\, t \geq 0\}$ on the collection of continuous functions on Λ, and $\{S(t),\, t \geq 0\}$ generates a Feller process $\{P^\eta,\, \eta \in \Lambda\}$ in Λ. The Feller process and the Feller semigroup are related by

$$S(t)f(\eta) = E^\eta\left[f(\eta_t)\right],$$

for all continuous functions $f(\eta)$ on Λ, $\eta \in \Lambda$, and $t \geq 0$.

Throughout this chapter we assume that the $c_T(\eta, \zeta)$ are translation invariant. That is, for all x,

$$P\Big(\eta_t^\zeta \in \cdot\Big) = P\Big(\eta_t^{\zeta+x} \in \cdot + x\Big). \tag{4.2}$$

The regularity conditions for $c_T(\eta, \zeta)$, which guarantees that Ψ is a Markov generator, are very simple in the case $c_T(\eta, \zeta) = 0$ unless $|T| = 1$ (spin systems). In this case we write $c_T(\eta, \zeta) = c(x, \eta)$. Then the collection of functions $c(x, \eta)$ generates a Markov generator whenever

$$\sup_{x \in Z^d} \sum_{u \in Z^d} \sup_{\eta \in \Lambda} \Big| c(x, \eta) - c(x, \eta_u) \Big| < +\infty, \tag{4.3}$$

where

$$\eta_u(x) = \begin{cases} \eta(x) & \text{if } x \neq u \\ 1 - \eta(x) & \text{if } x = u \end{cases} \tag{4.4}$$

(see [99], page 122). Let $\mathcal{M}(\Lambda)$ denote the collection of all probability measures (distributions) on Λ. By δ_A we denote the probability measure concentrated at A, and by ν_θ the bernoulli probability measure with density θ. That is, $\nu_\theta\{\eta : \eta(x) = 1\} = \theta$, for all x. A $\mu \in \mathcal{M}(\Lambda)$ is said to be invariant for the process with Feller semigroup $\{S(t), t \geq 0\}$ if $\mu S(t) = \mu$ for all $t \geq 0$. The class of all invariant distributions is denoted by $\mathcal{I}$.

It is well known that $\mathcal{I}$ is a nonempty compact convex subset of $\mathcal{M}(\Lambda)$, and that $\nu \in \mathcal{I}$ whenever $\mu S(t) \to \nu$, for some $\mu \in \mathcal{M}(\Lambda)$. Here $\mu S(t) \to \nu$ means that $\mu_n = \mu S(t_n) \to \nu$, for all sequences t_n such that $t_n \to +\infty$, and $\to$ denotes weak convergence. The following characterization of $\mathcal{I}$ will be used in Section 4.3.

$$\mathcal{I} = \left\{\nu \in \mathcal{M}(\Lambda) : \int \Psi f_A d\nu = 0 \text{ for all } f_A \in D(\Lambda)\right\}. \tag{4.5}$$

An IPS η_t is said to be ergodic if $\mathcal{I} = \{\nu\}$, for some ν in $\mathcal{M}(\Lambda)$.

4.2 Models and dual systems

We start this section by introducing several models of IPS, which will be studied in Section 4.3. For all these models the functions

$c_T(\eta, \zeta)$ satisfy the regularity conditions mentioned in Section 4.1. Thus the dynamics of these processes guarantee their existence and uniqueness as a Feller process.

Example 4.1 Biased voter model ($\lambda > 0$). Suppose $p(x, y)$ is the transition function of an irreducible Markov chain. If $|T| \geq 2$, $c_T(\eta, \zeta) = 0$. If $T = \{x\}$,

$$c(x, \eta) = \begin{cases} (1 - \eta(x))\,(1 + \lambda)\,\sum_y p(x, y)\eta(y) \\ \eta(x)\,\sum_y p(x, y)(1 - \eta(y)), \end{cases}$$

An interpretation of this model views the sites as individuals and the values of 0 and 1 represent the opinion of the individuals with regard to a particular issue. Individuals are influenced to change their opinion by individuals with which their position differs, and the number (weight) $p(x, y)$ is a measure of the strength of the influence that individual y has on individual x. Individual x changes his opinion at a rate equal to ε times the sum of the weights assigned to all individuals of different opinion, where $\varepsilon = 1 + \lambda$, if his opinion is 0, and $\varepsilon = 1$ otherwise.

Example 4.2 Contact process. If $|T| \geq 2$, $c_T(\eta, \zeta) = 0$. If $T = \{x\}$,

$$c(x, \eta) = \begin{cases} \eta(x) \\ \lambda(1 - \eta(x))\,\sum_{|y-x|=1} \eta(y), \end{cases}$$

where $\lambda \geq 0$. The contact process can be used to model the spread of infection among a population. The sites represent individuals and the values 1 or 0 in a site indicates the correspondent individual is infected or non infected, respectively. Infected individuals recover at rate 1, whereas healthy individuals are infected at a rate proportional to the number of infected neighbors.

Example 4.3 Annihilating branching process with death rate $\delta \geq 0$ (BAP). If $|T| \geq 2$, $c_T(\eta, \zeta) = 0$. If $T = \{x\}$,

$$c(x, \eta) = \delta\,\eta(x) + \sum_{|y-x|=1} \eta(y).$$

If we view $\eta_t(x) = 1$ as a particle being at site x, this process describes the evolution of particles in Z^d which obeys the following rules:

(i) each particle gives birth to a new particle on a neighboring site at rate 1;

(ii) if there is a birth on a site which is already occupied, annihilation occurs;

(iii) each particle dies at rate δ, $\delta \geq 0$.

Example 4.4 Branching annihilating random walk (BARW). Suppose $p(x, y)$ is the transition function of an irreducible Markov chain. If $|T| > 2$, $c_T(\eta, \zeta) = 0$. Otherwise

$$c_T(\eta, \zeta) = \begin{cases} \sum_y p(y, x)\eta(y) & \text{if } T = \{x\} \\ \alpha p(x, y)\eta(x) \text{ or } \alpha p(y, x)\eta(y) & \text{if } T = \{x, y\}. \end{cases}$$

If we view $\eta_t(x) = 1$ as a particle being at site x, this process describes the evolution of particles in Z^d which obeys the following rules:

(i) a particle at site x jumps to a site y with rate $\alpha p(x, y)$;

(ii) a particle at site x gives birth to a particle on site y at rate $p(x, y)$;

(iii) if two particles occupy the same site annihilation occurs.

Example 4.5 Recurrent exclusion process with variable birth rate. Suppose $p(x, y)$ and $q(x, y)$ are the transition functions of a symmetric irreducible and a double stochastic Markov chain, respectively. If $|T| > 2$, $c_T(\eta, \zeta) = 0$. If $T = \{x, y\}$,

$$c_T(\eta, \zeta) = c(x, y, \eta) = p(y, x)\,(1 - \eta(x))\eta(y) \quad \text{or} \quad p(x, y)\,\eta(x)(1 - \eta(y)),$$

whereas if $T = \{x\}$,

$$c(x, \eta) = (1 - \eta(x)) \sum_y \eta(y) q(x, y).$$

If we view $\eta_t(x) = 1$ as a particle being at site x, this process also describes the evolution of particles in Z^d. However the rules governing the evolution are now given by:

(i) at most one particle is allowed in each site;

(ii) the particle at x chooses site y at rate $p(x, y)$. If the site is not occupied by another particle, the particle at x moves to y. Otherwise the particle remains at x.

(iii) the particle at y chooses site x at rate $q(y, x)$. If the site is not occupied by another particle, the particle at y gives birth to a particle at x. Otherwise no birth takes place.

For the models introduced above our aim is to describe their limiting behavior starting from translation invariant distributions. To make this problem precise we need to introduce a few more definitions and related results.

It is easy to see that if $\mu \in \mathcal{M}(\Lambda)$, then μ is completely determined by its values on the sets $\{\eta | \eta \cap A \neq \emptyset\}$, or alternatively on the sets $\{\eta | \, |\eta \cap A| \text{odd}\}$, where $A \subset Z^d$, $|A| < +\infty$. Hence these sets also determine convergence. That is, if μ_n and μ are in $\mathcal{M}(\Lambda)$, $n = 1, 2, \cdots$, then μ_n converges weakly to μ ($\mu_n \to \mu$) if, and only if, for all finite $A \subset Z^d$,

$$\lim_{n \to +\infty} \mu_n[\eta : \eta \cap A \neq \emptyset] = \mu[\eta : \eta \cap A \neq \emptyset], \tag{4.6}$$

or, alternatively, if, and only if, for all finite $A \subset Z^d$,

$$\lim_{n \to +\infty} \mu_n S(t)[\eta : |\eta \cap A| \text{ odd}] = \mu S(t)[\eta : |\eta \cap A| \text{ odd}]. \tag{4.7}$$

In the next section we prove convergence of a large collection of IPS, which covers Examples 1-5, starting from arbitrary translation invariant distributions. We also characterize their limiting distributions, thus giving a complete description of all translation invariant elements of $\mathcal{I}$. To accomplish this we are going to investigate (4.6) and (4.7) via products of random variables. Note that

$$\begin{aligned} \mu S(t)[\eta : \eta \cap A \neq \emptyset] &= 1 - \mu S(t)[\eta : \eta \cap A = \emptyset] \\ &= 1 - \int_\Lambda \prod_{x \in A} (1 - \eta(x)) d\mu S(t). \end{aligned}$$

and hence (4.6) can be investigated by analyzing the limiting behavior of the product of the random variables $(1 - \eta_t(x))$. In dealing with (4.7) products are introduced by considering the quantity

$$\begin{aligned} f_{\mu_t}(A) &= \mu S(t)[\eta : |\eta \cap A| \text{ even}] - \mu S(t)[\eta : |\eta \cap A| \text{ odd}] \\ &= \int_\Lambda (1_{|\eta\cap A|\text{o}} - 1_{|\eta\cap A|\text{e}})\, d\mu S(t), \end{aligned}$$

where o=odd and e = even. Let $A = \{x_1, \cdots, x_n\}$. For $1 \le i \le n$ we define $A_o = A$ and $A_i = A_{i-1} \backslash \{x_i\}$. Then by recursively applying

$$\begin{aligned} 1_{|\eta\cap A_i|\text{o}} - 1_{|\eta\cap A_i|\text{e}} &= (1-\eta(x_i))1_{|\eta\cap A_{i-1}|\text{o}} + \eta(x_i)1_{|\eta\cap A_{i-1}|\text{e}} \\ &\quad -(1-\eta(x_i))1_{|\eta\cap A_{i-1}|\text{e}} - \eta(x_i)1_{|\eta\cap A_{i-1}|\text{o}} \\ &= (1-2\eta(x_i))(1_{|\eta\cap A_{i-1}|\text{o}} - 1_{|\eta\cap A_{i-1}|\text{e}}), \end{aligned}$$

we obtain

$$f_{\mu_t}(A) = -\int_\Lambda \prod_{i=1}^{n} (1 - 2\eta(X_i)) d\mu S(t).$$

Hence (4.7) can be investigated by analyzing the limiting behavior of the product of the random variables $(1 - 2\eta_t(x_i))$.

To be able to exploit the product representation of (4.6) and (4.7) we introduce another important tool in the study of IPS, which is the dual system and the related duality equation.

Let η_t be an IPS on Λ, and $\hat{\eta}$ an IPS on the collection of all finite subsets, say $\hat{\Lambda}$, of Z^d. We say that η_t and $\hat{\eta}$ are dual of each other if

$$P(\eta_t^\zeta \cap A \neq \emptyset) = \hat{P}(\hat{\eta}_t^A \cap \zeta \neq \emptyset) \tag{4.8}$$

for all $\zeta \in \Lambda$ and all $A \in \hat{\Lambda}$, or

$$P(|\eta_t^\zeta \cap A| \text{ odd}) = \hat{P}(|\hat{\eta}_t^A \cap \zeta| \text{ odd}) \tag{4.9}$$

for all $\zeta \in \Lambda$ and all $A \in \hat{\Lambda}$. Since the functions

$$f_A = \left(1_{|\eta\cap A|\text{o}}\right) \quad \text{and} \quad g_A = \left(1_{\eta\cap A\neq\emptyset}\right)$$

are a determining class of functions, the validity of (4.8) or (4.9) implies that one can study the process η_t via the process $\hat{\eta}_t$. IPS which satisfy (4.8) are called additive, whereas those which satisfy (4.9) are called cancellative.

Given an IPS η_t, there are several techniques available to determine the existence and form of a dual process. We describe two of them, which can be used to derive the dual process for all the IPS considered in this chapter.

The first method is based on constructing η_t using a graphical representation, from which the existence of a dual can be easily derived. The following approach is based on defining a percolation structure invented by Harris [78] and developed by Griffeath [74]. We show this by deriving a dual for the BARW.

Let us consider $Z^d \times [0, \infty)$, and think of this as assigning a time line to each $x \in Z^d$. For each $x, y \in Z^d$ with $p(x, y) > 0$, let $\left\{T_i^{(x,y)}(t),\, t \geq 0\right\}$, $i = 1, 2$, be two independent Poisson processes with rate $p(x, y)$ and $\alpha p(x, y)$, respectively, and $\left\{T_{i,n}^{(x,y)},\, n \geq 1\right\}$ their respective arrival times. At each $T_{1,n}^{(x,y)}$ we draw an arrow from $(x, T_{1,n}^{(x,y)})$ to $\left(y, T_{1,n}^{(x,y)}\right)$, and at each $T_{2,n}^{(x,y)}$ we draw an arrow from $\left(x, T_{2,n}^{(x,y)}\right)$ to $\left(y, T_{2,n}^{(x,y)}\right)$ and put a δ at $\left(x, T_{2,n}^{(x,y)}\right)$. We say that there is a path from (x, s) to (y, t), $s < t$, if there is a chain of upward vertical and directed horizontal segments leading from (x, s) to (y, t) that does not go through any δ 's.

Let $N_t^x(y)$ be the number of paths from $(x, 0)$ to (y, t), and

$$N_t^A(y) = \sum_{x \in A} N_t^x(y).$$

Define

$$\eta_t^A = \left\{y \in Z^d : \; N_t^A(y) \equiv 1 \mod 2\right\}.$$

Since the dynamics of the BARW guarantees its uniqueness as an interacting particle system, we need only to check that the effect of the arrows is consistent with the dynamics of the BARW. Let $\eta(x) = 1$ if $x \in \eta$, $\eta(x) = 0$ otherwise.

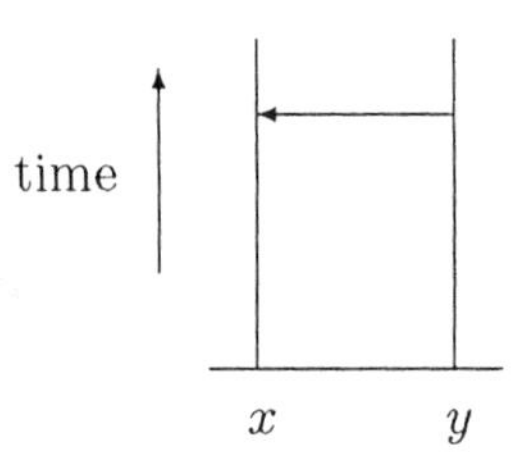

before arrow	after arrow
$(\eta(x), \eta(y))$	$(\eta(x), \eta(y))$
$(0,0)$	$(0,0)$
$(0,1)$	$(1,1)$
$(1,0)$	$(1,0)$
$(1,1)$	$(0,1)$

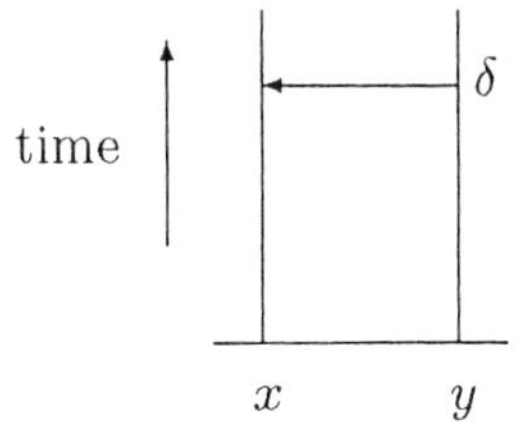

before arrow	after arrow
$(\eta(x), \eta(y))$	$(\eta(x), \eta(y))$
$(0, 0)$	$(0, 0)$
$(0, 1)$	$(1, 0)$
$(1, 0)$	$(1, 0)$
$(1, 1)$	$(0, 0)$

Hence, the process construct via the above percolation structure is a BARW. To derive a dual process we proceed as follows. First we reverse time and the direction of the arrows in the percolation structure, and then define in the same probability space a new process $\hat{\eta}$, with the property that $\forall \zeta$, $A \subset Z^d$, with A finite,

$$P\left[|\eta_t^\zeta \cap A|\,\text{odd}\right] = \hat{P}\left[|\hat{\eta}_t^A \cap \zeta|\,\text{odd}\right],$$

where $|\hat{\eta}_t^A \cap \zeta|$ denotes the number of paths in the new percolation structure from (A, t) down to $(\zeta, 0)$ (Griffeath [74], Chapter 3). By analyzing the effect of the reversed arrows, it is easy to see that $\tilde{\eta}$ is a spin system. That is, change in the configuration only occurs at one site at a time, and $\hat{\eta}(x)$ changes to $1 - \hat{\eta}(x)$ at a rate given by

$$c(x, \hat{\eta}) = \begin{cases} \alpha \sum_y p(x, y)\hat{\eta}(y) + \sum_y p(x, y)\hat{\eta}(y), & \text{if } \hat{\eta}(x) = 0 \\ \alpha \sum_y p(x, y)(1 - \hat{\eta}(y)) + \sum_y p(x, y)\hat{\eta}(y) & \text{if } \hat{\eta}(x) = 1. \end{cases}$$

The next procedure to construct a dual system is based on the uniqueness property of Markov generators (see Liggett [99]), and we describe this method by constructing a dual for the BWD. Let $c_T(x, \eta)$ be as in Model 3, and for $A \subset Z^d$, $|A| < +\infty$, define

$$H(\eta, A) = \prod_{x \in A} (1 - 2\eta(x)).$$

Then if Ψ is the generator for the BWD, and η_x is as in (4.4), we have

$$\begin{aligned}\Psi H(\eta, A) &= \sum_{x\in A} c_T(x,\eta)\left[H(\eta_x, A) - H(\eta, A)\right] \\ &= \delta \sum_{\substack{x\in A \\ \eta(x)=1}} \left[H(\eta_x, A) - H(\eta, A)\right] \\ &\quad + \sum_{x\in A} \sum_{\substack{|x-y|=1 \\ \eta(y)=1}} \left[H(\eta_x, A) - H(\eta, A)\right].\end{aligned}$$

Let $\eta^A = \hat{\eta}^A$. Then

$$H(\eta, A) = \prod_{x\in A}(1 - 2\eta(x)) = \prod_{x\in\eta}(1 - 2\hat{\eta}^A(x)) = H(\hat{\eta}^A, A)$$

which immediately gives

$$\begin{aligned}\Omega H(\eta, A) &= \delta \sum_{\substack{x\in\eta \\ \hat{\eta}^A(x)=1}} \left[H(\hat{\eta}^A_x, \eta) - H(\hat{\eta}^A, \eta)\right] + \\ &\quad \sum_{x\in\eta} \sum_{\substack{|x-y|=1 \\ \hat{\eta}^A(y)=1}} \left[H(\hat{\eta}^A_x, \eta) - H(\hat{\eta}^A, \eta)\right] \\ &= \sum_{x\in\eta} c_T(x, \hat{\eta}^A)\left[H(\hat{\eta}^A_x, \eta) - H(\hat{\eta}^A, \eta)\right] = \hat{\Psi} H(\hat{\eta}^A, \eta).\end{aligned}$$

Hence, $\hat{\Psi}$ is the generator of a BDW on the collection of finite subsets of Z^d, and

$$\begin{aligned}S(t)H(\eta, A) &= \hat{S}(t)H(\hat{\eta}^A, \eta) \\ E^\eta\left[\prod_{x\in A}(1 - 2\eta(x))\right] &= E^{\hat{\eta}^A}\left[\prod_{x\in\eta}(1 - 2\hat{\eta}^A(x))\right].\end{aligned}$$

This implies

$$\begin{aligned}&P^\eta[|\eta_t \cap A| \text{ even}] - P^\eta[|\eta_t \cap A| \text{ odd}] \\ &= P^{\hat{\eta}}[|\hat{\eta}^A_t \cap \eta| \text{ even}] - P^{\hat{\eta}}[|\hat{\eta}^A_t \cap \eta| \text{ odd}]\end{aligned}$$

and hence

$$P^{\eta}[|\eta_t \cap A| \text{ odd}] = P^{\hat{\eta}}[|\hat{\eta}_t^A \cap \eta| \text{ odd}].$$

The following two inequalities will be needed.

Lemma 4.1 *Let $X_1, \cdots, X_n$ be random variables with $0 \le Y_i \le 1$ and*

$$P(X_i > c) \le \epsilon \qquad \textit{for some } \epsilon > 0,\ c > 0,\ 1 \le i \le n.$$

Then

$$E[X_1 X_2 \cdots X_n] \le c^n + \epsilon.$$

PROOF. The proof is based on applying Hölder inequality $k-1$ times, each time with different conjugates p_i and q_i $(\frac{1}{p_i} + \frac{1}{q_i} = 1)$. Precisely, $p_1 = k, p_2 = k-1, \cdots, p_{k-1} = 2$. Using these conjugates we obtain

$$\begin{aligned} E[X_1 \cdots X_k] &\le (EX_1^k)^{1/k} \left(E(X_2 \cdots X_k)^{k/(k-1)}\right)^{(k-1)/k} \\ &\le (EX_1^k)^{1/k}(EX_2^k)^{1/k} \left(E(X_3 \cdots X_k)^{k/(k-2)}\right)^{(k-2)/k} \\ &\quad \vdots \\ &\le (EX_1^k)^{1/k}(EX_2^k)^{1/k} \cdots (EX_k^k)^{1/k} \\ &\le \max_{1 \le i \le k} E(X_i^k). \end{aligned}$$

Since for arbitrary i, $1 \le i \le k$, we have

$$E[X_i^k] = \int_{\{X_i > c\}} X_i^k \, dP + \int_{\{X_i \le c\}} X_i^k \, dP \le c^k + \epsilon,$$

we obtain $E[X_1 \cdots X_k] \le c^k + \epsilon$ as desired, and the lemma is proved. ∎

Lemma 4.2 *Let $Y_1, \cdots, Y_n$ be random variables with $0 \leq Y_i \leq 1$ and*

$$P\left(|X_i - \frac{1}{2}| \geq \delta\right) \leq \epsilon \qquad \text{for some } \epsilon > 0,\ c > 0,\ 1 \leq i \leq n.$$

Then

$$E\left[(1-2Y_1)(1-2Y_2)\cdots(1-2Y_n)\right] \leq (2\delta)^n + \epsilon.$$

PROOF. Let $X_i = |1 - 2Y_i|$. Then $0 \leq X_i \leq 1$ and

$$P(X_i > 2\delta) \leq \epsilon,$$

and Lemma 2 immediately follows from Lemma 4.1. ∎

Definition *For any distribution μ, $A \subset Z^d$ finite and $t \geq 0$, let*

$$\psi_t^\mu(A) = \mu S(t)[\eta : |\eta \cap A| \ odd]. \tag{4.10}$$

Definition *For any distribution μ, $A \subset Z^d$ finite and $t \geq 0$, let*

$$\phi_t^\mu(A) = \mu S(t)[\eta : \eta \cap A \neq \emptyset]. \tag{4.11}$$

From the preceding definitions we can derive the following relations.

$$\begin{aligned}
\psi_{t+s}^\mu(A) &= \int_\Lambda 1_{|\eta\cap A|\,\mathrm{odd}} d\mu S(t+s)(\zeta) = \int_\Lambda S(s) 1_{|\eta\cap A|\,\mathrm{odd}} d\mu S(t)(\eta) \\
&= \int_\Lambda P\left[|\zeta_s^\eta \cap A|\,\mathrm{odd}\right]\ d\mu S(t)(\eta) = \int_\Lambda \tilde{P}\left[|\tilde{\eta}_s^A \cap \eta|\,\mathrm{odd}\right]\ d\mu S(t)(\eta) \\
&= \int_\Lambda \sum_{\substack{C \subset Z^d \\ |C| < +\infty}} \tilde{P}\left[\tilde{\eta}_s^A = C\right] 1_{|C\cap\eta|\,\mathrm{odd}}\ d\mu S(t)(\eta).
\end{aligned}$$

Thus we have

$$\begin{aligned} \psi^{\mu}_{t+s}(A) &= \sum_{\substack{C\subset Z^d \\ |C|<+\infty}} \tilde{P}\left[\tilde{\eta}^A_s = C\right] \int_{\Lambda} 1_{|C\cap\eta|\,\text{odd}}\, d\mu S(t)(\eta) \\ &= \sum_{\substack{C\subset Z^d \\ |C|<+\infty}} \tilde{P}\left[\tilde{\eta}^A_s = C\right] P\left[|\eta^{\mu}_t \cap C|\,\text{odd}\right] \\ &= \sum_{\substack{C\subset Z^d \\ |C|<+\infty}} \tilde{P}\left[\tilde{\eta}^A_s = C\right] \psi^{\mu}_t(C). \end{aligned} \tag{4.12}$$

Similarly one obtains

$$\phi^{\mu}_{t+s}(A) = \sum_{\substack{C\subset Z^d \\ |C|<+\infty}} \tilde{P}\left[\tilde{\eta}^A_s = C\right] \phi^{\mu}_t(C). \tag{4.13}$$

4.3 IPS with dual of increasing size

We start by proving the following theorem, which is due to Harris [79].

Theorem 4.1 *Let η_t be such that for all $t > 0$, and for all $x \in Z^d$, for all $\zeta \in \Lambda$, $\zeta \neq \emptyset$,*

$$P\left(\eta^{\zeta}_t(x) = 1\right) > 0, \tag{4.14}$$

and let μ be a translation invariant distribution such that $\mu(\{\emptyset\}) = 0$. Then for all $t > 0$ and all $\epsilon > 0$ there is an N, $N = N(\epsilon, t)$, such that whenever $|A| \geq N$,

$$\mu S(t)\left[\eta :\ \eta \cap A = \emptyset\right] < \epsilon. \tag{4.15}$$

PROOF. We follow the proof of Harris [79]. Let $t > 0$ and $\epsilon > 0$ be given. By condition (4.14) gives that

$$\mu\left[\zeta :\ P(\eta^{\zeta}_t(x) = 0) = 1\right] = 0,$$

and therefore there is a δ, $0 < \delta < 1$ such that

$$\mu\left[\zeta :\ P(\eta^{\zeta}_t(x) = 0) > \delta\right] < \frac{\epsilon}{3}. \tag{4.16}$$

Our initial assumption expressed by in (4.2) extends the validity of (4.16) to any $x \in Z^d$. Let N be chosen such that every subset A of Z^d of size greater than or equal to N contains a subset of size greater than or equal to t, say $\{x_{i_1}, x_{i_2} \cdots, x_{i_k}\}$, $k \geq t$, where $d(x_{i_j}, x_{i_l}) \geq M(\epsilon/(3n))$, and t is chosen to satisfy $(1-\delta)^t \leq \epsilon/(3n)$.

Let $A = \{x_1, x_2, \cdots, x_n\}$, $n \geq N(\epsilon/(3n))$, be an arbitrary subset of Z^d, and denote by $\{x_{i_1}, x_{i_2}, \cdots, x_{i_k}\}$ the subset of A with $d(x_{i_j}, x_{i_l}) \geq M(\epsilon/(3n))$. Then

$$
\begin{aligned}
\mu S(t)\,[\eta : \eta \cap A = \emptyset] &= \int_\Lambda \prod_{i=1}^{n} (1 - \eta(x_i)) d\mu S(t)(\eta) \\
&\leq \int_\Lambda \prod_{j=1}^{k} (1 - \eta(x_{i_j})) d\mu S(t)(\eta) \\
&= \int_\Lambda S(t) \prod_{j=1}^{t} (1 - \eta(x_{i_j})) d\mu S(t)(\eta) \\
&\leq \int_\Lambda \prod_{j=1}^{k} S(t)\,(1 - \eta(x_{i_j})) d\mu(\eta) + \frac{\epsilon}{3t} \qquad (4.17) \\
&\leq (1-\delta)^t + \frac{2\epsilon}{3} \leq \epsilon \qquad (4.18)
\end{aligned}
$$

where the second equation follows from properties of Feller semigroups, $S(t+s) = S(t)\,S(s)$, the second inequality from a mixing property of IPS (see Liggett [99], page 40), and our choice of the x_{i_j},and the first inequality in (4.18) follows from (4.16) and Lemma 4.1. The last inequality follows from the defining property of t. The proof is completed. ∎

The next result deals with the parity of the number of particles in η_t. The result is due to Simonelli [138].

Theorem 4.2 *Let η_t be such that $\forall t > 0$, $\forall x \in Z^d$, $\forall \zeta \in X$, $\zeta \neq \emptyset$*

$$
0 < P\left(\eta_t^\zeta(x) = 1\right) < 1, \qquad (4.19)
$$

and let μ be a translation invariant distribution such that $\mu(\{\emptyset\}) = 0$. Then $\forall t > 0$, $\forall \epsilon > 0$ there is an N, $N = N(\epsilon, t)$, such that whenever

$|A| \geq N$,

$$\left| P\left(|\eta_t^\mu \cap A| \ odd \right) - \frac{1}{2} \right| < \epsilon$$

and

$$\left| P\left(|\eta_t^\mu \cap A| \neq 0 \ and \ even \ \right) - \frac{1}{2} \right| < \epsilon.$$

PROOF. We start our proof by observing that the validity of (4.19) implies the validity of (4.14), and hence, by Theorem 4.1, we only need to show that under the conditions of Theorem 4.2, $\forall t > 0$, $\forall \epsilon > 0$ there is an N, $N = N(\epsilon, t)$, such that whenever $|A| \geq N$,

$$\left| P\left(|\eta_t^\mu \cap A| \text{ odd} \right) - P\left(|\eta_t^\mu \cap A| \text{ even } \right) \right| < \epsilon.$$

Let $\epsilon > 0$, $t > 0$ be given. For every $s \in [0, \epsilon]$, assumption (4.19) implies that

$$\mu S(t) \left[\eta : \ P^\eta \big(\eta_s(0) = 1 \big) = 1 \right] = 0.$$

Therefore $\exists \delta_1 \in (\frac{1}{2}, 1)$ such that $\forall s \in [0, \epsilon]$

$$\mu S(t) \left[\eta : \ P^\eta \big(\eta_s(0) = 1 \big) > \delta_1 \right] < \frac{\epsilon}{6}.$$

A similar argument, combined with the assumption $\mu(\{\emptyset\}) = 0$, gives the existence of $\delta_2 \in (\frac{1}{2}, 1)$ such that $\forall s \in [0, \epsilon]$

$$\mu S(t) \left[\eta : \ P^\eta \big(\eta_s(0) = 0 \right] > \delta_2 \big] < \frac{\epsilon}{6}.$$

Let $\delta = \max\{\delta_1, \delta_2\}$. Then $\forall s \in [0, \epsilon]$ we have

$$\mu S(t) \left[\eta : \ 1 - \delta < P^\eta \big(\eta_s(0) = 1 \big) < \delta \right] > 1 - \frac{\epsilon}{3}. \tag{4.20}$$

The validity of (4.2) extends (4.20) to any x in place of 0. Let $A = \{x_1, x_2, \cdots, x_n\}$, n to be specified later, be an arbitrary subset of Z^d, and consider

$$\begin{aligned} f_{\mu_t}(A) &= \left| P\left(|\eta_t^\mu \cap A| \text{ odd} \right) - P\left(|\eta_t^\mu \cap A| \text{ even } \right) \right| \\ &= -\int_\Lambda \prod_{i=1}^n (1 - 2\eta(x_i)) \, d\mu S(t)(\eta). \end{aligned}$$

Since $g(\eta) = 1 - \eta(x) \in D(\Lambda)$, and the map $s \mapsto S(s)f$ is continuous, there exists an $r > 0$ such that

$$\| S(r)g - g \| \leq \frac{\epsilon}{3n}.$$

Assumption (4.2) further implies that r does not depend on the value of x. Let $g_i(\eta) = 1 - \eta(x_i)$. Then for arbitrary η,

$$\begin{aligned} g_1(\eta)\cdots g_n(\eta) &\leq (1 - 2S(r)\eta(x_1))g_2(\eta)\cdots g_n(\eta) + \frac{\epsilon}{3\cdot n} \\ &\leq (1 - 2S(r)\eta(x_1))(1 - 2S(r)\eta(x_2))\cdots g_n(\eta) + \frac{2\epsilon}{3\cdot n} \\ &\ \ \vdots \\ &\leq \prod_{i=1}^{n}(1 - 2S(r)\eta(x_i)) + \frac{n\epsilon}{3\cdot n}, \end{aligned}$$

which gives

$$g_1(\eta)\cdots g_n(\eta) \leq |1 - 2S(r)\eta(x_1)|\cdots|1 - 2S(r)\eta(x_n)| + \frac{\epsilon}{3}.$$

Since $-(g_1(\eta)g_2(\eta)\cdots g_n(\eta))$ also satisfies the above inequality, we have

$$|f_{\mu_t}(A)| \leq \int_\Lambda |1 - 2S(r)\eta(x_1)|\cdots|1 - 2S(r)\eta(x_n)|\ d\mu S(t) + \frac{\epsilon}{3}.$$

Next let $Y_i(\eta) = S(r)\eta(x_i) = P[\eta_r^\eta(x_i) = 1]$. Then Y_i is a random variable on X, $0 \leq Y_i \leq 1$, and by (4.20) we have that $\mu S(t)[\eta : (1-\delta) \leq Y_i \leq \delta] \geq 1 - \frac{\epsilon}{3}$. Therefore the $Y_i's$ satisfy the conditions of Lemma 4.2, and we have

$$|f_{\mu_t}(A)| \leq \frac{\epsilon}{3} + (2\delta - 1)^n + \frac{\epsilon}{3}.$$

Hence, if we let $N(\epsilon, t) = N$ be such that $(2\delta - 1)^N < \frac{\epsilon}{3}$, then $|f_{\mu_t}(A)| \leq \epsilon$ whenever $|A| \geq N$ and the proof of Theorem 4.2 is complete. ∎

It is important to stress the role of Theorems 4.1 and 4.2 in the next two results. They contain the essential conditions that allow to unify the study of the asymptotics of a large class of interacting particle systems. We clarify this and motivate the next two results with the following example, which also has the advantage to make transparent the importance that products have in this investigation.

Example 4.6 Let us consider the evolution of an interacting particle system η_t with initial distribution given by a product measure ν_θ with variable density θ. That is, $\nu_\theta[\eta : \eta(x) = 1] = \theta(x)$, $0 < \theta(x) < 1$. We start by assuming the existence of a dual system $\tilde{\eta}_t$ which satisfies (4.9). Then for any finite $A \subset Z^d$, (4.12) gives, with $t = 0$,

$$\psi_s^{\nu_\theta}(A) = \sum_{\substack{C \subset Z^d \\ |C| < +\infty}} \tilde{P}\left[\tilde{\eta}_s^A = C\right] \psi^{\nu_\theta}(C). \tag{4.21}$$

Let $C = \{x_1, x_2, \cdots, x_t\}$, and define $X_i(\eta) = X_i = \eta(x_i)$. Then X_i, $i = 1, 2, \cdots, t$ can be viewed as random variables in Λ, $X_i = 0, 1$, $P(X_i = 1) = \theta(x_i)$, and $\psi_t^{\nu_\theta}(C) = P(m_t = \text{ odd })$, and, by using results developed in Chapter 1, conditions on η_t or $\tilde{\eta}_t$ can now be given to assure the existence of a limiting distribution. For example, if the size of the dual process is known to increase with no bound as a function of s, and the $\theta(x_i)$ are chosen such that for any infinite collection of sites x_{i_j}, $\sum_j \theta(x_{i_j}) = +\infty$, then (4.21) and Theorem 1.23 would immediately prove the existence of a symmetric limiting distribution.

Next we assume $\tilde{\eta}_t$ satisfies (4.8). Let ν_θ and A be as before. Then (4.13) gives, with $t = 0$,

$$\phi_s^{\nu_\theta}(A) = \sum_{\substack{C \subset Z^d \\ |C| < +\infty}} \tilde{P}\left[\tilde{\eta}_s^A = C\right] \phi^{\nu_\theta}(C).$$

If C and X_i are as before, then $\phi^{\nu_\theta}(C) = 1 - \prod_{i=1}^t (1 - \theta(x_i))$ and we obtain the interesting result that the sufficient conditions needed for cancellative systems to have a symmetric limiting distribution are also sufficient for additive systems to have a limiting distribution. In the additive case the limiting distribution ν_1 is given by

$$\nu_1(A) = \lim_{s \to +\infty} \phi_s^{\nu_\theta}(A) = \tilde{P}\left(\tilde{\eta}_t^A \neq \emptyset \, \forall t > 0\right).$$

This surprising similarity between these two classes of interacting particle systems, considered and analyzed separately in most of the present literature, can be further explained with the help of Theorems 4.1 and 4.2. In fact, under our assumptions, these two theorems justify to write

$$\begin{aligned}\lim_{s\to+\infty}\phi_s^{\nu_\theta}(A) &= \lim_{s\to+\infty}\sum_{\substack{C\subset Z^d\\|C|<+\infty}}\tilde{P}\left[\tilde{\eta}_s^A = C\right]\phi^{\nu_\theta}(C)\\ &= \lim_{s\to+\infty}\sum_{\substack{C\subset Z^d\\|C|<+\infty}}\tilde{P}\left[\tilde{\eta}_s^A = C\right]2\psi^{\nu_\theta}(C),\end{aligned}$$

which directly shows the interdependence between the existence of the two described limiting distributions.

The next two results generalize the above example to all translation invariant initial distribution. Example 4.6 also justifies to start by considering cancellative systems. The following theorem is due to Simonelli [138]

Theorem 4.3 *Assume η_t and $\hat{\eta}$ satisfy the duality equation (4.18) and the following conditions:*

(a) for each $t>0$, $x\in S$, $\zeta\subset Z^d$, $\zeta\neq\emptyset$, $0<P\left[\eta_t^\zeta(x)=1\right]<1$;

(b) on $\widetilde{\mathcal{A}}_\infty$, $\widetilde{\mathcal{A}}_\infty=\{\hat{\eta}:\ \hat{\eta}\neq\emptyset,\ \forall t\geq 0\}$, $|\hat{\eta}_t|\to+\infty$ with t a.s..

Then

(i) there exists a measure $\nu_{\frac{1}{2}}$ such that for all translation invariant distributions ν,

$$\lim_{t\to+\infty}\nu S(t)=\nu(\{\emptyset\})\delta_\emptyset+(1-\nu(\{\emptyset\}))\nu_{\frac{1}{2}},$$

where

$$\phi^{\nu_{\frac{1}{2}}}(A)=\lim_{t\to+\infty}\phi_t^{\delta_{Z^d}}(A)=\frac{1}{2}\widetilde{P}^A\left[\hat{\eta}_t\neq\emptyset\,\forall t\geq 0\right];$$

(ii) if $\nu_{\frac{1}{2}}=\delta_\emptyset$ then $\delta_\emptyset$ is the unique invariant distribution;

(iii) If $\nu_{\frac{1}{2}}\neq\delta_\emptyset$ then $\nu_{\frac{1}{2}}(\{\emptyset\})=0$.

PROOF. First we note that the existence of a dual system $\hat{\eta}_t$ that satisfies (4.18) implies that $\delta_\emptyset$ is an invariant measure, and hence $\delta_\emptyset S(t) = \delta_\emptyset \ \forall t \geq 0$. Next let ν be an arbitrary translation invariant distribution. Then ν can be written as

$$\begin{aligned} \nu &= \nu(\{\emptyset\})\delta_\emptyset + (1-\nu(\{\emptyset\}))\left[\frac{\nu - \nu(\{\emptyset\})\delta_\emptyset}{1-\nu(\{\emptyset\})}\right] \\ &= \nu(\{\emptyset\})\delta_\emptyset + (1-\nu(\{\emptyset\}))\mu \end{aligned}$$

where μ is translation invariant and $\mu(\{\emptyset\}) = 0$. Hence

$$\begin{aligned} \lim_{t\to+\infty} \nu S(t) &= \lim_{t\to+\infty}\left[\nu(\{\emptyset\})\delta_\emptyset + (1-\nu(\{\emptyset\}))\left[\frac{\nu - \nu(\{\emptyset\})\delta_\emptyset}{1-\nu(\{\emptyset\})}\right]\right] S(t) \\ &= \nu(\{\emptyset\})\delta_\emptyset + (1-\nu(\{\emptyset\}))\lim_{t\to+\infty}\mu S(t)\,. \end{aligned}$$

Therefore to prove part (i) and (ii) of the theorem it suffices that for any translation invariant distribution μ, $\mu(\{\emptyset\}) = 0$,

$$\lim_{t\to+\infty} \psi_t^\mu(A) = \frac{1}{2}\hat{P}^A\left[\hat{\eta}_t \neq \emptyset \forall t \geq 0\right]$$

for all finite subsets A of Z^d.

First let us assume that $\hat{P}^A[\widetilde{\mathcal{A}}_\infty] = 0$, which implies that $\nu_{\frac{1}{2}} = \delta_\emptyset$, and note that the validity of (4.18) also implies that $\{\emptyset\}$ is an absorbing state for $\hat{\eta}_t$. Then for any initial distribution ν,

$$\begin{aligned} \lim_{t\to+\infty} \psi_t^\nu(A) &= \lim_{t\to+\infty} P\left[|\eta_t^\mu \cap A| \text{ odd }\right] \\ &= \lim_{t\to+\infty}\int_\Lambda \hat{P}\left[|\hat{\eta}_t^A \cap \zeta| \text{ odd }\right] \, d\mu(\zeta) \\ &\leq \lim_{t\to+\infty} \hat{P}\left[\hat{\eta}_t^A \neq \emptyset\right] = \hat{P}^A[\widetilde{\mathcal{A}}_\infty] = 0. \end{aligned}$$

Next we assume that $\hat{P}^A[\widetilde{\mathcal{A}}_\infty] \neq 0$, and let μ be an arbitrary translation invariant distribution with $\mu(\{\emptyset\}) = 0$. Then

$$\begin{aligned} \lim_{t\to+\infty} \psi_t^\mu(A) &= \lim_{t\to+\infty}\left[\sum_{\substack{B\subset Z^d \\ |B|<+\infty}} \hat{P}\left[\hat{\eta}_{t-s}^A = B\right]\cdot\psi_s^\mu(B)\right] \\ &= \lim_{t\to+\infty}\left[\sum_{\substack{B\subset Z^d \\ |B|<+\infty}} \hat{P}\left[\hat{\eta}_{t-s}^A = B|\widetilde{\mathcal{A}}_\infty\right]\cdot\hat{P}^A\left[\widetilde{\mathcal{A}}_\infty\right]\cdot\psi_s^\mu(B)\right], \end{aligned}$$

where in the first equation we have used (4.12), and the second equation holds because

$$\lim_{t\to+\infty}\left[\sum_{\substack{B\subset Z^d\\|B|<+\infty}} \hat{P}\left[\hat{\eta}^A_{t-s}=B,\,\widetilde{\mathcal{A}}^c_\infty\right]\cdot\psi^\mu_s(B)\right]=0.$$

Let $\epsilon>0$ be given. By assumption (b) of Theorem 4.3, on $\widetilde{\mathcal{A}}_\infty$, $|\hat{\eta}_t|\to+\infty$ as $t\to+\infty$ $a.s.$, that implies

$$\exists t_1(\epsilon) \text{ such that on } \widetilde{\mathcal{A}}_\infty,\quad |\hat{\eta}_t|\geq N(\epsilon,t_1-s)\quad \forall t\geq t_1(\epsilon),$$

where $N(\epsilon,t_1-s)$ is chosen as in Theorem 4.2. Hence, by Theorem 4.2, $\forall t\geq t_1(\epsilon)$ we have

$$\left|\sum_{\substack{B\subset Z^d\\|B|<+\infty}} \hat{P}\left[\hat{\eta}^A_{t-s}=B|\widetilde{\mathcal{A}}_\infty\right]\cdot\hat{P}^A\left[\widetilde{\mathcal{A}}_\infty\right]\cdot\psi^\mu_s(B)-\frac{\hat{P}^A[\widetilde{\mathcal{A}}_\infty]}{2}\right|\leq\epsilon.$$

Therefore

$$\lim_{t\to+\infty}\psi^\mu_t(A)=\frac{1}{2}\hat{P}^A\left[\hat{\eta}_t\neq\emptyset,\ \forall t\geq 0\right].$$

It is left to show the validity of part (iii) of the theorem. So let us assume that $\nu_{\frac{1}{2}}\neq\delta_{\delta_\emptyset}$. We are going to show that $\nu_{\frac{1}{2}}(\{\emptyset\})=0$. Let $\epsilon>$ 0 be given. We are going to show $\nu_{\frac{1}{2}}(\{\emptyset\})<\epsilon$, for any given $\epsilon>0$, which would then imply $\nu_{\frac{1}{2}}(\{\emptyset\})=0$. By definition,

$$\nu_{\frac{1}{2}}(\{\emptyset\})=\lim_{t\to+\infty}\int_\Lambda\prod_{x\in S}(1-\eta(x))\ d\mu S(t)=\lim_{t\to+\infty}h(t),$$

where μ is any translation invariant distribution with $\mu(\{\emptyset\})=0$, and $h(t)$ is defined by the equation above. Let $\epsilon>0$ be given. Then $\exists t_1$ such that if $t\geq t_1$,

$$|\nu_{\frac{1}{2}}(\{\emptyset\})-h(t)|\leq\frac{\epsilon}{2}.$$

Moreover,

$$\nu_{\frac{1}{2}}(\{\emptyset\})\leq\int_\Lambda\prod_{x\in z^d}(1-\eta(x))\ d\mu S(t)+\frac{\epsilon'}{3}\leq\int_\Lambda\prod_{i=1}^{n}(1-\eta(x_i))\ d\mu S(t)+\frac{\epsilon'}{3}$$

for all integers $n > 0$ and for all $\{x_1, x_2, \cdots, x_n\} \subset S$. Let $t \geq t_1$, and note that condition (a) of Theorem 4.3 satisfies condition (4.14) of Theorem 4.1, and therefore n can be chosen such that

$$\int_\Lambda \prod_{i=1}^{n}(1 - \eta(x_i))\, d\mu S(t) \leq \frac{\epsilon}{2}\,.$$

Hence

$$\nu_{1/2}(\{\emptyset\}) \leq \epsilon,$$

and since ϵ was arbitrary, the validity of part (iii) immediately follows. The proof of the theorem is completed. ∎

Next we turn to additive systems.

Theorem 4.4 *Assume η_t and $\hat{\eta}$ satisfy the duality equation (4.8) and the following conditions:*

(a) for each $t > 0$, $x \in Z^d$, $\zeta \subset S$, $\zeta \neq \emptyset$, $P\left[\eta_t^\zeta(x) = 1\right] > 0$,

(b) let $\widetilde{A}_\infty = \{\tilde{\eta} : \tilde{\eta} \neq \emptyset, \forall t \geq 0\}$. Then on $\widetilde{A}_\infty$, $|\hat{\eta}| \to +\infty$ with t a.s..

Then we have,

(i) for all translation invariant distributions ν,

$$\lim_{t\to+\infty} \nu S(t) = \nu(\{\emptyset\})\delta_\emptyset + (1 - \nu(\{\emptyset\}))\nu_1,$$

where

$$\phi^{\nu_1}(A) = \lim_{t\to+\infty} \phi_t^{\delta_{Z^d}}(A) = \hat{P}^A\left[\hat{\eta}_t \neq \emptyset\, \forall t \geq 0\right];$$

(ii) if $\nu_1 = \delta_\emptyset$ then $\delta_\emptyset$ is the unique invariant distribution;

(iii) If $\nu_1 \neq \delta_\emptyset$ then $\nu_1(\{\emptyset\}) = 0$.

Proof. Let μ be an arbitrary translation invariant initial distribution. By arguing as at the beginning of the previous proof, we can assume $\mu\{\emptyset\} = 0$. To prove that for arbitrary finite $A \subset Z^d$,

$$\lim_{t\to+\infty} \phi_t^\mu(A) = \hat{P}^A\left[\hat{\eta}_t \neq \emptyset\, \forall t \geq 0\right],$$

we again consider the two cases $\hat{P}(\widetilde{\mathcal{A}}_\infty) = 0$ and $\hat{P}(\widetilde{\mathcal{A}}_\infty) \neq 0$. If $\hat{P}(\widetilde{\mathcal{A}}_\infty) = 0$, the validity of parts (ii) and (iii) immediately follows by repeating the same calculations as used in the proof of part (ii) and (iii) of Theorem 4.3. To prove part (i) in the case $\hat{P}(\widetilde{\mathcal{A}}_\infty) \neq 0$, we proceed as in the last part of Example 4.6, and for arbitrary t and s we write

$$\phi^\mu_{t+s}(A) = \sum_{\substack{B \subset Z^d \\ |B| < +\infty}} \hat{P}\left[\hat{\eta}^A_s = C\right] 2\psi^\nu_t(B).$$

This, Theorem 4.2, and assumption (b) of Theorem 4.4 give that for any $\epsilon > 0$ we can choose t to be sufficiently large such that

$$\lim_{s \to +\infty} \left| \phi^\mu_{t+s}(A) - \hat{P}^A\left[\hat{\eta}_t \neq \emptyset \, \forall t \geq 0\right] \right| \leq \epsilon,$$

which proves the validity of part (i) in the case $\hat{P}(\widetilde{\mathcal{A}}_\infty) \neq 0$. The proof of the theorem is now complete. ∎

Theorems 4.2, 4.3, and 4.4 cover a large class of interacting particle systems. In particular we are going to show that the limiting behavior of all the systems introduced in Examples 1-5 can be recovered via these theorems. Before we show this, it is convenient to make the following remark. We have already mentioned that the existence of a dual system $\hat{\eta}$ which satisfies (4.8) or (4.9) implies that $\emptyset$ is an absorbing state for both η_t and $\hat{\eta}_t$. From classical results on Markov chain (see Bramson et al. [24], for details) this implies that if

$$\inf_{|B| \leq k} P\big(\hat{\eta}^B_t = \emptyset, \quad \text{for some } t > 0\big) < 1, \tag{4.22}$$

then

$$P\Big(\lim_{t \to +\infty} |\hat{\eta}^B_t| = +\infty \Big| \widetilde{\mathcal{A}}_\infty\Big) = 1.$$

Therefore, to prove the validity of condition (b) in Theorems 4.3 and 4.4, it suffices to show the validity of (4.22).

In what follows, $\nu_{1/2}$ and ν_1 are the probability measures defined in Theorems 4.3 and 4.4, respectively.

We start by considering the system described in Example 4.3.

Theorem 4.5 *Let η_t be a BAP ($\delta \geq 0$). If δ is small, then the set of extreme translation invariant distributions is $\{\delta_\emptyset, \nu_{1/2}\}$. Moreover, if ν is any translation invariant distribution, then*

$$\lim_{t\to+\infty} \nu S(t) = \nu(\{\emptyset\})\delta_\emptyset + (1-\nu(\{\emptyset\}))\nu_{1/2}.$$

The above theorem was first proved by Bramson, Ding Wanding and Durrett [24] , and for the case $\delta = 0$ independently by A. Sudbury [145].

PROOF OF THEOREM 4.5. Let η_t be a BAP. To prove the theorem it suffices to show that η_t satisfies Theorem 4.3. It is clear from the dynamics of the system that condition (a) of Theorem 4.3 is satisfied, and we have already shown that $\hat{\eta}_t$ exists, and is a BAP (see computation preceding Lemma 4.1). Bramson et al. [24] showed that if δ is small, then $P(\eta_t^0 \neq \emptyset$ for all $t) > 0$. So it is left to show that on $\widetilde{\mathcal{A}}_\infty$, $|\hat{\eta}_t| \to +\infty$ *a.s.*. For the case $\delta > 0$ this follows from

$$\inf_{0<|B|\leq n} P\big[\hat{\eta}_t^B = \emptyset \text{ for some } t \geq 0\big] \geq \Big(\frac{\delta}{2d+\delta}\Big)^n > 0, \qquad (4.23)$$

which follows from the dynamics of the process (see Bramson et al. [24] for the details, or the proof of the next theorem for similar arguments). For the case $\delta = 0$ the above argument cannot be used because the right hand side of (4.23) equals 0. In this case the desired result follows from showing that the $\hat{\eta}_t$ dominates an oriented percolation process with parameter $1-\epsilon$, with the property that if ϵ is small enough,

$$\text{on } \ \Omega_\infty = \{\, W_n^o \neq \emptyset \text{ for all n}\,\}, \quad |W_n^o| \to +\infty \ \text{ a.s. as } n \to +\infty,$$

where $W_n^o = \{y \in Z^d : \text{there is an open path from } (0,0) \text{ to } (y,n)\}$ (see [24] for details). ∎

Next we consider the particle system in Example 4.4, and we prove the following result (Simonelli [138]).

Theorem 4.6 *Let η_t be a BARW. If α is small, but $\alpha > 0$, then the set of extreme translation invariant distributions is $\{\delta_\emptyset, \nu_{1/2}\}$. Moreover, if ν is any translation invariant distribution, then*

$$\lim_{t\to+\infty} \nu S(t) = \nu(\{\emptyset\})\delta_\emptyset + (1-\nu(\{\emptyset\}))\nu_{1/2}.$$

PROOF. Let η_t be a BARW. It is clear from the dynamics of the system (and our assumption on $p(x, y)$) that condition (a) of Theorem 3 is satisfied, and that $\emptyset$ is an absorbing state. Bramson et al. [24] showed that if α is small, then the process starting from any finite and fixed configuration will survive with positive probability, and if α is large enough, then extinction is certain (see also Bramson and Gray [25]). The existence of a dual system $\hat{\eta}_t$ was shown earlier by using a percolation structure. Next we are going to show that (4.22) holds. We only consider the case $p(x, y)$ is a symmetric random walk on Z^d, the same type of calculation applies to the general case. Let $|B| \leq n$, be arbitrary, and suppose first that $\alpha \leq 1$. From the rate of the dual system we have

$$\begin{aligned} &P[\hat{\eta}_t^B = \emptyset \text{ for some } t \geq 0] \\ &\quad \geq \hat{P}[\hat{\eta}_t^B = \emptyset \text{ for some } t \geq 0 \,|\, |\hat{\eta}_s^B| \geq |\hat{\eta}_{s+u}^B| \,\forall\, s, u \geq 0] \\ &\quad \geq \Big(\frac{\alpha}{\alpha+1}\Big)^n > 0. \end{aligned}$$

Next suppose $\alpha > 1$. let

$$c_1(x, \eta) = \begin{cases} c(x, \eta), & \text{if } \eta(x) = 0, \\ 1, & \text{if } \eta(x) = 1, \end{cases}$$

and denote by ζ_t the process defined by $c_1(x, \eta)$. Hence, for $\alpha > 1$,

$$\begin{aligned} &P[\hat{\eta}_t^B = \emptyset \text{ for some } t \geq 0] \\ &\quad \geq \hat{P}[\hat{\eta}_t^B = \emptyset \text{ for some } t \geq 0 \,|\, |\hat{\eta}_s^B| \geq |\hat{\eta}_{s+u}^B| \,\forall\, s, u \geq 0] \\ &\quad \geq \hat{P}[\hat{\zeta}_t^B = \emptyset \text{ for some } t \geq 0 \,|\, |\hat{\zeta}_s^B| \geq |\hat{\zeta}_{s+u}^B| \,\forall\, s, u \geq 0] \\ &\quad \geq \Big(\frac{1}{\alpha+2}\Big)^n > 0. \end{aligned}$$

Therefore for arbitrary $\alpha > 0$,

$$P[\hat{\eta}_t^B = \emptyset \text{ for some } t \geq 0] \geq \min\left\{\Big(\frac{\alpha}{\alpha+1}\Big)^n, \Big(\frac{1}{\alpha+2}\Big)^n\right\} > 0,$$

thus implying the validity of condition (b) of Theorem 4.3. ∎

We now turn to additive systems. The next two results are due to Schwartz [135].

Theorem 4.7 *Let η_t be a biased voter model with parameter $\lambda > 0$. The set of extreme translation invariant distributions is $\{\delta_\emptyset, \nu_1\}$. Moreover, if ν is any translation invariant distribution, then*

$$\lim_{t \to +\infty} \nu S(t) = \nu(\{\emptyset\})\delta_\emptyset + (1 - \nu(\{\emptyset\}))\nu_1.$$

PROOF. Let η_t be a biased voter model. We claim that η_t satisfies the conditions of Theorem 4.4. Condition (a) of the theorem immediately follows from the dynamics of the process. Next we need to show the existence of a dual system which satisfies condition (b). To this end we construct η_t using a percolation structure.

For each (x, y) in $Z^d \times Z^d$ with $p(x, y) > 0$ we associate two independent Poisson processes, $\{T_i^{(x,y)}(t), t \geq 0\}$, $i = 1, 2$, with rates $p(x, y)$ and $\lambda p(x, y)$, respectively. Let $\{T_{i,n}^{(x,y)}, n \geq 0\}$, $i = 1, 2$, be the corresponding arrival times. At each $T_{1,n}^{(x,y)}$ we draw an arrow from $(x, T_{1,n}^{(x,y)})$ to $(y, T_{1,n}^{(x,y)})$, and put a δ at $(y, T_{1,n}^{(x,y)})$, whereas at each $T_{2,n}^{(x,y)}$ we simply draw an arrow from $(x, T_{1,n}^{(x,y)})$ to $(y, T_{1,n}^{(x,y)})$. We say that there is a path from (x, s) to (y, t), $s < t$, if there is a chain of upwards vertical and directed segments leading from (x, s) to (y, t) that does not go through any δ's. Let $N_t^x(y)$ denote the number of paths from (x, y) to (y, t). We define

$$\xi_t^A = \{y \in Z^d : N_t^x(y) \neq 0 \text{ for some } x \in A\}.$$

Let us check the dynamics of the process we have just constructed.

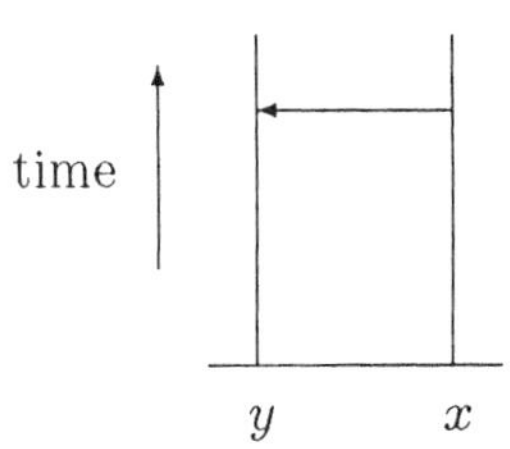

before arrow	after arrow
$(\eta(x), \eta(y))$	$(\eta(x), \eta(y))$
$(0, 0)$	$(0, 0)$
$(0, 1)$	$(1, 1)$
$(1, 0)$	$(1, 0)$
$(1, 1)$	$(1, 1)$

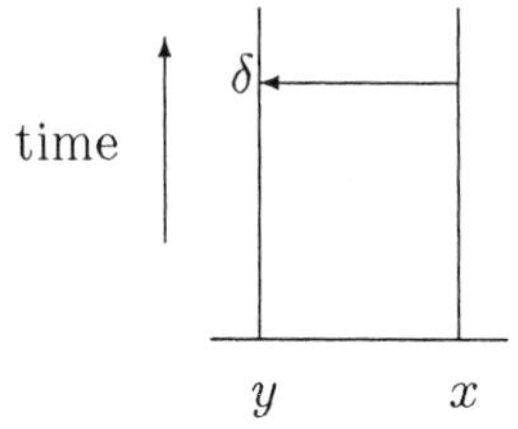

before arrow	after arrow
$(\eta(x), \eta(y))$	$(\eta(x), \eta(y))$
$(0,0)$	$(0,0)$
$(0,1)$	$(1,1)$
$(1,0)$	$(0,0)$
$(1,1)$	$(1,1)$

Hence, ξ_t is a biased voter model. To derive a dual process we reverse time and direction of the arrows in the percolation structure, and define a new process $\hat{\xi}_t$ with the property that for all η, $A \subset Z^d$, with A finite,

$$P[\xi_t^\eta \cap A \neq \emptyset] = \hat{P}[\hat{\xi}_t^A \cap \eta \neq \emptyset].$$

The rate of the dual system can be recovered by analyzing the effect of reversing the arrows, and the dual rate is given by the following rules:

(a) $c_T(\eta, d\zeta) = 0$ if $|T| > 2$,

(b) $\eta(x)$ changes to $1 - \eta(x)$ at a rate given by

$$\begin{cases} \lambda p(y,x) & \text{if } \eta(x) = 0,\ \eta(y) = 1 \\ p(x,y) & \text{if } \eta(x) = 1, \eta(y) = 1, \end{cases}$$

(c) $\eta(x)$ and $\eta(y)$ change to $1 - \eta(x)$ and $1 - \eta(y)$, respectively, at a rate given by

$$c(x,y,\eta) = p(x,y) \quad \text{if } \eta(x) = 1,\ \eta(y) = 0.$$

That is $\hat{\xi}_t$ is a coalescing branching random walk. To show that the dual satisfies property (b) of Theorem 3.4, we replace the coalescing property with the annihilating property, thus obtaining a BARW, say ζ_t. Clearly for every finite A, $|\hat{\xi}_t^A| \geq |\zeta_t^A|$, and the result now follows by arguing as in Theorems 4.5 and 4.6 (via (4.22)) (see the proof of Theorem 4.10 for details). ∎

Next we consider the system in Example 4.5.

Theorem 4.8 *Let η_t be a recurrent exclusion process with variable birth rate. Then the set of extreme translation invariant distributions of η_t is $\{\delta_\emptyset, \nu_1\}$. Moreover, if ν is any translation invariant distribution, then*

$$\lim_{t\to+\infty} \nu S(t) = \nu(\{\emptyset\})\delta_\emptyset + (1-\nu(\{\emptyset\}))\nu_1.$$

PROOF. Let η_t be a simple exclusion process which satisfies the conditions of the theorem. From the dynamics of the system we immediately have that property (a) of Theorem 4.4 is satisfied. We next prove the existence of a system $\hat{\eta}_t$ which satisfies the duality equation (4.6).

Let $A \subset Z^d$ be finite, and let $f(\eta) = \prod_{z\in A}(1-\eta(z))$. Then

$$\begin{aligned}
\Omega f(\eta) &= \sum_T c_T(\eta,\zeta)\Big[f_T(\eta) - f(\eta)\Big] \\
&= \sum_{\eta(x)=1,\,\eta(y)=0} p(x,y)\Big[\prod_{z\in A}(1-\eta_{xy}(z)) - \prod_{z\in A}(1-\eta(z))\Big] \\
&\quad + \sum_{\eta(x)=0,\,\eta(y)=1} q(x,y)\Big[\prod_{z\in A}(1-\eta_x(z)) - \prod_{z\in A}(1-\eta(z))\Big].
\end{aligned}$$

Let us consider the first sum above. By considering the cases obtained by assuming x and y are or are not in A, we obtain that the first sum can be written as

$$\begin{aligned}
&\sum_{\substack{\eta(x)=1,\,\eta(y)=0\\ A(x)=1,\,A(y)=0}} p(x,y)\prod_{z\in A}(1-\eta_{xy}(z)) - \sum_{\substack{\eta(x)=1,\,\eta(y)=0\\ A(x)=0,\,A(y)=1}} p(x,y)\prod_{z\in A}(1-\eta(z)) \\
&= \sum_{\substack{\eta(x)=0,\,\eta(y)=1\\ A(x)=1,\,A(y)=0}} p(x,y)\prod_{z\in\eta}(1-A(z)) - \sum_{\substack{\eta(x)=1,\,\eta(y)=0\\ A(x)=0,\,A(y)=1}} p(x,y)\prod_{z\in\eta}(1-A(z)) \\
&= \sum_{A(y)=1,\,A(x)=0} p(x,y)\Big[\prod_{z\in\eta}(1-A_{xy}(z)) - \prod_{z\in\eta}(1-A(z))\Big].
\end{aligned}$$

By proceeding similarly, the second sum in the definition $\Omega f(\eta)$ can be written as

$$\sum_{A(x)=0,\,A(y)=1} q(x,y)\Big[\prod_{z\in\eta}(1-A_x(z)) - \prod_{z\in\eta}(1-A(z))\Big].$$

Therefore

$$
\begin{aligned}
\Omega f(\eta) = & \sum_{A(y)=1,\, A(x)=0} p(x,y)\Big[\prod_{z\in\eta}(1 - A_{xy}(z)) - \prod_{z\in\eta}(1 - A(z))\Big] \\
& + \sum_{A(x)=0,\, A(y)=1} q(y,x)\Big[\prod_{z\in\eta}(1 - A_x(z)) - \prod_{z\in\eta}(1 - A(z))\Big] \\
& = \hat{\Omega}\hat{f}(A),
\end{aligned}
$$

where $\hat{\Omega}\hat{f}(A) = \prod_{z\in\eta}(1 - A(z))$. Hence, $\hat{\Omega}$ is the generator of a process $\hat{\eta}_t$ which differs from η_t only in the branching rates: in $\hat{\eta}_t$ particles are generated in the empty site y at a rate given by $\sum_x \hat{\eta}(x)q(x,y)$. Therefore, for all finite $A \subset Z^d$,

$$
P\Big[\eta_t^{\zeta} \cap A \neq \emptyset\Big] = \hat{P}\Big[\hat{\eta}_t^{A} \cap \zeta \neq \emptyset\Big].
$$

It is only left to verify the validity of part (b) of Theorem 4.4, which is certainly true due to the branching part of the process (and the absence of any size reduction rule). ∎

The contact process (Example 4.2) also satisfies Theorem 4.4 (we leave for the reader to verify the validity of properties (a) and (b)). That is, the following result holds (see [79]).

Theorem 4.9 *Let η_t be a contact process. Then the set of extreme translation invariant distributions of η_t is $\{\delta_\emptyset, \nu_1\}$. Moreover, if ν is any translation invariant distribution, then*

$$
\lim_{t\to+\infty} \nu S(t) = \nu(\{\emptyset\})\delta_\emptyset + (1 - \nu(\{\emptyset\}))\nu_1.
$$

Next we exploit the relationship between η_t and $\hat{\eta}_t$ in Example 4.4, to prove extinction of the BARW in Z.

Theorem 4.10 *Let $\hat{\eta}_t$ be the dual of a BARW in Z^d. Then*

(i) $\hat{\eta}_t$ satisfies Theorem 4.3.

(ii) If $d = 1$, then $\hat{\eta}_t$ is ergodic for all $\alpha \geq 1$.

(iii) If $d = 1$, *and* $\alpha \geq 1$, *then for all finite A subset Z,*

$$P\big[\hat{\eta}_t^A = \emptyset \text{ for some } t \geq 0\big] = 1.$$

Theorem 4.11 *Let* η_t *be a BARW in* Z. *Then*

(i) η_t *is ergodic for all* $\alpha \geq 1$.

(ii) For all $\alpha \geq 1$ *and* $A \subset Z$, *A finite,*

$$P\big[\eta_t^A = \emptyset \text{ for some } t \geq 0\big] = 1.$$

We start by proving Theorem 4.10.

PROOF OF THEOREM 4.10. Let $\hat{\eta}_t$ be a dual of a BARW. We start our proof by showing that $\hat{\eta}_t$ is a well defined process for any arbitrary initial configuration $\zeta \subset Z^d$, not necessarily finite. Since $\hat{\eta}_t$ is a spin system it is enough to check the validity of (4.3). From the rate of the system we have that

$$\sup_{x \in Z^d} \sum_{u \in Z^d} \sup_{\eta \subset Z^d} |c(x,\eta) - c(x,\eta_u)| < 1 + 3\beta < +\infty.$$

Hence the operator $\hat{\Psi}$ defined by

$$\hat{\Psi} f(\eta) = \sum_{x \in Z^d} c(x,\eta)\left[f(\eta_x) - f(\eta)\right], \quad \text{for } f \in D(\Lambda) \tag{4.24}$$

is the generator of a Markov semigroup $\hat{S}(t)$ that we identify with $\hat{\eta}_t$.

Now that we have proved that $\hat{\eta}_t$ is a well defined process in Z^d, we turn to the proof of the theorem.

For part (i) we only need to verify that conditions (a) and (b) of Theorem 4.3 are satisfied. Clearly, if η_t is a BARW, then η_t is a dual of $\hat{\eta}_t$. The validity of (a) again follows from the dynamics of the process. We claim that

$$\inf_{|B| \leq n} P\big[\eta_t^B = \emptyset \text{ for some } t \geq 0\big] > 0.$$

Assume that $|B| = n$. From the dynamics of the BARW we have that if we are interested in the specific event that $|\eta_t^B| = n - 1$, and

$|\eta_s^B| \le n+1$, $0 \le s \le t$, then the configuration B^*, $|B| = |B^*|$, which minimizes the probability of this event is given by isolated particles. That is, if $B^*(x) = 1$, then $B^*(y) = 0$ whenever $|x - y| = 1$. Then

$$\begin{aligned} & P\Big[|\eta_t^B| = n-1, \text{ and } |\eta_s^B| \le |B| + 1, 0 \le s \le t\Big] \\ \ge\ & P\left[\begin{array}{l} \text{a particle in } B^* \text{ generates an offspring and then the} \\ \text{offspring and the parent both die before any of} \\ \text{the other particles has generated any particle} \end{array}\right] \\ \ge\ & \frac{\alpha}{(n+1)(4cd(\alpha+1))}, \end{aligned}$$

where $c = \max\{\alpha, 1\}$. By repeating the same calculation n times we obtain

$$\inf_{0<|B|\le n} P\big[\eta_t^B = \emptyset \text{ for some } t \ge 0\big] \ge \frac{1}{(n+1)!}\left[\frac{\alpha}{4cd(\alpha+1)}\right]^n > 0.$$

Hence (i) holds. For part (ii), since we are assuming that $\alpha \ge 1$, it is convenient to write the rate of $\hat{\eta}_t$ as

$$c(x,\eta) = \begin{cases} 2\sum_y p(x,y)\eta(y) + \beta \sum_y p(x,y)\eta(y) & \text{if } \eta(x) = 0 \\ 1 + \beta \sum_y p(x,y)(1-\eta(y)) & \text{if } \eta(x) = 1, \end{cases}$$

where $\beta = \alpha - 1 \ge 0$.

From the rate $c(x,\eta)$ it is clear that if $\hat{\eta}_t$ is ergodic the only invariant measure is $\delta_\emptyset$. Let

$$H(\eta, A) = \prod_{x \in A} \eta(x), \quad A \subset Z^d,\ |A| < +\infty$$

and let us consider $\hat{\Omega} H(\eta, A)$. Then by 4.24),

$$\begin{aligned} \hat{\Psi} H(\eta, A) &= \sum_x c(x,\eta)\,[H(\eta_x, A) - H(\eta, A)] \\ &= \sum_{x \in A} c(x,\eta)\,[H(\eta_x, A) - H(\eta, A)] \end{aligned}$$

where the last equality follows because if $x \notin A$ then $H(\eta_x, A) - H(\eta, A) = 0$.

If we replace $c(x,\eta)$ by its definition, we have

$$\begin{aligned}
\hat{\Psi}H(\eta,A) &= (2+\beta)\sum_{x\in A}\sum_{y}p(x,y)\eta(y)(1-\eta(x))H(\eta,A\setminus\{x\}) \\
&\quad -\sum_{x\in A}H(\eta,A)-\beta\sum_{x\in A}\sum_{y}p(x,y)(1-\eta(y))H(\eta,A) \\
&= (2+\beta)\sum_{x\in A}\sum_{y}p(x,y)H(\eta,(A\setminus\{x\})\cup\{y\}) \\
&\quad -(2+\beta)\sum_{x\in A}\sum_{y}p(x,y)H(\eta,A\cup\{y\}) \\
&\quad -\sum_{x\in A}H(\eta,A)-\beta\sum_{x\in A}\sum_{y}p(x,y)H(\eta,A) \\
&\quad +\beta\sum_{x\in A}\sum_{y}p(x,y)H\left(\eta,A\cup\{y\}\right).
\end{aligned}$$

If we let

$$p(x,y)=p(0,y-x)=\begin{cases}\frac{1}{2}, & |x-y|=1\\ 0, & \text{otherwise}\end{cases}$$

then

$$\begin{aligned}
\hat{\Psi}H(\eta,A) &= \frac{(2+\beta)}{2}\sum_{x\in A}\sum_{|x-y|=1}H(\eta,(A\backslash\{x\})\cup\{y\}) \\
&\quad -\sum_{x\in A}\sum_{|y-x|=1}H(\eta,A\cup\{y\})-(1+\beta)|A|H(\eta,A).
\end{aligned} \tag{4.25}$$

We remind the reader that by part **(i)**, the proof of part **(ii)** will be complete if we can show that

$$\lim_{t\to+\infty}\delta_Z\hat{S}(t)=\delta_{\emptyset}.$$

To be able to prove this we use the characterization $\mathcal{I}$ in (4.5).

Since $H(\eta,A)=\prod_{x\in A}\eta(x)\in D(\Lambda)$, and $\lim_{t\to+\infty}\delta_{Z^d}S(t)=\mu$ is in $\mathcal{I}$, we must have

$$\int_{\Lambda}\hat{\Psi}H(\eta,A)\,d\mu=0. \tag{4.26}$$

If we replace $\hat{\Psi} H(\eta, A)$ by equation (4.25), and if we let

$$\phi(A) = \int_{\Lambda} H(\eta, A)\, d\mu$$

then (4.26) becomes

$$\begin{aligned} 0 &= -(1+\beta)|A|\phi(A) + \frac{(2+\beta)}{2} \sum_{x \in A} \sum_{|y-x|=1} \phi((A \backslash \{x\}) \cup \{y\}) \\ &\quad - \sum_{x \in A} \sum_{|y-x|=1} \phi(A \cup \{y\}). \end{aligned} \tag{4.27}$$

Case 1. $A = \{0\}$. Then (4.27) gives

$$\begin{aligned} 0 &= -(1+\beta)\phi(\{0\}) + \frac{(2+\beta)}{2} [\phi(\{-1\}) + \phi(\{1\})] \\ &\quad - [\phi(\{0,1\}) + \phi(\{-1,0\})] \\ 0 &= +\phi(\{0\}) - 2\phi(\{0,1\}) \end{aligned}$$

where the above equality follows from the translation invariant property of μ. Therefore

$$\mu\{\eta : \eta(0) = 1\} = 2\mu\,\{\eta : \eta(0) = \eta(1) = 1\}\,. \tag{4.28}$$

Case 2. $A = \{0, 1\}$. Then (4.27) gives

$$\begin{aligned} 0 &= -(1+\beta)2\phi(\{0,1\}) + \frac{(2+\beta)}{2} [2\phi(\{0\}) + 2\phi(\{0,2\})] \\ &\quad - [2\phi(\{0,1\}) + 2\phi(\{0,1,2\})] \end{aligned}$$

and after applying (4.28)

$$\begin{aligned} 0 &= \phi(\{0\}) - \beta\phi(\{0\}) + 2\phi(\{0\}) + \beta\phi(\{0\}) + (2+\beta)\phi(\{0,2\}) \\ &\quad -\phi(\{0\}) - 2\phi(\{0,1,2\}) \\ 0 &= 2\phi(\{0,1,2\}) + \beta\phi(\{0,2\}). \end{aligned}$$

Since $\hat{\eta}_t$ is attractive μ has positive correlation (see [99], Chapter 2), and this implies that the right side is not zero unless $\phi(\{0\}) = 0$. Hence we have

$$\lim_{t \to +\infty} \delta^Z \hat{S}(t) = \delta_\emptyset$$

and the proof of part **(ii)** is complete.

It is left to prove part **(iii)** of the theorem. Let $\alpha \geq 1$. We want to show that for all $A \subset Z$ finite,

$$P\big[\hat{\eta}_t^A = \emptyset \text{ for some } t \geq 0\big] = 1.$$

To prove this we proceed as follows. Let $Z(t)$ be the distance between the two furthest particles in $\hat{\eta}_t^A$, if $|\hat{\eta}_t^A| = 1$ we let $Z(t) = 0$. Then $Z(t) = 0, 1, 2, \cdots$. Furthermore, from the dynamics of the process and the definition of $Z(t)$ we have:

(a) $Z(t) = 0 \to 1$ at rate $\alpha + 1$ and $Z(t) = 0 \to |\hat{\eta}_t^A| = 0$ at rate α,

(b) $Z(t) > 0 \to Z(t) + 1$ at rate $\alpha + 1$,

(c) $Z(t) > 0 \to Z(t) - i, \quad i = 1, \cdots, Z(t)$, at rate at least $\alpha + 1$.

Therefore the walk $Z(t)$, $Z(t) > 0$, can be bounded from above by a continuous time symmetric random walk with rate at least 2. Hence $Z(t) = 0$ an infinite number of times and $\eta_t^A = \emptyset$ with probability one. ∎

We are now ready to prove Theorem 4.11.

PROOF OF THEOREM 4.11. To prove part (i), we want to show that for all $A \subset Z$ finite and for all $\mu \in \mathcal{M}(Z^d)$,

$$\lim_{t \to +\infty} P\big[|\eta_t^\mu \cap A| \text{ odd}\big] = 0.$$

From Theorem 4.10, it suffices to consider $\mu = \delta_Z$. But from Theorem 4.3,

$$\lim_{t \to +\infty} P\big[|\eta_t^Z \cap A| \text{ odd}\big] = \frac{1}{2} P\big[\hat{\eta}_t^A \neq \emptyset\, \forall t \geq 0\big]$$

and the result follows from part (iii) of Theorem 4.11.

For part (ii), by Theorem 4.11 part (ii), we know that for all $A \subset Z$ finite,

$$\lim_{t \to +\infty} P\big[|\hat{\eta}_t^Z \cap A| \text{ odd}\big] = 0.$$

But by part (i) of Theorem 4.11 and by Theorem 4.3, we also have

$$\lim_{t \to +\infty} P\big[|\hat{\eta}_t^Z \cap A| \,\mathrm{odd}\big] = \frac{1}{2} P\big[\eta_t^A \neq \emptyset \ \forall t \geq 0\big]$$

that implies

$$P\big[\eta_t^A \neq \emptyset \ \forall t \geq 0\big] = 0$$

and the proof is complete. ∎

4.4 Annihilating and coalescing random walks

In this section we consider the evolution of an arbitrary collection of particles that perform a continuous time random walk on Z^d. That is, each particle at a site x waits an exponential time with mean 1 and then jumps into a site y with probability $p(x, y)$, where $p(x, y)$'s are the transition probabilities associated with an irreducible Markov chain. Two simple collision rules are introduced: annihilation and coalescence. In the annihilation case, each particle moves independently, and whenever one jumps into a site that is already occupied both particles disappear. In the coalescence case, the particles again move independently, but whenever one jumps into a site that is already occupied the two particles merge into one. We will call these two processes the annihilating particle model, a.p.m., and the coalescing particle model, c.p.m., respectively. From the above dynamics we recover the rates $c_T(\eta, \zeta)$ of the two systems. For the a.p.m., $c_T(\eta, \zeta) = 0$ if $|T| \neq 2$, and if $T = \{x, y\}$, then $c_T(\eta, \zeta) = c(x, t, \eta)$ puts mass

(i) $p(x, y)$ on $\eta(x) = 0$, $\eta(y) = 1$ if $\eta(x) = 1$ and $\eta(y) = 0$, or

$p(y, x)$ on $\eta(y) = 0$, $\eta(x) = 1$, if $\eta(y) = 1$ and $\eta(x) = 0$;

(ii) $p(x, y)$ on $\eta(x) = \eta(y) = 0$ if $\eta(x) = \eta(y) = 1$, or

$p(y, x)$ on $\eta(x) = \eta(y) = 0$ if $\eta(x) = \eta(y) = 1$.

For the c.p.m., $c_T(\eta, \zeta) = 0$, if $|T| > 2$. If $T = \{x\}$, then $c(x, \eta)$ puts mass $p(x, y)$ on $\eta(x) = 0$, if $\eta(x) = 1$ and $\eta(y) = 1$, whereas if $T = \{x, y\}$, $c(x, y, \eta)$ puts mass $p(x, y)$ on $1 - \eta(x)$ and $1 - \eta(y)$ if

$\eta(x) = 1$ and $\eta(y) = 0$, or puts mass $p(y,x)$ on $1-\eta(y)$ and $1-\eta(x)$ if $\eta(y) = 1$ and $\eta(x) = 0$.

It is well known (see Griffeath [73]) that for both models

$$\mathcal{I} = \{\delta_\emptyset\}.$$

Even if this result answers all questions about the invariant measures of the a.p.m. and the c.p.m., there are many other interesting properties that one can investigate. We start by considering a property which deals with the relationship between the a.p.m. η_t and the c.p.m. ξ_t.

It is easy to show that for all common initial configurations B, there is a coupling that for all $t \geq 0$

$$\eta_t^B \subset \xi_t^B,$$

and this relation motivates the study of ratios of the type

$$\frac{P[\eta_t^B(0) = 1]}{P[\xi_t^B(0) = 1]}.$$

We are going to prove the following theorem, which is due to Simonelli [138]. For related work see also Arratia [12].

Theorem 4.12 *Let η_t be an a.p.m. and ξ_t a c.p.m. with common irreducible probability density $p(x,y)$. Then for all finite $A \subset Z^d$,*

$$\lim_{t\to+\infty} \frac{P[|\eta_t^{Z^d} \cap A| \text{ odd}]}{P[\xi_t^{Z^d} \cap A \neq \emptyset]} = \frac{1}{2}. \tag{4.29}$$

Before we go into the proof of Theorem 4.12, we need to characterize a dual system for the a.p.m and the c.p.m.. This will be done by constructing the systems on a percolation structure. We start with the latter one.

For each (x,y) in $Z^d \times Z^d$ with $p(x,y) > 0$ we associate a Poisson process $\{T^{(x,y)}(t), t \geq 0\}$, and denote by $\{T_n^{(x,y)}, n \geq 1\}$ its arrival times. At each $T_n^{(x,y)}$ we draw an arrow from $(x, T_n^{(x,y)})$ to $(y, T_n^{(x,y)})$, and put a delta at $(x, T_n^{(x,y)})$. We say that there is a path from (x,s) to (y,t), $s < t$, if there is a chain of upwards vertical and directed segments leading from (x,s) to (y,t) that does not go through any

δ's. Let $N_t^x(y)$ denote the number of paths from (x, y) to (y, t). We define

$$\xi_t^A = \{y \in Z^d : N_t^x(y) \neq 0 \text{ for some } x \in A\}.$$

Next we verify that the effect of the arrows is consistent with the dynamics of the c.p.m..

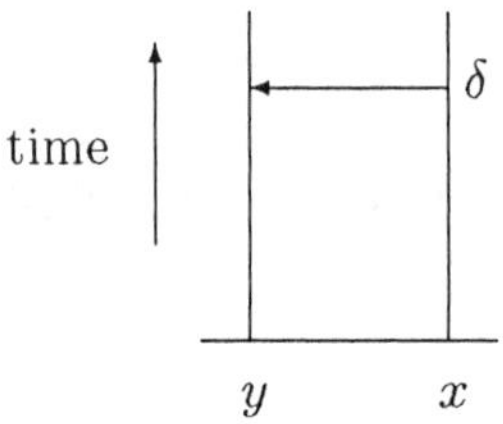

before arrow	after arrow
$(\eta(x), \eta(y))$	$(\eta(x), \eta(y))$
$(0,0)$	$(0,0)$
$(0,1)$	$(1,0)$
$(1,0)$	$(1,0)$
$(1,1)$	$(1,0)$

We construct a dual system again by reversing the direction of the arrows in the percolation structure, and obtain a system $\hat{\xi}_t = \zeta_t$ defined by the following rates. If $|T| > 1$, $c_T(\eta, \zeta) = 0$. If $T = \{x\}$,

$$c_T(\eta, \zeta) = \begin{cases} \sum_y p(x,y)\eta(y) & \text{if } \eta(x) = 0 \\ \sum_y p(x,y)(1 - \eta(y)) & \text{if } \eta(x) = 1. \end{cases}$$

That is, ζ_t is a voter model (the model in Example 4.1 with $\lambda = 0$).

The similarity between the a.p.m. and the c.p.m. allows us to define the a.p.m. by using the same percolation structure used for the c.p.m.. We define

$$\eta_t^A = \{y \in Z^d : \sum_{x \in A} N_t^x(y) \text{ is odd for some } x \in A\}.$$

In this case the effect of the arrows gives

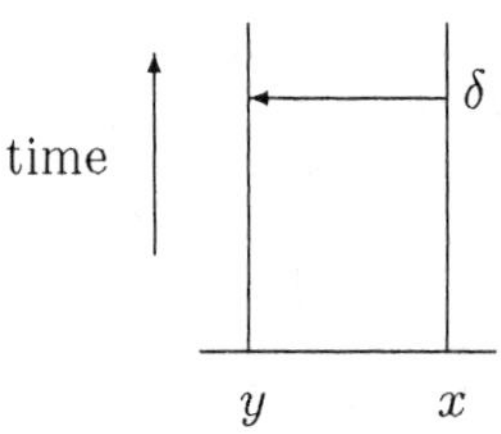

before arrow	after arrow
$(\eta(x), \eta(y))$	$(\eta(x), \eta(y))$
$(0,0)$	$(0,0)$
$(0,1)$	$(1,0)$
$(1,0)$	$(1,0)$
$(1,1)$	$(0,0)$

Hence, η_t is an a.p.m.. It is easy to see that the dual system of η_t is also a voter model ζ_t. Therefore we have proved that for all finite $A \subset Z^d$,

$$P\Big[\xi_t^B \cap A \neq \emptyset\Big] = \hat{P}\Big[\zeta_t^A \cap B \neq \emptyset\Big]$$

and

$$P\Big[|\xi_t^B \cap A| \text{ odd } \neq \emptyset\Big] = \hat{P}\Big[|\zeta_t^A \cap B| \text{ odd }\Big].$$

In what follows we need the following result due to Arratia [12].

Lemma 4.3 *Let ζ_t be a voter model with $p(x, y)$ denoting a nontrivial random walk. For arbitrary positive integer k,*

$$\lim_{t\to+\infty} P\big[|\zeta_t^0| \geq k | \zeta_t^0 \neq \emptyset\big] = 1.$$

The additive property of the voter model, $\zeta_t^{A\cup B} = \zeta_t^A \cup \zeta_t^B$, immediately implies the validity of Lemma 4.3 in the case the initial configuration $\{0\}$ is replaced by any finite configuration.

It is interesting to notice that the coalescing and annihilating particle systems satisfy Theorem 4.4 and Theorem 4.3, respectively. However, as we have mentioned before, these systems are ergodic, with $\mathcal{I} = \{\delta_\emptyset\}$. This follows from the result that for a voter model η_t

$$\lim_{t\to+\infty} P\big[\eta_t^A \neq \emptyset\big] = 0,$$

for any finite set $A \subset Z^d$.

We are now ready to prove Theorem 4.12.

PROOF OF THEOREM 4.12. Let $A \subset Z^d$ be finite. We want to show that

$$\lim_{t\to+\infty} \frac{P\big[|\eta_t^{Z^d} \cap A| \text{ odd }\big]}{P\big[\xi_t^{Z^d} \cap A \neq \emptyset\big]} = \frac{1}{2}.$$

Let $s > 0$ be arbitrary, and define

$$g(s,t) = \frac{P\big[|\eta_{t+s}^{Z^d} \cap A| \text{ odd }\big]}{P\big[\xi_t^{Z^d} \cap A \neq \emptyset\big]}.$$

Then by utilizing (4.12) and (4.13) we obtain

$$\begin{aligned}
g(s,t) &= \frac{P[|\eta_s^{Z^d} \cap \zeta_t^A| \text{ odd }]}{P[\zeta_t^A \neq \emptyset]} \\
&= \frac{\sum_{\substack{C\subset Z^d \\ |C|<+\infty}} P[\zeta_t^A = C]P[|\eta_s^{Z^d} \cap C| \text{ odd }]}{P[\zeta_t^A \neq \emptyset]} \\
&= \sum_{\substack{C\subset Z^d \\ |C|<+\infty}} P[\zeta_t^A = C \,|\zeta_t^A \neq \emptyset\,]P[|\eta_s^{Z^d} \cap C| \text{ odd }].
\end{aligned}$$

Hence,

$$\begin{aligned}
\lim_{t\to+\infty} g(s,t) &= \lim_{t\to+\infty} \sum_{\substack{C\subset Z^d \\ |C|<+\infty}} P[\zeta_t^A = C \,|\zeta_t^A \neq \emptyset\,]P[|\eta_s^{Z^d} \cap C| \text{ odd }] \\
&= \frac{1}{2},
\end{aligned}$$

where the second equality follows from the remarks after Lemma 4.3, and Theorem 4.2. Next, let $\{t_k\}_{k=1}^{+\infty}$ be a sequence such that $t_k \to +\infty$ as $k \to +\infty$ and

$$\lim_{k\to+\infty} \frac{P[|\eta_{t_k}^{Z^d} \cap A| \text{ odd }]}{P[\xi_{t_k}^{Z^d} \cap A \neq \emptyset]} = \alpha,$$

for some $\alpha \geq 0$. Choose an arbitrary $k > 0$, and let $0 < s < t_k$. Then

$$\begin{aligned}
\frac{P[|\eta_{t_k}^{Z^d} \cap A| \text{ odd }]}{P[\xi_{t_k}^{Z^d} \cap A \neq \emptyset]} &= \frac{P[|\zeta_{t_k-s}^A \cap \eta_s^{Z^d}| \text{ odd }]}{P[\zeta_{t_k}^A \neq \emptyset]} \\
&= \frac{P[|\zeta_{t_k-s}^A \cap \eta_s^{Z^d}| \text{ odd }]}{P[\zeta_{t_k}^A \neq \emptyset|\zeta_{t_k-s}^A \neq \emptyset]P[\zeta_{t_k-s}^A \neq \emptyset]} \\
&= \frac{g(s,t_k-s)}{P[\zeta_{t_k}^A \neq \emptyset|\zeta_{t_k-s}^A \neq \emptyset]},
\end{aligned}$$

where the second equality holds because $\{\emptyset\}$ is an absorbing state for ζ_t. Hence,

$$\alpha = \lim_{k\to+\infty} \frac{g(s,t_k-s)}{P[\zeta_{t_k}^A \neq \emptyset|\zeta_{t_k-s}^A \neq \emptyset]}.$$

Since $\lim_{k\to+\infty} g(s,t_k-s)=\frac{1}{2}$ for all $s>0$, then

$$\lim_{k\to+\infty} P\big[\,\zeta_{t_k}^A\neq\emptyset|\zeta_{t_k-s}^A\neq\emptyset\,\big]$$

also exists and is independent of s. By Lemma 4.3 this limit is equal to one. Therefore all convergent subsequences will converge to the same limit, and

$$\lim_{t\to+\infty}\frac{P\big[|\eta_t^{Z^d}\cap A|\text{ odd }\big]}{P\big[\xi_t^{Z^d}\cap A\neq\emptyset\big]}=\frac{1}{2}.$$

■

In our next results we apply Theorem 2.8 to find the limiting distribution of a two particle coalescing and annihilating random walks after proper normalization. We limit our discussion to the discrete time version of these processes, but our results generalize to continuous time.

Let $\overline{X}_n=(\overline{X}_n^1,\overline{X}_n^2)$ and $\overline{Y}_n=(\overline{Y}_n^1,\overline{Y}_n^2)$ be a two particle coalescing and annihilating random walk on $Z^d\times Z^d$, respectively, with $p(x,y)=p(0,y-x)=1/(2d)$ if $|y-x|=1$, and 0 otherwise. That is, $\overline{X}_n^1$, $\overline{X}_n^2$, $\overline{Y}_n^1$, and $\overline{Y}_n^2$, are symmetric random walks on Z^d that move independently until they meet. When $\overline{X}_n^1$ and $\overline{X}_n^2$ meet they coalesce, whereas when $\overline{Y}_n^1$ and $\overline{Y}_n^2$ meet they annihilate each other.

Let Z be a normal random variable on R^d with mean 0 and covariance (1/d)I, i.e., if A is an open ball in R^d,

$$P(Z\in A)=(\frac{2}{2\pi})^{d/2}\int_A e^{-d|x|^2/2}\,dx_1\,dx_2\cdots dx_d.$$

Set $\phi(A)=P(Z\in A)$. We prove the following results.

Theorem 4.13 *Let $\overline{X}_n=(\overline{X}_n^1,\overline{X}_n^2)$ be a two particle coalescing random walk on $Z^d\times Z^d$, $\overline{X}_0^1=0$, $\overline{X}_0^2=z$, and $\tau=\inf\{n:\overline{X}_n^1=\overline{X}_n^2\}$. Then if $B=B_1\times B_2$, where B_i is an open ball in R^d, $i=1,2$,*

$$\lim_{n\to+\infty} P\Big(\frac{\overline{X}_n}{\sqrt{n}}\in B\Big)$$
$$=P\big(\tau=+\infty\big)\phi(B_1)\,\phi(B_2)+P\big(\tau<+\infty\big)\phi(B_1\cap B_2).$$

Theorem 4.14 *Let $\overline{Y}_n = (\overline{Y}^1_n, \overline{Y}^2_n)$ be a two particle annihilating random walk on $Z^d \times Z^d$, $\overline{Y}^1_0 = 0$, $\overline{Y}^2_0 = z$, and $\tau = \inf\{n : \overline{Y}^1_n = \overline{Y}^2_n\}$. Then if $B = B_1 \times B_2$, where B_i is an open ball in R^d, $i = 1, 2$,*

$$\lim_{n \to +\infty} P\Big(\frac{\overline{Y}_n}{\sqrt{n}} \in B\Big) = P(\tau = +\infty)\phi(B_1)\,\phi(B_2).$$

PROOF. Let $X_n = (X^1_n, X^2_n)$, where X^1_n and X^2_n are independent symmetric random walks on Z^d, $X^1_0 = 0$, $X^2_0 = z$. From the dynamics of the process, $\tau \stackrel{d}{=} \hat{\tau} = \inf\{n : X^1_n = X^2_n\}$. Let $A_{+\infty} = \{\hat{\tau} = +\infty\}$, and write

$$\begin{aligned} &P\Big(\frac{\overline{X}_n}{\sqrt{n}} \in B\Big) \\ &= P\Big(\{\frac{\overline{X}_n}{\sqrt{n}} \in B\} \cap \{\tau < +\infty\}\Big) + P\Big(\{\frac{\overline{X}_n}{\sqrt{n}} \in B\} \cap \{\tau = +\infty\}\Big) \\ &= P\Big(\{\frac{X^1_n}{\sqrt{n}} \in B_1 \cap B_2\} \cap A^c_{+\infty}\Big) + P\Big(\{\frac{X_n}{\sqrt{n}} \in B\} \cap A_{+\infty}\Big) \qquad (4.30) \\ &P\Big(\{\frac{X^1_n}{\sqrt{n}} \in B_1 \cap B_2\} | A^c_{+\infty}\Big) P\Big(A^c_{+\infty}\Big) + P\Big(\{\frac{X_n}{\sqrt{n}} \in B\} | A_{+\infty}\Big) P\Big(A_{+\infty}\Big). \end{aligned}$$

From classical results from the theory of random walks we have that

$$\lim_{n \to +\infty} P\Big(\{\frac{X^1_n}{\sqrt{n}} \in B_1 \cap B_2\}\Big) = \phi(B_1 \cap B_2)$$

and

$$\lim_{n \to +\infty} P\Big(\{\frac{X_n}{\sqrt{n}} \in B\}\Big) = \phi(B_1)\phi(B_2).$$

Now if $d \geq 3$, $0 < P(A_{+\infty}) < 1$, and an appeal to Theorem 2.8 completes the proof of the theorem. If $d \leq 2$, then $P(A_{+\infty}) = 0$, and the second term in (4.30) goes to zero as n goes to infinity. Then an appeal to Theorem 2.8 for the first limit in line below (4.30) completes the prove in this case as well. ∎

The proof of Theorem 4.14 is similar, and we leave the details for the reader.

We end this section by describing an application of products to a model of cellular automata. We define the model on Z, but the discussion and results can be generalized to Z^d.

Let $\{Y_n(x)\}$, $x \in Z$, $n \geq 1$, be a collection of independent random variables with $Y_n(x) = -1$ or 1, and $P(Y_n(x) = -1) = p_n$. We consider a discrete time stochastic process ξ_n on $\{-1, 1\}^Z$ by the following rule:

for $n \geq 1$, $x \in Z$,

$$\xi_n(x) = \xi_{n-1}(x-1)\xi_{n-1}(x+1)Y_n(x). \tag{4.31}$$

We further assume that ξ_0 is independent of the $Y_n(x)$'s.

Several authors have studied this model, see Miyamoto [112], and Lind [100], for example.

Our aim is to investigate the asymptotic of $\xi_n^{\nu_\theta}$, for arbitrary Bernoulli measure ν_θ as initial distribution. To this end we are going to iterate (4.31) and exploit the multiplicative property of the process. The following definition will be useful.

For each $x \in Z$, and $n \geq 1$, we define a path of length n starting from x, say $\tau_{n,j}^x$, to be a collection of sites in Z, say $x_{0,j}, x_{1,j}, \cdots, x_{n-1,j}$, such that $x_{n-1,j} = x+1$ or $x-1$, and for $0 \leq k < n-1$, $x_{k,j} = x_{k+1,j}+1$ or $x_{k+1,j}-1$.

The definition of paths, and (4.31), allow us to write $\xi_n(x)$ as

$$\begin{aligned}
\xi_n(x) &= \xi_{n-1}(x-1)\,\xi_{n-1}(x+1)Y_n(x) \\
&= \xi_{n-2}(x-2)\xi_{n-2}(x)Y_{n-1}(x-1) \\
&\quad \cdot\xi_{n-2}(x)\xi_{n-2}(x+2)Y_{n-1}(x+1)Y_n(x) \\
&= \cdots \\
&= \cdots \\
&= \prod_{\tau_{n,j}^x} \xi_0(x_{0,j}) \prod_{\tau_{n,j}^x} \left(\prod_{k=1}^{n-1} Y_k(x_{k,j}) \right) Y_n(x) \\
&= \prod_{\tau_{n,j}^x} \xi_0(x_{0,j}) \prod_{k=1}^{n-1} \left(\prod_{\tau_{n,j}^x} Y_k(x_{k,j}) \right) Y_n(x),
\end{aligned} \tag{4.32}$$

where $\prod_{\tau_{n,j}^x}$ denotes product over all paths of length n from x. Let $A = \{x^{(1)}, x^{(2)}, \cdots, x^{(t)}\}$ be an arbitrary finite subset of Z. The above derivation immediately extends to $\xi_n(A) = \prod_{x^{(i)} \in A} \xi_n(x^{(i)})$, giving

$$\xi_n(A) \quad = \Big(\prod_{x^{(i)} \in A,\, \tau^x_{n,j}} \xi_0(x^{(i)}_{0,j}) \prod_{k=1}^{n-1} \prod_{\tau^{x^{(i)}}_{n,j}} Y_k(x^{(i)}_{k,j}) \Big) \prod_{x^{(i)} \in A} Y_n(x^{(i)})$$

As in the case of IPS, the asymptotics of ξ_n can be obtained by investigating

$$\lim_{n \to +\infty} P\Big(\xi_n(A) = -1\Big),$$

for $A \subset Z$ finite.

The reader has certainly recognized the similarity between ξ_n and some of the interacting particle systems we have discussed in the previous section. In fact ξ_n could have been defined on $\{0,1\}^Z$, call this equivalent system ζ_n, in such a way that $P(\xi_n(A) = -1) = P(|\zeta_n \cap A| \text{ is odd })$.

We are now ready to discuss the ergodicity of ξ_n. We first consider the case $p_n = 0$ for $n \geq 1$. In this case, with $A = \{x\}$,

$$\lim_{n \to +\infty} P\Big(\xi_n^{\nu_\theta}(x) = -1\Big) = \lim_{n \to +\infty} P\Big(\prod_{\tau^x_{n,j}} \xi_0^{\nu_\theta}(x_{0,j}) = -1\Big), \quad (4.33)$$

where the $\xi_0^{\nu_\theta}(x_{0,j})$ are independent random variables with

$$P(\xi_0^{\nu_\theta}(x_{0,j}) = -1) = \theta.$$

It is clear from the above expression that the existence of the limit in (4.33) is determined by the structure of the product in $P(\ )$ and the value of θ. From the derivation of (4.33) we obtain

$$\prod_{\tau^x_{n,j}} \xi_0^{\nu_\theta}(x_{0,j}) = \prod_{j=-n}^{n} \left(\xi_0^{\nu_\theta}(x+j)\right)^{\delta_{n,j}},$$

where

$$\delta_{n,|j|} = \binom{n}{j} \qquad \text{and} \qquad \sum_{j=-n}^{n} \delta_{n,j} = 2^n.$$

From the properties of binomial coefficients (see Wolfram [157], for example), we have that

$$\delta_{n,j} \equiv 0 \text{ modulo } 2, \ -n < j < n, \qquad \text{if } n = 2^k,$$

whereas

$$\delta_{n,j} \equiv 1 \text{ modulo } 2, \ -n \leq j \leq n, \qquad \text{if } n = 2^k - 1.$$

Hence the limit in (4.33) cannot exist unless $\theta = 0, 1, 1/2$. We have proved the following theorem.

Theorem 4.15 *The cellular automaton $\xi_t^{\nu_\theta}$ with $p_n = 0$, $n \geq 1$, has a limiting distribution if, and only if, $\theta = 0, 1/2, 1$.*

Next we consider the case $0 < p_n < 1$. Let

$$X_{n,0}(A) = \prod_{x^{(i)} \in A, \, \tau_{n,j}^{x}} \xi_0(x_{0,j}^{(i)})$$

and

$$Y_n(A) = \prod_{x^{(i)} \in A} Y_n(x^{(i)}).$$

Then by using the representation of $\xi_n(A)$ we derived previously (see displayed formula after (4.32)), we can write

$$\lim_{n \to +\infty} P\Big(\xi_n(A) = -1\Big)$$
$$= \lim_{n \to +\infty} P\left(X_{n,0}(A) \, Y_n(A) \prod_{k=1}^{n-1} \prod_{\tau_{n,j}^{x^{(i)}}} Y_k(x_{k,j}^{(i)}) = -1 \right).$$

Note that unless $\theta = 0, 1/2, 1$, $X_{n,0}(A)$ does not have a limiting distribution, and therefore the above limit exists if, and only if, the limiting distribution of the remaining factors in the product is invariant with respect to convergent subsequences of the distribution of $X_{n,0}(A)$. This is possible if, and only if, the limiting distribution of

$$Y_n(A) \prod_{k=1}^{n-1} \prod_{\tau_{n,j}^{x^{(i)}}} Y_k(x_{k,j}^{(i)})$$

is symmetric, which can happen (see Theorem 1.21) if, and only if, $p_n = 1/2$ for some n, or

$$\lim_{n\to+\infty} \sum_{k=1}^{n-1} \sum_{\tau^{x^{(i)}}_{n,j}} \min\{P\big(Y_k(x^{(i)}_{k,j}) = -1\big),\, P\big(Y_k(x^{(i)}_{k,j}) = 1\big)\} = +\infty.$$

By investigating the growth of ξ_n, Miyamoto [112] proved that the above limit holds if, and only if, the number of n for which $0 < p_n < 1$ is infinite, thus obtaining the following result.

Theorem 4.16 *Assume that $\theta \neq 0, 1/2, 1$. Then the cellular automaton $\xi_t^{\nu_\theta}$ is ergodic if, only if, $p_n = 1/2$ for some n, or if the number of n for which $0 < p_n < 1$ is infinite.*

Chapter 5

Applications to arithmetical functions

5.1 Introduction

Throughout this chapter we put $Z = \{1, 2, \cdots\}$. An *arithmetical function* $f(n)$ is a real or complex valued function whose domain is either a subset of, or the whole set Z. Thus, there is no difference between the arithmetical function $f(n)$ or the sequence f_n, $n \geq 1$, except that when we call a sequence a function then the emphasis on $f(n)$ is the values of $f(n)$ based on the arithmetical structure of n.

Let us put $2 = p_1 < p_2 < \cdots$ for the sequence of the consecutive prime numbers, and p will always denote a generic prime. We use the notation $p|n$, or, in general, $a|b$, $a, b \in Z$, for p dividing n and a dividing b, respectively. Furthermore, $p^k \parallel n$ denotes that $p^k|n$ but p^{k+1} does not divide n (in notation, $p^{k+1} \not| n$). The fundamental theorem of arithmetic states that every $n \in Z$ has a unique representation in the form of

$$n = \prod_{j=1}^{w} p_{i_j}^{a_{i_j}}, \quad w = w(n), \quad a_{i_j} = a_{i_j}(n),\ n \geq 2 \tag{5.1}$$

where $w(n)$ and $a_{i_j}(n)$ are positive integers. We can also write (5.1) into the infinite product

$$n = \prod_{j=1}^{+\infty} p_j^{s_j}, \quad s_j = s_j(n) \geq 0 \text{ integers.} \tag{5.2}$$

Clearly $s_j = a_{i_j}$ if $p_{i_j}|n$ and $s_j = 0$ if $p_{i_j} \not| n$.

The functions $w(n)$, $s_j(n)$ and $a_{i_j}(n)$ are arithmetical functions, and they indeed play an important role in the sequel. The function $w(n)$ counts the number of distinct prime divisors of n. We can write

$$w(n) = \sum_{p|n} 1,$$

and thus if n and m have no common prime divisors, in notation $(n, m) = 1$, which is referred to as n and m are relatively prime, then

$$w(nm) = w(n) + w(m), \ (n, m) = 1. \tag{5.3}$$

If we count the prime divisors of n with multiplicity,

$$W(n) = \sum_{p_{i_j}^{a_{i_j}} \parallel n} a_{i_j}$$

then

$$W(mn) = W(n) + W(m) \tag{5.4}$$

without the restriction $(n, m) = 1$. We call any function having the property (5.3) *additive*, and if (5.4) holds (no restrictions on n and m) *completely additive*. If an additive function $f(n)$ satisfies $f(p^k) = f(p)$ for all $k \geq 1$, then we call $f(n)$ *strongly additive*.

An additive function $f(n)$ can be written as

$$f(n) = \sum_{p^k \parallel n} f(p^k) = \sum_{j=2}^{+\infty} \sum_{k=0}^{+\infty} f(p_j^k) \epsilon_{k,p_j}(n) \tag{5.5}$$

where

$$\epsilon_{k,p}(n) = \begin{cases} 1 & \text{if } p^k \parallel n \\ 0 & \text{otherwise.} \end{cases} \tag{5.6}$$

The simple function $g_t(n) = n^t$, $n \in Z$, t a fixed real or complex number, exemplifies another important class of arithmetical functions. Clearly,

$$g_t(mn) = g_t(m) g_t(n) \tag{5.7}$$

a property that represents the family of *(completely) multiplicative* functions. If (5.7) is valid for $(m, n) = 1$ only, $g_t(n)$ is called *multiplicative.* A multiplicative function $g(n)$ is strongly multiplicative if $g(p^k) = g(p)$ for all $k \geq 1$.

There is a strong relation between additive ($f(n)$) and multiplicative ($g(n)$) functions. Indeed, if a multiplicative function $g(n) > 0$ for all n, then $f(n) = \log g(n)$ is additive. However, when $g(m)$ can be both positive and negative, the relation weakens. Such questions as the asymptotic properties of the arithmetical mean of $g(n)$ cannot be treated with any relation to additive functions.

Just as (5.3) and (5.4) extend to a representation of an infinite sum ((5.5) and (5.6)), the multiplicative property (5.7) leads to an infinite product representation. That is, if $g(n)$ is a multiplicative function,

$$g(n) = \prod_{j=2}^{+\infty} \prod_{k=1}^{+\infty} g(p_j^k)^{\epsilon_{k,p_j}(n)} \tag{5.8}$$

with $\epsilon_{k,p}(n)$ as at (5.6).

We say that n is square-free if $\epsilon_{k,p}(n) \equiv 0$ for all p and all $k \geq 2$. In other words, for a square-free integer $n \in Z$, every $a_{i_j} = 1$ in (5.1). An important multiplicative function is the so called Möbius-function $\mu(n)$: $\mu(n) = 1$ or -1, defined on square-free numbers, it is multiplicative and $\mu(p) = -1$ for every prime p. It can also be extended to all $n \in Z$ by defining $\mu(n) = 0$ if n is not square-free. Clearly, $\mu(n)$ remains multiplicative after such an extension. It is easily obtained that $\mu(n) = 1$ or -1 according as the square-free n has an even number or an odd number of prime divisors. At first glance it therefore appears that it should not be difficult to conclude that

$$\frac{1}{N} \sum_{n=2}^{N} \mu(n) \to 0 \qquad \text{as } N \to +\infty. \tag{5.9}$$

Note that only the square-free numbers n contribute to the sum above. The reader, who has not yet gone through an extensive course of number theory, is advised to make an attempt at (5.9) to appreciate the results that follow on the subsequent pages.

In this chapter we want to deal with additive and multiplicative arithmetical functions by using tools of probability theory, in particular, results of the earlier chapters of the present book. For such an aim we introduce the following probability space. Let $\Omega_N = \{1, 2, \cdots, N\}$, $\mathcal{A}_N$ the set of all subsets of Ω_N, and $P_N(A) = k/N$ for every $A \subset \Omega_N$ where k is the number of elements in A. In other words, we have a uniformly distributed random variable X on Ω_N and $P_N(A) = P_N(X \in A)$.

As an example for gaining an advantage from this probabilistic model, let us determine the number $\varphi(N)$ of those integers $1 \le k \le N$ which are relatively prime to N; that is, $(k, N) = 1$. Let $p|N$. Define A_p as the set of those integers in Ω_N which are also divisible by p. Then, with $[y]$ signifying the integer part of y,

$$P_N(A_p) = \frac{[\frac{N}{p}]}{N} = \frac{1}{p}$$

and for $p \neq p^*$, $p|N$ and $p^*|N$, $P_N(A_p \cap A_{p^*}) = [N/(pp^*)]/N = 1/(pp^*)$, and a similar rule obtains for any number of $p|N$. In other words, the A_p, $p|N$, are independent events. But then the A_p^c are also independent. We thus have that for $p^{(1)}, p^{(2)}, \cdots, p^{(w(N))}$, each $p^{(j)}|N$,

$$P_N\left(A^c_{p(1)} \cap A^c_{p(2)} \cap \cdots \cap A^c_{p(w(N))}\right) = \prod_{p|N} \left(1 - \frac{1}{p}\right).$$

On the other hand, by definition, the left hand side above equals $\varphi(N)/N$, so

$$\varphi(N) = N \prod_{p|N} \left(1 - \frac{1}{p}\right) = \prod_{\substack{p \\ p^k \,\|\, N}} p^k \left(1 - \frac{1}{p}\right). \qquad (5.10)$$

The formula (5.10) implies that $\varphi(N)$ is multiplicative, and it gives its exact value. The function $\varphi(N)$ is known as Euler's function.

For a real valued arithmetical function $f(n)$ we define

$$F_N(x) = P_N\left(n \in \Omega_N : f(n) \le x\right) = P_N\left(f(n) \le x\right). \qquad (5.11)$$

The function $F_N(x)$ is called the distribution function of $f(n)$ on $(\Omega_N, \mathcal{A}_N, P_N)$, and we seek conditions under which $F_N(x)$ converges completely to a proper distribution function $F(x)$. In general, for a set $A \subset Z$, with $A_N = A \cap \Omega_N$,

$$\lim_{N \to +\infty} P_N(A_N) = d(A) \tag{5.12}$$

is called *(the natural) density* of A. Hence our aim at (5.11) is to determine conditions guaranteeing that the set $A = A(x) = \{n \in Z : f(n) \leq x\}$ has a natural density $F(x)$ for given x (with the additional restriction that x is a point of continuity of $F(x)$). There is an easy solution of this for $f(n) = \epsilon_{k,p}(n)$ at (5.6). Clearly, $F(x) = 0$ for $x < 0$, and, for $0 < x < 1$,

$$\begin{aligned} 1 - F_N(x) &= P_N\left(n \in \Omega_N : p^k \parallel n\right) \\ &= \frac{[\frac{N}{p^k}] - [\frac{N}{p^{k+1}}]}{N} \to \frac{1}{p^k}(1 - \frac{1}{p}) \end{aligned} \tag{5.13}$$

and finally, $F_N(x) = 1$ for all $x > 1$ and all $N \geq 2$. It therefore suggests that, for additive arithmetical functions $f(n)$ and multiplicative functions $g(n)$, the representations (5.5) and (5.8), respectively, give an easy tool for finding the limits at (5.12). There are two main reasons why it does not follow easily. First, the functions at (5.6) are not independent, although they are independent asymptotically. However, the asymptotic independence does not translate to independent sums at (5.5) or products at (5.8). The error terms in such asymptotic independence very quickly become larger than the main terms. Indeed, let us just compute the values

$$\begin{aligned} & P_N\left(\epsilon_{1,p}(n) = 1,\, \epsilon_{1,p^*}(n) = 1\right) \\ &= \frac{[\frac{N}{pp^*}] - [\frac{N}{p^2p^*}] - [\frac{N}{p(p^*)^2}] + [\frac{N}{p^2(p^*)^2}]}{N} \\ &= \frac{1}{pp^*}\left(1 - \frac{1}{p}\right)\left(1 - \frac{1}{p^*}\right) + \mathrm{O}\left(\frac{1}{N}\right) \end{aligned} \tag{5.14}$$

where the constant in O( ) is bounded by 2. A comparison with (5.13) does give the asymptotic independence (for these two values). However, when one chooses p and p^* such that $pp^* > N$, then the exact value of $P_N(\cdots) = 0$ while the "main" term in the asymptotic value is a positive number.

A second difficulty arises from the fact that the density $d(A)$ is not a probability on the collection of sets A having natural density. As an example, take the sets

$$A_j = \{j^2+1, j^2+2, \cdots, (j+1)^2\},\ j \geq 1.$$

Then $\cup A_j = Z$, $P_N(A_{j,N}) \to d(A_j) = 0$, so

$$1 = d(Z) = d(\cup_{j=1}^{+\infty} A_j) \neq \sum_{j=1}^{+\infty} d(A_j) = 0.$$

In spite of these difficulties, we can proceed by comparing the sum at (5.5) and the product at (5.8) with sums and products of independent random variables, respectively, and obtain necessary and sufficient conditions for the convergence set out at (5.11). Our approach differs significantly from the traditional ones in probabilistic number theory.

The set $A_{k,p}$ of integers whose density is computed in (5.13) is not a typical set whose density does exist. In fact, in $A_{k,p}$ the integers have a very regular structure: they are all multiples of p^k with p and k fixed (and multiples of p^{k+1} are deleted). In Section 3 of the present chapter we shall show that gaps between two consecutive members of a set A with $d(A) > 0$ can become arbitrarily large.

5.2 The Kubilius-Turán inequality

Let $f(m)$ be a real valued additive arithmetical function. We evaluate, and then estimate, the variance $V_N(f)$ of $f(m)$ on the probability space $(\Omega_N, \mathcal{A}_N, P_N)$, introduced in the previous section. We use (5.5) in the finite form

$$f(m) = \sum_{p^k \leq m} f(p^k)\epsilon_{k,p}(m) \tag{5.15}$$

where $\epsilon_{k,p}(m)$ is defined at (5.6). For strongly additive functions (when $f(p^k) = f(p)$ for all prime p and $k \geq 2$), (5.15) becomes

$$f(m) = \sum_{p \leq m} f(p)\beta_p(m) \tag{5.16}$$

with

$$\beta_p(m) = \begin{cases} 1 & \text{if } p \mid m \\ 0 & \text{otherwise.} \end{cases} \tag{5.17}$$

We use the notation $E_N(g)$ for the expected value (mean value) of an arithmetical function $g(m)$ and $V_N(g)$ for its variance. That is,

$$E_N(g) = \frac{1}{N}\sum_{m=1}^{N} g(m) \tag{5.18}$$

and

$$V_N(g) = \frac{1}{N}\sum_{m=1}^{N}\Big(g(m) - E_N(g)\Big)^2 = \frac{1}{N}\sum_{m=1}^{N} g^2(m) - E_N^2(g). \tag{5.19}$$

Note that

$$\begin{aligned} E_N(\epsilon_{k,p}) &= E_N(\epsilon_{k,p}^2) = P_N(\epsilon_{k,p}(m) = 1) \\ &= \frac{1}{N}\left\{[\frac{N}{p^k}] - [\frac{N}{p^{k+1}}]\right\} \\ &= \frac{1}{p^k}\Big(1 - \frac{1}{p}\Big) + \frac{c(p,N)}{N}, \qquad \big|c(p,N)\big| \le 1 \end{aligned} \tag{5.20}$$

and

$$\begin{aligned} E_N(\beta_p) &= E_N(\beta_p^2) = P_N(\beta_p(m) = 1) \\ &= \frac{1}{N}\Big[\frac{N}{p}\Big] = \frac{1}{p} + \frac{c^*(p,N)}{N}, \end{aligned} \tag{5.21}$$

where $c^*(p,N)| \le 1$. Hence, by the linearity of the $E_N(\cdot)$ operation,

$$\begin{aligned} E_N(f) &= \sum_{p^k \le N} f(p^k) E_N(\epsilon_{k,p}) \\ &= \sum_{p^k \le N} \frac{f(p^k)}{p^k}(1 - \frac{1}{p}) + \frac{\eta(N)}{N}\sum_{p^k \le N} \big|f(p^k)\big|, \end{aligned} \tag{5.22}$$

with $|\eta(N)| \le 1$. Furthemore, from (5.16) and (5.22) (see the definition in (5.19)),

$$\begin{aligned} V_N(f) &= E_N\left\{\Big[\sum_{p^k\le N} f(p^k)\big(\epsilon_{k,p}(m) - E_N(\epsilon_{k,p})\big)\Big]^2\right\} \\ &= \sum_{p^k\le N} f^2(p^k)V_N(\epsilon_{k,p}) + 2\sum_{\substack{p^k\le N,\ q^t\le N\\ p\ne q}} f(p^k)f(q^t) \\ &\qquad \cdot E_N\left\{\big(\epsilon_{k,p} - E_N(\epsilon_{k,p})\big)\big(\epsilon_{k,q} - E_N(\epsilon_{k,q})\big)\right\}, \end{aligned}$$

where q also is running through the primes as indicated. We put

$$\begin{aligned} \gamma_N(f) &= 2\sum_{\substack{p^k\le N,\ q^t\le N\\ p\ne q}} f(p^k)f(q^t) \\ &\qquad \cdot E_N\left\{\big(\epsilon_{k,p} - E_N(\epsilon_{k,p})\big)\big(\epsilon_{k,q} - E_N(\epsilon_{k,q})\big)\right\}, \end{aligned} \tag{5.23}$$

so, we have

$$V_N(f) = \sum_{p^k\le N} f^2(p^k)V_N(\epsilon_{k,p}) + \gamma_N(f). \tag{5.24}$$

Now, we complete the exact computation of $V_N(f)$. By (5.19) and (5.20),

$$\begin{aligned} V_N(\epsilon_{k,p}) &= E_N(\epsilon_{k,p}^2) - E_N^2(\epsilon_{k,p}) = E_N(\epsilon_{k,p}) - E_N^2(\epsilon_{k,p}) \\ &= \frac{1}{p^k}(1-\frac{1}{p}) - \frac{1}{p^{2k}}(1-\frac{1}{p})^2 + \frac{c(p,N)}{N}\left(1 + \frac{2}{p^k}(1-\frac{1}{p})\right) \\ &\quad - \frac{c^2(p,N)}{N^2}. \end{aligned} \tag{5.25}$$

On the other hand, for $p \ne q$,

$$\begin{aligned} & E_N\left\{\Big(\epsilon_{k,p} - E_N(\epsilon_{k,p})\Big)\Big(\epsilon_{t,q} - E_N(\epsilon_{t,q})\Big)\right\} \\ &= E_N\big(\epsilon_{k,p}\,\epsilon_{t,q}\big) - E_N\big(\epsilon_{k,p}\big)E_N\big(\epsilon_{t,q}\big) \\ &= P_N\big(\epsilon_{k,p}(m) = 1,\ \epsilon_{t,q}(m) = 1\big) - E_N(\epsilon_{k,p})E_N(\epsilon_{t,q}) \\ &= \frac{1}{N}\left\{\left[\frac{N}{p^kq^t}\right] - \left[\frac{N}{p^{k+1}q^t}\right] - \left[\frac{N}{p^kq^{t+1}}\right] + \left[\frac{N}{p^{k+1}q^{t+1}}\right]\right\} \\ &\quad - E_N(\epsilon_{k,p})E_N(\epsilon_{t,q}). \end{aligned} \tag{5.26}$$

Upon collecting the expressions (5.24), (5.23), (5.25), and (5.26), we established a formula for $V_N(f)$ which is adequate for major limit theorems. We find that the main term in $V_N(f)$ appears to be

$$B_N^2(f) = \sum_{p^k \le N} \frac{f^2(p^k)}{p^k}(1 - \frac{1}{p}). \tag{5.27}$$

However, in order to see that the several other terms are indeed either comparable to, or smaller in magnitude than $B_N^2(f)$, some number theory must enter. At this stage we formulate the inequality that goes back to Turán [151] in some special cases of $f(m)$ and to Kubilius [96] for the general case, and which bounds $V_N(f)$ by a constant multiple of $B_N^2(f)$

Theorem 5.1 *(The Kubilius-Turán inequality). For every real or complex valued additive arithmetical function $f(m)$*

$$D_N^2(f) = \frac{1}{N}\sum_{n=1}^{N}\Big|f(n) - \sum_{p^k \le N} \frac{f(p^k)}{p^k}(1 - \frac{1}{p})\Big|^2 \le C\, B_N^2(f)$$

where $C > 0$ is a universal constant (not depending on f).

PROOF. Note that if $f(m)$ is complex valued, and if we prove the theorem for $f_1(m) = Re\, f(m)$ and $f_2 = Im\, f(m)$, both real valued and additive functions, then the theorem becomes valid for the comples valued $f(m)$ as well on account of

$$|z|^2 = |x + iy|^2 = x^2 + y^2$$

for any complex number $z = x + iy$, x, y real. Therefore, in the sequel, we assume that $f(m)$ is real valued.

A second remark for the start of the proof is to emphasize that we replaced $E_N(f)$ by its approximation

$$A_N(f) = \sum_{p^k \le N} \frac{f(p^k)}{p^k}(1 - \frac{1}{p}) \tag{5.28}$$

in the statement. However, since $(a+b)^2 \leq 2a^2 + 2b^2$ for any real number a and b,

$$\begin{aligned} V_N(f) &= \frac{1}{N}\sum_{n=1}^{N}\Big(f(n) - A_N(f) + A_N(f) - E_N(f)\Big)^2 \\ &\leq 2D_N^2(f) + 2\Big(E_N(f) - A_N(f)\Big)^2 \end{aligned} \tag{5.29}$$

and, in view of (5.22),

$$\begin{aligned} \Big(E_N(f) - A_N(f)\Big)^2 &= \Big(\frac{\eta(N)}{N}\sum_{p^k \leq N} |f(p^k)|\Big)^2 \\ &= \Big(\frac{\eta(N)}{N}\sum_{p^k \leq N} \frac{|f(p^k)|}{\sqrt{p^k}}\sqrt{p^k}\Big)^2 \\ &\leq \Big(\sum_{p^k \leq N} \frac{f^2(p^k)}{p^k}\Big)\Big(\frac{1}{N^2}\sum_{p^k \leq N} p^k\Big) \\ &< B_N^2(f)\Big(\frac{1}{N}\sum_{p^k \leq N} 1\Big) < B_N^2(f), \end{aligned}$$

where the first inequality is due to the Cauchy-Schwarz inequality, while in the last two inequalities we utilized that $p^k \leq N$. In summary, the theorem states that

$$V_N(f) \leq (2C+2)B_N^2(f). \tag{5.30}$$

We can reverse the role of $V_N(f)$ and $D_N^2(f)$ in (5.29) by simply switching the roles of $E_N(f)$ and $A_N(f)$. We get

$$D_N^2(f) \leq 2V_N(f) + 2B_N^2(f). \tag{5.31}$$

Therefore, we can estimate $D_N^2(f)$ by utilizing the earlier expressions for $V_N(f)$.

We go back to (5.24) and estimate the first term on the right hand side by an appeal to (5.20) and (5.25). We have, with some

$|\rho(N)| \le 3$,

$$\begin{aligned} 0 &< \sum_{p^k \le N} f^2(p^k) V_N(\epsilon_{k,p}) \\ &= B_N^2(f) - \sum_{p^k \le N} \frac{f^2(p^k)}{p^{2k}} (1 - \frac{1}{p}) + \frac{\rho(N)}{N} \sum_{p^k \le N} f^2(p^k) \quad (5.32) \\ &< B_N^2(f) + \frac{1}{2} B_N^2(f) + 3B_N^2(f) = 4.5 B_N^2(f), \end{aligned}$$

where we used in the middle term that $(1-1/p)/p^k \le 1/2$ for all p and k, and in the last term that $\rho(N)/N \le 3/p^k$. Next, for estimating $\gamma_N(f)$ in (5.23)-(5.24), we simplify (5.26). First, note that each of the integer parts on the extreme right hand side of (5.26) is zero if $p^k q^t > N$. Next, we drop the integer parts for $p^k q^t \le N$, and thus the extreme right hand side of (5.26) becomes

$$\begin{aligned} &(1 - \Delta(p,q)) \frac{\delta(N)}{N} + \frac{c(p,N)}{N} \frac{1}{q^t} (1 - \frac{1}{q}) \\ &+ \frac{c(q,N)}{N} \frac{1}{p^k} (1 - \frac{1}{p}) - \frac{1}{p^k} (1 - \frac{1}{p}) \frac{1}{q^t} (1 - \frac{1}{q}) \Delta(p,q) \end{aligned} \quad (5.33)$$

where $\delta(N)$ is some number with $|\delta(N)| \le 3$, $|c(p,N)| \le 1$, $|c(q,N)| \le 1$, just as at (5.20), and $\Delta(p,q) = 1$ if $p^k q^t > N$, and $\Delta(p,q) = 0$ if $p^k q^t \le N$ (the term $c(p,N)c(q,N)/N^2$ is absorbed into the term $1/(Nq^t)$, say). We substitute (5.33) into (5.23), and estimate each of the resulting four sums, denoted S_j, $1 \le j \le 4$, individually. We get:

$$|S_1| \le \frac{6}{N} \sum_{\substack{p^k q^t \le N \\ p \ne q}} |f(p^k) f(q^t)| = \frac{6}{N} \sum_{\substack{p^k q^t \le N \\ p \ne q}} \frac{|f(p^k)|}{\sqrt{p^k}} \frac{|f(q^t)|}{\sqrt{q^t}} \sqrt{p^k q^t}$$

which, by the Cauchy-Schwarz inequality, is not exceeding

$$\frac{6}{N} \left(\sum_{p^k \le N} \frac{f^2(p^k)}{p^k} \sum_{q^t \le N} \frac{f^2(q^t)}{q^t} \right)^{1/2} \left(\sum_{\substack{p^k q^t \le N \\ p \ne q}} p^k q^t \right)^{1/2} \le 6\, B_N^2(f).$$

The estimates for S_2 and S_3 are similar to each other, so we carry out the estimate only once. By the Cauchy-Schwarz inequality again

$$
\begin{aligned}
|S_2| &\le \frac{1}{N} \sum_{\substack{p^k \le N \\ q^t \le N \\ p \ne q}} |f(p^k)||f(q^t)| \frac{1}{q^t} \left(1 - \frac{1}{q}\right) \\
&\le \frac{1}{N} \Big(\sum_{\substack{p^k \le N \\ q^t \le N}} \frac{f^2(p^k)}{p^k} \frac{f^2(q^t)}{q^t} \Big)^{1/2} \Big(\sum_{\substack{p^k \le N \\ q^t \le N \\ p \ne q}} p^k \frac{1}{q^t} \left(1 - \frac{1}{q}\right)^2 \Big)^{1/2} \\
&\le B_N^2(f) \frac{1}{N} \Big(\sum_{p^k \le N} p^k \Big)^{1/2} \Big(\sum_{q^t \le N} \frac{1}{q^t} \left(1 - \frac{1}{q}\right)^2 \Big)^{1/2} \\
&\le B_N^2(f) \frac{1}{\sqrt{N}} \Big(\sum_{p^k \le N} 1 \Big)^{1/2} \Big(\sum_{q^t \le N} \frac{1}{q^t} \Big)^{1/2} \qquad (5.34)
\end{aligned}
$$

where we have to know only that the number of primes $p \le N$, denoted $\pi(N)$, satisfies $\pi(N)/N \to 0$ as $N \to +\infty$ and $\sum_{p \le N} 1/p$ is not as large as $N/\pi(N)$ (the terms $k \ge 2$ and $t \ge 2$ contribute only a bounded number; more accurate estimates on $\pi(N)$ and on $\sum_{p \le N} 1/p$ will be discussed later). We thus have, with some constant C^*,

$$|S_2| \le C^* B_N^2(f) \quad \text{and} \quad |S_3| \le C^* B_N^2(f).$$

Finally,

$$|S_4| = \sum_{\substack{p^k \le N \\ q^t \le N \\ p^k q^t > N}} |f(p^k)|\,|f(q^t)| \frac{1}{p^k}\left(1 - \frac{1}{p}\right) \frac{1}{q^t}\left(1 - \frac{1}{q}\right).$$

The first step in estimating $|S_4|$ is the Cauchy-Schwarz inequality again. We have

$$|S_4| \le \Big(\sum_{\substack{p^k \le N \\ q^t \le N}} \frac{f^2(p^k)}{p^k} \frac{f^2(q^t)}{q^t} \Big)^{1/2} \Big(\sum_{\substack{p^k \le N \\ q^t \le N \\ p^k q^t > N}} \frac{1}{p^k q^t} \Big)^{1/2}$$

where in the first factor we dropped the restriction $p^k q^t > N$, so it becomes $B_N^2(f)$. We just have to estimate the second factor. The terms corresponding to $k \geq 2$ and $t \geq 2$ are clearly contributing only a constant, so our main concern is the contributions from $k = 1$ or $t = 1$. By symmetry in p^k and q^t,

$$\sum_{\substack{p^k \leq N \\ q^t \leq N \\ p^k q^t \leq N}} \frac{1}{p^k q^t} = 2 \sum_{p^k \leq \sqrt{N}} \frac{1}{p^k} \Big(\sum_{\frac{N}{p} < q^t \leq N} \frac{1}{q^t} \Big) + \Big(\sum_{\sqrt{N} < p^k \leq N} \frac{1}{p^k} \Big)^2 \tag{5.35}$$

which turns out to be bounded, based on a more accurate formula for $\sum_{p \leq N} 1/p$. In fact, using

$$\sum_{p \leq N} \frac{1}{p} = \log\log(N) + C^{**} + o(1) \tag{5.36}$$

where C^{**} is a constant and $o(1)$ is an expression tending to zero as $N \to +\infty$, we get (dropping the terms $k \geq 2$ and $t \geq 2$, whose contributions add up to a constant)

$$\begin{aligned} & 2 \sum_{p \leq \sqrt{N}} \frac{1}{p} \big(\log\log N - \log\log \frac{N}{p} \big) + \big(\log\log N - \log\log \sqrt{N} \big)^2 \\ &= 2 \sum_{p \leq \sqrt{N}} \frac{1}{p} \log \frac{\log N}{\log N - \log p} + (\log 2)^2 \\ &\sim 2 \sum_{p \leq \sqrt{N}} \frac{\log p}{p} \frac{1}{\log N} + (\log 2)^2 \end{aligned} \tag{5.37}$$

which is indeed bounded by yet another formula from number theory:

$$c_2 \frac{n}{\log n} \leq \pi(n) \leq c_1 \frac{n}{\log n}, \quad c_1 > 0,\ c_2 > 0 \ \text{ constants.} \tag{5.38}$$

One just has to use the Abelian method

$$\begin{aligned} \sum_{p \leq N} \frac{\log p}{p} &= \sum_{n=1}^{N} \frac{\log n}{n} \{\pi(n) - \pi(n-1)\} \\ &= \sum_{n=2}^{N-1} \pi(n) \big(\frac{\log n}{n} - \frac{\log(n+1)}{n+1} \big) + \pi(N) \frac{\log N}{N} \end{aligned}$$

and then Taylor expansions

$$\begin{aligned}\frac{\log(n+1)}{n+1} &= \frac{\log n + \log(1+1/n)}{n(1+1/n)} \\ &= \frac{\log n}{n}\left(1-\frac{1}{n}+\frac{\vartheta}{n^2}\right) + \frac{\frac{1}{n}+\frac{\delta}{n^2}}{n}\left(1-\frac{1}{n}+\frac{\vartheta}{n^2}\right).\end{aligned}$$

This yields the major term $\sum_{n=2}^{N} 1/n$ for $\sum_{p\le N}(\log p)/p$, which suffices for the boundedness in (5.37). The proof is completed. ∎

Remark 5.1 In order to emphasize the extent to which results of number theory are involved in the proof, we intentionally used the same results in (5.36) and (5.38) in a weaker form at (5.34) than at (5.35). In Chapter 6 we shall deal with some number theoretic results including (5.36) and (5.38).

Remark 5.2 There are several extensions of the Kubilius-Turán inequality. One direction of extension is to investigate $D_N^2(f)$ by summing over subsequences of the consecutive integers, and other directions are to replace the square distance used in $D_N^2(f)$ by other "subadditive " functions. The reader can find a comprehensive list of reference for these extensions in a paper by Indlekofer and Kátai [84].

We reformulate Theorem 5.1 for strongly additive functions. It requires no new detailed proof, but an estimate like the one at (5.29) immediately yields the result. This version also is known as the Kubilius-Turán inequality.

Theorem 5.1a *(An Inequality of Kubilius and Turán). Let $f(m)$ be a strongly additive arithmetical function. Then*

$$\frac{1}{N}\sum_{n=1}^{N}\left|f(m)-\sum_{p\le N}\frac{f(p)}{p}\right|^2 \le C_1\sum_{p\le n}\frac{f^2(p)}{p} \qquad (5.39)$$

where $C_1 > 0$ is a universal constant.

The seeds of all variants of the Kubilius-Turán inequalities are found in the paper [152] of Turán, who first proved (5.39) for the number $\omega(m)$ of prime divisors of m. Since $\omega(m)$ is strongly additive and $w(p) = 1$, (5.39), in view of (5.36), states that $E_N(\omega(m)) = \log\log N + C^{**}$ and $V_N(\omega(m)) \leq C_1 \log\log N$. Hence, by Chebyshev's inequality

$$P_N\left(\left|\omega(m) - \log\log N\right| \geq \psi(N)\sqrt{\log\log N}\right) \leq \frac{1}{\psi^2(N)}$$

and thus, by choosing $\psi(N) \to +\infty$ with N but slower that $\sqrt{\log\log N}$ we get (note that $\log\log n$ grows so slowly that $\log\log N \sim \log\log n$ for "almost all $n \leq N$"):

Corollary 5.1 *The natural density of those integers n whose number of prime divisors is asymptotically* $\log\log n$ *is one.*

The surprising result ($\omega(m)$ varies very wildly over the integers in general) of Corollary 5.1 is due to Hardy and Ramanujan [77]. It was Turán's proof of Corollary 5.1 via (5.39) that moved the result to probabilistic number theory (Turán himself did not use probabilistic language and did not refer to Chebyshev's inequality, but established (5.39) for $\omega(m)$ and deduced from it Corollary 5.1 in a manner that is usual for proving Chebyshev's inequality.)

Just as for $\omega(m)$, (5.39) can be used "for laws of large numbers" for a large class of additive functions. The inequalities of Theorems 5.1 and 5.1a serve as a more important tool in this book: in the subsequent sections of the present chapter we transform probabilistic results to results on additive functions by utilizing the Kubilius-Turán inequalities.

5.3 A theorem of Bakštys and the Erdös-Wintner theorem

Let $g(m)$ be a multiplicative function with distribution function $F_N(x)$ on $(\Omega_N, \mathcal{A}_N, P_N)$. We further assume that $g(1) = 1$. The aim of this section is to determine conditions that guarantee the existence of a nonsymmetric weak limit of $F_N(x)$. This will be done by introducing a probabilistic model in which the s_{p_j} are independent. To this end we introduce an abstract probability space $(\Omega, \mathcal{A}, P)$, on which there are random variables $e_{p_j} = e_{p_j}(\omega)$ which take the nonnegative integers $k_j \geq 0$ with distribution

$$P\left(e_{p_j} \geq k_j,\, 1 \leq j \leq t\right) = \frac{1}{p_1^{k_1} p_2^{k_2} \cdots p_t^{k_t}}.$$

Hence, the random variables e_{p_j}, $j \geq 1$, are stochastically independent. Let

$$G_N(\omega) = \prod_{p \leq N} g(p^{e_p}). \tag{5.40}$$

Note that since (5.40) consists of a product of independent random variables, results from Sections 1.4 and 1.5 provide a complete solution to the existence of limiting distribution of $G_N(\omega)$. Our aim is to reduce the limiting distributional properties of $g(m)$ to those of $G_N(\omega)$, thus obtaining results on $g(m)$ by purely probabilistic arguments. We are going to prove the following result.

Theorem 5.2 *(Bakštys) Let $g(m) \neq 0$ and assume*

$$\sum_{g(p)<0} \frac{1}{p} < +\infty.$$

Then $g(m)$ has a limiting distribution function $F(x)$, continuous at 0, if, and only if, $g^*(m) = |g(m)|$ does. Equivalently, $g(m)$ has a limiting distribution function continuous at 0 if, and only if, for arbitrary $0 < M < +\infty$, the three series

$$(i) \sum{}^{*} \frac{1}{p} \qquad (ii) \sum{}^{**} \frac{\log|g(p)|}{p} \qquad (iii) \sum{}^{**} \frac{\log^2|g(p)|}{p} \tag{5.41}$$

converge, where $\sum^*$ and $\sum^{**}$ denote summation over primes p such that $|\log|g(p)|| > M$ and $|\log|g(p)|| < M$, respectively. The limiting distribution function $F(x)$ is symmetric if, and only if, $g(2^k) = -1$ for all k.

From (5.41) it is clear that $g(m)$ and $g^*(m)$ have limiting distribution functions at the same time.

The special case $g(m) > 0$, for all $m \geq 1$, allows one to take logarithm of $g(m)$. We have from (5.8)

$$
\begin{aligned}
f(m) &= \log g(m) = \sum_{j=2}^{+\infty} \sum_{k=0}^{+\infty} \left[\log g(p^k)\right] \epsilon_{k,p_j}(m) \\
&= \sum_p \log g(p^{s_p(m)}) = \sum_p f(p^{s_p(m)}).
\end{aligned}
$$

Hence, Theorem 5.2 implies the celebrated Erdös-Wintner theorem.

Theorem 5.3 *An additive arithmetical function $f(m)$ has a limiting distribution if, and only if, the three series at (5.41) converge (upon replacing* $\log|g(p)|$ *by $f(p)$).*

Theorem 5.3 resembles the three-series theorem of Kolmogorov from probability theory (Theorem 1.16). It came up in the literature several times, and both P. Erdös and A. Rényi suggested to one of us (Galambos) in personal discussions to reduce Theorem 5.3 from the Kolmogorov three-series theorem rather than proceeding with a dependent model (see Theorem 5.4 later in this section) and extending the three series theorem to such model in order to prove Theorem 5.3 on probabilistic lines. Such a solution is now given via proving the more general Theorem 5.2. The following proof is due to Galambos and Simonelli [63].

PROOF OF THEOREM 5.2. First let us assume the validity of the three series in (5.41). This, Kolmogorov three series theorem, and Theorem 1.26 immediately imply that $G_N(\omega)$ converges weakly to a random variable continuous at 0. Theorem 1.23 provides conditions for this limit to be symmetric. We claim that $G_N(\omega)$ and $g(m)$ have the same limiting distribution. Since for each T, $\prod_{p\leq T} g(p^{s_p(m)})$ converges weakly to $G_T(\omega)$, as $N \to +\infty$, to prove our claim it suffices to show that for any $\delta > 0$,

$$
R = \lim_{T\to+\infty} \limsup_{N\to+\infty} P_N\Big(\Big|\prod_{p\leq N} g(p^{s_p(m)}) - \prod_{p\leq T} g(p^{s_p(m)})\Big| > \delta\Big) = 0.
$$

Now, the following estimates

$$P_N\Big(\prod_{T<p\le N} g(p^{s_p(m)})<0\Big)$$
$$\le\left(\sum_{p>T:g(p)<0}\frac{1}{p}+\sum_{p>T}\frac{1}{p^2}\right)\to 0 \quad \text{as}\, T\to+\infty$$

and

$$P_N\Big(s_p(m)\neq\beta_p(m),\, p>T\Big)\le\sum_{p>T}\frac{1}{p^2}\to 0 \quad \text{as}\, T\to+\infty,$$

where $\beta_p(m)$ is as in (5.17), give

$$\begin{aligned} R &= \lim_{T\to+\infty}\limsup_{N\to+\infty} P_N\Big(\Big|\prod_{p\le T} g(p^{s_p(m)})\Big|\Big|\prod_{T<p\le N} g(p^{s_p(m)})-1\Big|>\delta\Big)\\ &\le \lim_{T\to+\infty}\limsup_{N\to+\infty} P_N\Big(\Big|\prod_{p\le T} g(p^{s_p(m)})\Big|\Big|\prod_{T<p\le N} \Big|g(p^{\beta_p(m)})\Big|-1\Big|>\delta\Big).\end{aligned}$$

This last estimate and the weak convergence of $G_T(\omega)$ to a proper distribution imply that $R=0$ holds whenever for arbitrary $\epsilon>0$

$$\lim_{T\to+\infty}\limsup_{N\to+\infty} P_N\Big(\Big|\prod_{T<p\le N}\Big|g(p^{\beta_p(m)})\Big|-1\Big|>\epsilon\Big)=0. \tag{5.42}$$

The convergence of the first series in (5.41) and the inequality

$$P_N\Big(\Big|\prod_{T<p\le N}\Big|g(p^{\beta_p(m)})\Big|-1\Big|>\epsilon\Big)$$
$$\le\sum_{T<p\le N}^{*}\frac{1}{p}+P_N\Big(\Big|\prod_{T<p\le N}^{**}|g(p^{\beta_p(m)})|-1\Big|>\epsilon\Big),$$

where $\prod^{**}$ is over all primes p such that $|\log|g(p)||<M$, give that (5.42) is equivalent to

$$\lim_{T\to+\infty}\limsup_{N\to+\infty} P_N\Big(\Big|\prod_{T<p\le N}^{**}|g(p^{\beta_p(m)})|-1\Big|>\epsilon\Big)$$
$$=\lim_{T\to+\infty}\limsup_{N\to+\infty} R_{N,T}=0.$$

Let $0 < \epsilon_1 < \epsilon$. The convergence of the second series in (5.41) implies we can find a T_o such that whenever $N > T > T_o$,

$$\Big| \prod_{T<p\le N}^{**} |g(p)|^{\frac{1}{p}} - 1 \Big| < \epsilon_1.$$

Hence,

$$\begin{aligned}
&\Big\{ \Big| \prod_{T<p\le N}^{**} |g(p^{\beta_p(m)})| - 1 \Big| > \epsilon \Big\} \\
= &\Big\{ \prod_{T<p\le N}^{**} |g(p^{\beta_p(m)})| > 1+\epsilon \ \text{ or } \prod_{T<p\le N}^{**} |g(p^{\beta_p(m)})| < 1-\epsilon \Big\} \\
\subseteq &\Big\{ \prod_{T<p\le N}^{**} \frac{|g(p^{\beta_p(m)})|}{|g(p)|^{\frac{1}{p}}}(1+\epsilon_1) > 1+\epsilon \ \text{ or} \\
&\prod_{T<p\le N}^{**} \frac{|g(p^{\beta_p(m)})|}{|g(p)|^{\frac{1}{p}}}(1-\epsilon_1) < 1-\epsilon \Big\} \\
= &\Big\{ \ln\Big(\prod_{T<p\le N}^{**} \frac{|g(p^{\beta_p(m)})|}{|g(p)|^{\frac{1}{p}}} \Big) > \ln(\frac{1+\epsilon}{1+\epsilon_1}) \ \text{ or} \\
&\ln\Big(\prod_{T<p\le N}^{**} \frac{|g(p^{\beta_p(m)})|}{|g(p)|^{\frac{1}{p}}} \Big) < \ln(\frac{1-\epsilon}{1-\epsilon_1}) \Big\} \\
\subseteq &\Big\{ \Big| \sum_{T<p\le N}^{**} \ln|g(p^{\beta_p(m)})| - \sum_{T<p\le N}^{**} \frac{\ln|g(p)|}{p} \Big| > \epsilon' \Big\},
\end{aligned}$$

where $\epsilon' = \min\{\ln(\frac{1+\epsilon}{1+\epsilon_1}), -\ln(\frac{1-\epsilon}{1-\epsilon_1})\}$. Hence

$$R_{N,T} \le P_N\Big(\Big| \sum_{T<p\le N}^{**} \ln|g(p^{\beta_p(m)})| - \sum_{T<p\le N}^{**} \frac{\ln|g(p)|}{p} \Big| > \epsilon' \Big).$$

Now, by Chebyshev's inequality, after estimating the variance by the Kubilius-Turán inequality, $P_N()$ above is at most

$$\frac{C}{\epsilon'^2} \sum_{p>T}^{**} \frac{\ln^2|g(p)|}{p},$$

and by assumption this bound goes to 0 as $T \to +\infty$. Hence $R = 0$ and the weak convergence of $g(m)$ and $|g(m)|$ is established.

Next we assume $|g(m)|$ has a limiting distribution continuous at zero. Let $|G_{T_k}(\omega)|$ be an arbitrary subsequence of $|G_N(\omega)|$ that converges weakly, and denote by $H(x)$ its limiting distribution. We claim that $H(x)$ is a proper distribution continuous at zero. If this is indeed the case, the convergence of $|g(m)|$ and the first part of the proof would imply that $H(x)$ must coincide with the limiting distribution of $|g(m)|$, and since $|G_{T_k}(\omega)|$ was arbitrary this would give the convergence of $|G_N(\omega)|$. The Kolmogorov three series theorem and the first part of our proof would further imply the convergence of the three series in (5.41) and of $g(m)$. So let us first assume that $H(x)$ is not continuous at zero. Then from Theorem 1.26 $|G_{T_k}(\omega)|$ converges to zero almost surely. We show that this is not possible since it would imply that the limiting distribution of $|g(m)|$ is also degenerate at 0, contradicting our initial assumption. To prove our claim it suffices to show that for arbitrary $\epsilon > 0$,

$$\lim_{k\to+\infty} \limsup_{n\to+\infty} P_{T_n}\left[\prod_{T_k<p\le T_n} |g(p^{\beta_p(m)})| > \epsilon\right] = 0. \tag{5.43}$$

In fact the convergence of $|g(m)|$ and (5.43) immediately give that for arbitrary k

$$\begin{aligned}
\lim_{N\to+\infty} P_N\left[\prod_{p\le N} |g(p^{s_p})| > \epsilon\right] &= \lim_{n\to+\infty} P_{T_n}\left[\prod_{p\le T_n} |g(p^{s_p})| > \epsilon\right] \\
\le \limsup_{n\to+\infty} \Bigg(P_{T_n}&\left[\prod_{T_k<p\le T_n} |g(p^{\beta_p(m)})| > \epsilon\right] \\
+P_{T_n}\left[\prod_{p\le T_k} |g(p^{s_p(m)})| \ge 1\right] &+ \sum_{p>T_k} \frac{1}{p^2}\Bigg),
\end{aligned}$$

and the right hand side of the above inequality goes to zero as $k \to +\infty$. By Markov's inequality

$$\begin{aligned} &\limsup_{n\to+\infty} P_{T_n}\left[\prod_{T_k<p\le T_n} |g(p^{\beta_p(m)})| > \epsilon\right] \\ &\le \frac{1}{\epsilon}\limsup_{n\to+\infty} E_{T_n}\left[\prod_{T_k<p\le T_n} |g(p^{\beta_p(m)})|\right], \end{aligned}$$

and by Cauchy-Schwarz's inequality the above expectation is bounded by

$$\left(E_{T_n}\left[\prod_{T_k<p\le T_{k+t}} |g(p^{\beta_p(m)})|^2\right] E_{T_n}\left[\prod_{T_{k+t}<p\le T_n} |g(p^{\beta_p(m)})|^2\right]\right)^{\frac{1}{2}}. \tag{5.44}$$

Since the first expectation in (5.44) goes to zero as n and then t go to infinity, the validity of (5.43) will follow if we can show that

$$\alpha = \lim_{t\to+\infty}\limsup_{n\to+\infty} E_{T_n}\left[\prod_{T_{k+t}<p\le T_n} |g(p^{\beta_p(m)})|^2\right]$$

is bounded. Let t_n be the number of primes p, $T_{k+t} < p \le T_n$. By applying Hölder's inequality $t_n - 1$ times one obtains that α is at most

$$\lim_{t\to+\infty}\limsup_{n\to+\infty}\left(\prod_{T_{k+t}<p\le T_n} E_{T_n}\left[|g(p^{\beta_p(m)})|^{2t_n}\right]\right)^{\frac{1}{t_n}},$$

and hence to show that α is bounded it suffices to show that

$$\lim_{t\to+\infty}\limsup_{n\to+\infty}\prod_{T_{k+t}<p\le T_n} E_{T_n}\left[|g(p^{\beta_p(m)})|^{2t_n}\right]$$

is bounded.

$$\begin{aligned} E_{T_n}\left[|g(p^{\beta_p(m)})|^{2t_n}\right] &= 1-(1-|g(p)|^{2t_n})\frac{\lfloor T_n/p\rfloor}{T_n} \\ &\le \begin{cases} 1-\frac{1-|g(p)|^{2t_n}}{p} & \text{if } |g(p)|\ge 1 \\ 1-\frac{1-|g(p)|^{2t_n}}{p}+\frac{1-|g(p)|^{2t_n}}{T_n} & \text{if } |g(p)|<1. \end{cases} \end{aligned}$$

Let $\prod^{**}$ (and $\sum^{**}$) denote product (sum) over all p's with $|g(p)| < 1$, and $b_p(\omega) = 1$ if $e_p(\omega) \geq 1$, 0 otherwise. Then

$$\prod_{T_{k+t}<p\leq T_n} E_{T_n}\left[|g(p^{\beta_p(m)})|^{2t_n}\right]$$

$$\leq \prod_{T_{k+t}<p\leq T_n} E\left[|g(p^{b_p(\omega)})|^{2t_n}\right] \tag{5.45}$$

$$\cdot \prod_{T_{k+t}<p\leq T_n}^{**} \left[\frac{1-\frac{1-|g(p)|^{2t_n}}{p}+\frac{1-|g(p)|^{2t_n}}{T_n}}{1-\frac{1-|g(p)|^{2t_n}}{p}}\right]$$

$$\leq \prod_{T_{k+t}<p\leq T_n} E\left[|g(p^{b_p(\omega)})|^{2t_n}\right] \prod_{T_{k+t}<p\leq T_n}^{**} \left[1+\frac{p}{p-1}\frac{1}{T_n}\right]$$

$$\leq \prod_{T_{k+t}<p\leq T_n} E\left[|g(p^{b_p(\omega)})|^{2t_n}\right] e^{\sum^{**}_{T_{k+t}<p\leq T_n}\frac{2}{T_n}}$$

$$= E\left[\prod_{T_{k+t}<p\leq T_n} |g(p^{b_p(\omega)})|^{2t_n}\right] e^{\sum^{**}_{T_{k+t}<p\leq T_n}\frac{2}{T_n}}. \tag{5.46}$$

Since $|G_{T_k}(\omega)|$ converges to zero almost surely, the first term in (5.46) goes to zero as $n \to +\infty$, whereas the second term goes to one since the exponent of e is at most t_n/T_n that also goes to zero as $n \to +\infty$. Hence

$$\lim_{t\to+\infty} \limsup_{n\to+\infty} \prod_{T_{k+t}<p\leq T_n} E_{T_n}\left[|g(p^{\beta_p(m)})|^{2t_n}\right] = 0,$$

that implies α is bounded and consequently (5.44) holds, completing the proof of our claim.

Next we assume that $H(x)$, the limiting distribution of $|G_{T_k}|$, is continuous at 0 but not proper. In this case we consider $|G_{T_k}|^{-1}$. Since $g(m) \neq 0$, this new random variable is well defined. Since $H(x)$ is not proper, by Theorem 1.26 $|G_{T_k}|^{-1}$ converges to 0 almost surely, and as in the previous case this implies that the limiting distribution of the multiplicative function $|g(m)|^{-1}$ is also degenerate at 0 which contradicts our initial assumption on $g(m)$. Hence weak convergence of $|g(m)|$ implies weak convergence of $|G_N(\omega)|$, the convergence of the three series in (5.41), and by the first part of our proof weak convergence of $g(m)$.

Clearly, if $g(m)$ has a limiting distribution so does $|g(m)|$, reducing this case to the one we have just proved. The proof of Theorem 5.2 is now complete. ∎

We have alluded to a probabilistic method of Galambos [52] that provides a probabilistic proof for the sufficiency part of Theorem 5.3. We quote this result below.

Theorem 5.4 *Let $(S_N, \mathcal{B}_N, Q_N)$ be a sequence of probability spaces. Let $\delta_{k,N}$, $1 \le k \le n = n(N)$, be random variables on $(S_N, \mathcal{B}_N, Q_N)$, which take the values 0 and 1 only, with the properties*

(a) $Q_N(\delta_{k,N} = 1) = r_k + O(1/N)$

(b) with some integer $T = T(N) \to +\infty$ with N,

$$Q_N(\delta_{i_1,N} = \delta_{i_2,N} = \cdots = \delta_{i_t,N} = 1)$$
$$= r_{i_1} r_{i_2} \cdots r_{i_t} + R(T, N, i_1, \cdots, i_t)$$

such that, with some $0 < R^*(T,N) \to 0$ with $N \to +\infty$,

$$|R(T, N, i_1, i_2, \cdots, i_t)| \le R^*(T, N)$$

for all $1 \le i_1 < i_2 < \cdots < i_t \le T$,

(c) there is a subset Γ of sequences $\{c_j\}$ such that the variance

$$V_N\Big(\sum_{k=1}^{n} c_k \delta_{k,N}\Big) \le A \sum_{k=1}^{n} c_k^2 r_k + o(1) \qquad \text{as } N \to +\infty$$

where A is a universal constant on Γ.

Then

$$f_{n,N} = \sum_{k=1}^{N} c_k \delta_{k,n} \tag{5.47}$$

has a proper weak limit distribution $F(x)$, if $n = o(N)$, and each of the series

$$\sum_{\substack{k \\ |c_k|<1}} c_k r_k; \qquad \sum_{\substack{k \\ |c_k|<1}} c_k^2 r_k; \qquad \sum_{\substack{k \\ |c_k|\geq 1}} r_k \tag{5.48}$$

converges. The distribution function $F(x)$ is determined by the characteristic function

$$\varphi(t) = \prod_{k=1}^{+\infty} \left[1 + r_k\left(e^{itc_k} - 1\right)\right]. \tag{5.49}$$

Remark 5.3 While the conditions of the theorem seem to imitate the behavior of an additive function $f(m)$ on $(\Omega_N, \mathcal{A}_N, P_N)$, Theorem 5.4 is a significant generalization of the sufficiency part of the Erdös-Wintner theorem: it does not restrict the choice of S_N to the set Ω_N. Furthermore, the proof, whose outline is given below, is purely probabilistic and therefore it does not require tools from number theory beyond those required to establish conditions (a), (b), (c). One recognizes in (c) the Kubilius-Turán inequality, which can be the most crucial requirement. However, no new idea is needed to show that the conditions (a), (b), and (c) are valid if $S_N = \{N, N+1, \cdots, N+M\}$ with $M \geq N$, and Q_N is the uniform distribution on the set $\mathcal{B}_N$ of all subsets of S_N.

OUTLINE OF THE PROOF OF THEOREM 5.4. By an obvious extension of the method of inclusion and exclusion, for all subscripts $1 \leq i_1 < \cdots < i_s \leq m$ and $1 \leq j_1 < \cdots < j_k \leq m$, $i_u \neq j_v$, $m \leq T$,

$$\begin{aligned} &Q_N\left(\delta_{i_1,N} = \cdots = \delta_{i_s,N} = 0,\ \delta_{j_1,N} = \cdots = \delta_{j_k,N} = 1\right) \\ &\quad = \prod_{u=1}^{s}(1 - r_{i_u}) \prod_{v=1}^{k} r_{j_k} + O\left(2^m R^*(T,N)\right). \end{aligned}$$

Consequently ($E_N(\cdot)$ denotes expectation on $(S_N, \mathcal{B}_N, Q_N)$), the characteristic function

$$\begin{aligned} \varphi_{m,N}(t) &= E_N\left[e^{it\sum_{k=1}^{m} c_k\delta_{k,N}}\right] \\ &= \prod_{k=1}^{m}\left(1 - r_k + r_k e^{itc_k}\right) + O\left(2^{2m} R^*(T,N)\right). \end{aligned} \tag{5.50}$$

Let $m \leq n < N$, $m \to +\infty$, $n = o(N)$, and let the conditions (5.48) be satisfied. Then with

$$\varphi_{n,N}(t) = E_N\left[e^{it\sum_{k=1}^{n} c_k\delta_{k,N}}\right],$$

the difference

$$\left|\varphi_{n,N}(t) - \varphi_{m,N}(t)\right| \leq E_N\left(\left|e^{it\sum_{k=m+1}^{n} c_k\delta_{k,N}} - 1\right|\right] \to 0 \qquad (5.51)$$

as $m \to +\infty$. Indeed, by first using the Cauchy-Schwarz inequality, and then utilizing the fact that, for α and β real,

$$\left|e^{it(\alpha+\beta)} - 1\right|^2 \leq 2\min(4, t^2\alpha^2) + 2\min(4, t^2\beta^2) \qquad (5.52)$$

one can split the sum $\sum_{k=m+1}^{n} c_k\delta_{k,N}$ into "$\alpha + \beta$" by including in α those k for which $|c_k| \geq 1$, and the rest in β. For any fixed t, the terms in α by $\sum_{k,\,|c_k|\geq 1} 2\delta_{k,N}$, $m \leq k \leq n$, which, after the expectation, become bounded by a constant times the tail of the last sum of (5.48) + m/N, both of which tend to zero. In β, write (summations are for $m \leq k \leq n$, and $|c_k| < 1$)

$$\sum c_k\delta_{k,N} = \sum c_k r_k + \sum c_k(\delta_{k,N} - r_k)$$

and apply (5.52) again. One more appeal to (5.48) leads to (5.51). The combination of (5.50) and (5.51) enatil (5.49), which completes the proof in view of the continuity theorem of characteristic functions. ∎

The next theorem, due to Galambos and Kátai [59], gives an example for a sequence of numbers n_j, $j \geq 1$ such that the natural density $d(\{n_j\})$ exists and is positive, and the gaps $n_{j+1} - n_j = g_j$ become arbitrarily large. That is, the existence of the natural density does not imply a regular behavior. We need some notation and assumptions.

Let $f(m)$ be a strongly additive function such that $f(p)$ is decreasing and tends to zero. Furthermore, let

$$\sum f(p) = +\infty; \qquad \sum \frac{f(p)}{p} < +\infty; \qquad \sum_{y\leq p\leq 2y} f(p) \to 0 \quad (5.53)$$

as $y \to +\infty$.

Theorem 5.5 *Let $C > 0$ be an arbitrary fixed constant. Let $n_j = n_j(C)$ be the j-th integer such that $f(m) \leq C$, and set $g_j = n_{j+1} - n_j$. Define $\psi(y) = \sum_{p \leq y} f(p)$. Then, under condition (5.53),*

$$0 < \limsup_{j=+\infty} \frac{n_{j+1} - n_j}{\psi(\log n_j)} < +\infty. \tag{5.54}$$

PROOF. By (5.53), and by the assumptions on $f(p)$, $\sum_{|f(p)| \geq 1} 1/p$ is finite, and $0 < \sum f^2(p)/p \leq \sum f(p)/p < +\infty$, where we may have to exclude a finite number of terms. Hence, Theorem 5.3 applies, that is, whatever $C > 0$, the density $d(C) = F(C)$ of the sequence $\{n : f(n) \leq C\}$ exists (in fact $F(C)$ is continuous in C, so weak convergence coincides with pointwise convergence of $F_N(C) = P_N(f(m) \leq C)$– see Chapter 6 for a further discussion). We now show that $1 > F(C) > 0$ for all $C > 0$. Consequently, the sequence $n_j = n_j(C)$ is indeed such that its natural density exists and satisfies $0 < F(C) < 1$. In order to prove that $F(C) < 1$, choose an integer B such $\sum_{j=1}^{B} f(p_j) > C$. This can be done in view of (5.53). Define the integer $\alpha = p_1 p_2 \cdots p_B$. Then, for all integers $j\alpha$, $f(j\alpha) > C$, implying $F(C) \leq 1 - 1/\alpha < 1$. For proving that $F(C) > 0$, consider all integers n relatively prime to α. Either by the inclusion and exclusion method, or by the asymptotic independence of divisibility by a finite number of primes p_j, $1 \leq j \leq B$, whose complements also are independent, we get

$$d((n, \alpha) = 1) = \prod_{j=1}^{B} (1 - \frac{1}{p_j}) > 0.$$

For simplicity of writing, put $\gamma(B) = \{n : (n, \alpha) = 1\}$. We want to show that B can be chosen so that $F(C) \geq (1/2) d(\gamma(B)) > 0$. For this purpose, we introduce a new additive function

$$f_B(p) = \begin{cases} 0 & \text{if } p = p_j,\ 1 \leq j \leq B \\ f(p) & \text{otherwise.} \end{cases}$$

With no change in our proof of the Kubilius-Turán inequality we get

$$\frac{1}{N} \sum_{\substack{u \in \gamma(B) \\ u \leq N}} \Big(f_B(u) - \sum_{p_B \leq p \leq N} \frac{f(p)}{p} \Big)^2 \leq C_1 \sum_{p_B \leq p \leq N} \frac{f^2(p)}{p}$$

with some universal constant $C_1 > 0$. Furthermore, $C_1 f^2(p) \leq f(p)$ for $p \geq p_B$ if B is large enough (recall $f(p) \to 0$ and decreases), so, by Chebyshev's inequality, for $u \in \gamma(B)$, since $f(u) = f_B(u)$ on $\gamma(B)$,

$$f(u) \leq \sum_{p_B \leq p} \frac{f(p)}{p} + \Big(\sum_{p_B \leq p} \frac{f(p)}{p}\Big)^{1/2}$$

for at least 50% of $u \in \gamma(B)$. If B is large enough, the right hand side above $\leq C$ on account of (5.53), so $F(C) \geq (1/2)d(\gamma(B)) > 0$.

We have seen at the beginning of the proof that $F(C)$ always exists. We now show that, if we put $\nu_N\{\cdots\}$ for the number of integers for which the property specified in the dotted space holds,

$$F(C) = \lim_{N=+\infty} \frac{1}{M(N)} \nu_N\{n : n \leq M(N), f(N+n) \leq C\}, \tag{5.55}$$

where $M(N) = m(N)\psi(\log N)$, with $m(N) \to +\infty$ with N and $\psi(y) = \sum_{p \leq y} f(p)$. As mentioned in Remark 5.3, (5.55) would follow from Theorem 5.4 if $M(N)$ could be guaranteed to satisfy $M(N) \geq N$. However, we want to proceed with arbitrary $m(N)$ in (5.55), and in some cases, it may not be large enough to make Theorem 5.4 applicable. But we add here that, in view of Remark 5.3, we may assume that $M(N) < N$ in the present argument.

For proving (5.55), we use characteristic functions. Set

$$\varphi(t, M(N)) = \frac{1}{M(N)} \sum_{j=1}^{M(N)} e^{itf(N+j)}. \tag{5.56}$$

Our aim is to show that

$$\varphi(t, M(N)) \to \varphi(t) = \prod_p \Big(1 + \frac{e^{itf(p)} - 1}{p}\Big), \tag{5.57}$$

which is known to be the characteristic function of $F(C)$. We first simplify the argument by truncating $f(n)$, show that the limit in the previous sentence is valid with the truncated $f(n)$, and then we establish that the truncation had no effect on the limit. So, we define $f(n; y)$ as the strongly additive function in n such that $f(p; y) = f(p)$ for primes $p \leq y$, and $f(p; y) = 0$ if $p > y$. Let $g(n; y) = e^{itf(n;y)}$

and $h(p;y) = e^{itf(p)} - 1$ if $p \le y$ and $h(p;y) = 0$ if $p > y$ as well as $h(p^a;y) = 0$ if $a \ge 2$; furthermore, $h(n,y)$ is multiplicative. Then

$$g(n;y) = \sum_{d|n} h(d;y)$$

and

$$\begin{aligned}\varphi(t;M,y) &= \frac{1}{M}\sum_{j=1}^{M} g(N+j;y) = \frac{1}{M}\sum_{j=1}^{M}\sum_{d|N+j} h(d;y) \\ &= \sum_{d} \frac{h(d;y)}{d} + O\Big(\frac{1}{M}\sum_{d} |h(d;y)|\Big).\end{aligned}$$

Since

$$\sum_{d} \frac{h(d;y)}{d} = \prod_{p\le y}\Big(1 + \frac{e^{itf(p)} - 1}{p}\Big)$$

and

$$\begin{aligned}\sum_{d} |h(d;y)| &\le \prod_{p\le y}(1 + |h(p)|) \le \prod_{p\le y}(1 + |t|f(p)) \\ &\le e^{\sum_{p\le y}(|t|f(p))} \le 2^{cy/\log y}, \quad |t| \text{ bounded},\end{aligned}$$

where $c > 0$ is a constant, and $f(p) \le c^*$, so we can use (5.38) for $\pi(y) = \sum_{p\le y} 1$. Hence, $\varphi(t;M,y) \to \varphi(t)$ of (5.57) as $y \to +\infty$ if we choose $y = \log M$. On the other hand, with $y = \log M$,

$$\Big|\varphi(t;M,y) - \varphi(t;M,M)\Big| \le \frac{1}{M}\sum_{j=1}^{M} \Big|e^{it\sum_{\substack{y<p\le M\\ p|N+j}} f(p)} - 1\Big|$$

$$\le \frac{|t|}{M}\sum_{j=1}^{M}\sum_{\substack{y<p\le M\\ p|N+j}} f(p) \le \frac{|t|}{M}\sum_{y<p\le M} f(p)\Big(\frac{M}{p} + 1\Big)$$

$$\le |t|\sum_{y<p\le M} \frac{f(p)}{p} + |t|\frac{\pi(M)}{M} \to 0$$

on any finite interval of t, as M (and $y = \log M$) go to $+\infty$. In order to complete the proof of (5.57), we have to eliminate the truncation.

In the estimates above, M was arbitrary, except that $M \to +\infty$. The role of y disappeared since truncation at y or at M resulted in the same limit $\varphi(t)$ for $\varphi(t, M, y)$. From this point on, $M = M(N) = m(N)\psi(\log N)$. We show that, with this $M(N) = M$,

$$\begin{aligned} &\Big|\varphi(t,M) - \varphi(t,M,N)\Big| \\ &\le \frac{1}{M}\Big|\sum_{j=1}^{M}\left(e^{itf(N+j)} - e^{itf(N+j,M)}\right)\Big| \qquad (5.58) \\ &\le \frac{|t|}{M}\sum_{j=1}^{M}\big|f(N+j) - f(N+j,M)\big| = \frac{|t|}{M}\sum_{j=1}^{M}\sum_{\substack{p\mid N+j\\ p>M}} f(p) \to 0 \end{aligned}$$

on any finite interval of t, as $M(N) \to +\infty$. First, note that, on the extreme right hand side of (5.58), every $p > M$ and the same $f(p)$ can occur only once. Also, by the assumed monotonicity of $f(p)$, we can use consecutive primes $q_1 < q_2 < \cdots < q_R$, exceeding M, for estimating in (5.58). We thus have that the extreme right hand side of (5.58) does not exceed

$$\frac{|t|}{M}\sum_{j=1}^{R} f(q_j) \le \frac{|t|}{M}\psi(q_R), \qquad (5.59)$$

where we have quite a freedom in choosing R. If we choose R by the obviously permissible inequality

$$q_1 q_2 \cdots q_R \le (N + M(N))^{M(N)} < q_1 q_2 \cdots q_R q_{R+1},$$

then we get $q_R \le cM(N)\log(N + M(N))$ with some constant $c > 0$, from which, the last assumption of (5.53) implies $\psi(q_R) \le \psi(\log(N + M(N))) + O(\log M(N))$ and thus (5.59) tends to zero with our choice of $M(N)$, proving (5.58).

We finally turn to the establishment of the claimed magnitude of the gaps $n_{j+1}(C) - n_j(C)$. Fix $C > 0$. Let $\alpha > 0$ be a constant whose value we determine later, and set $K = [\alpha\psi(\log n_j(C))]$, where $[\cdot]$ signifies integer part. We show that, for all large j,

$$n_{j+1}(C) - n_j(C) \le K < \alpha\psi(\log n_j(C)), \qquad (5.60)$$

proving the upper inequality in (5.54). The basic idea is this: we start with the truncated version $f(N+t;K)$, $1 \le t \le K$, with $N = n_{j+1}(C)-1-K$ and show that among the values $f(N+t)$, $1 \le t \le K$, at least one $f(N+t) \le C$, so $n_j(C) \ge N+1 = n_{j+1}(C) - K$. In order to achieve this estimate, we utilize that $f(N+t;K) - f(N+t)$ is small for a majority of t, on account of our earlier result that the asymptotic distribution of the truncated version and the untruncated one is the same $F(C)$ (see (5.58)). So, let $\epsilon > 0$ be an arbitrary fixed number, and let $C_1 = C - \epsilon$. Then

$$\frac{1}{K}\nu_N\{j : j \le K,\ f(N+j;K) \le C_1\} \ge F(C_1) + o(1), \quad F(C_1) > 0.$$

On the other hand, by (5.58) and the estimates following (5.59),

$$\begin{aligned}
&\frac{1}{K}\nu_N\{j : j \le K,\ |f(N+j) - f(N+j;K)| \ge \epsilon\} \\
&\le \frac{1}{K\epsilon}\psi(\log(N+K)) + O(\frac{\log K}{K}) \\
&= \frac{1}{K\epsilon}\psi(\log(n_{j+1}(C)-1)) + O(\frac{\log K}{K}) \\
&\le \frac{1}{K\epsilon}\psi(\log n_j(C)) + O(\frac{\log K}{K}) \\
&\le \frac{1}{\alpha\epsilon} + O(\frac{\log K}{K}),
\end{aligned}$$

where, in the last but one inequality, we used $n_{j+1}(C) \le 2n_j(C)$, for large j, justified by (5.57), which implies that between $(1/2)n_{j+1}(C)$ and $n_{j+1}(C)$ there are as many as $(1/2)F(C)(1+o(1))n_{j+1}(C)$ members of integers satisfying $f(n) \le C$, and $F(C) > 0$. In addition, the change from $\log n_{j+1}(C)$ to $\log n_j(C)$ is an error that can be included in $O((\log K)/K)$ on account of (5.53). Hence, if we choose $\alpha > 2/\epsilon F(C_1)$, then at least one $1 \le j \le K$ has the property that if $f(N+j;K) \le C_1$ then $f(N+j) \le C$. The proof of (5.60) is concluded.

The reverse inequality in (5.60) with a different $\alpha > 0$ is a bit more sophisticated but it is also implicit in the properties established earlier. We show that, as $x \to +\infty$, there is an N between $x/2$ and x such that none of $N, N+1, \cdots, N+K$ belongs to the sequence $n_j(C)$, where, as previously, $K = [\alpha\psi(\log x)] = [\alpha\psi(\log n_J(C)](1 + o(1))$ where J is chosen so that $n_J(C) \le N < n_{J+1}(C)$, and α is to be chosen later. Note that, just as in the preceding case, $n_J(C) < n_{J+1}(C) \le 2n_J(C)$ and thus either $n_J(C)$ or $n_{J+1}(C)$ is between $x/2$ and x, hence (5.53) guarantees that $\psi(\log x)$ or $\psi(\log n_J(C))$ or $\psi(\log n_{J+1}(C))$ can freely be interchanged in asymptotic formulas. Such a 'construction' of $N+j$, $0 \le j \le K$, therefore proves the lower inequality of (5.54).

We fix x. Let $p_1 < p_2 < \cdots < p_R$ be the consecutive primes, where R is defined by

$$P(R) = p_1 p_2 \cdots p_R \le \frac{1}{2}x < P(R)p_{R+1}.$$

We place the primes p_j, $1 \le j \le R$, into K mutually exclusive groups, each non-empty, and the product of the primes in the t-th group is denoted P_t. So, $P(R) = P_1 P_2 \cdots P_K$, and we choose the subscripts t so that $f(P_1) \le f(P_2) \le \cdots \le f(P_K)$. Clearly, $f(P(R)) = f(P_1) + f(P_2) + \cdots + f(P_k)$, $f(P(R)) = \psi(p_R)$ and $p_R = (1 + o(1)) \log x$ (due to a result from number theory, to be discussed in Chapter 6), so $f(P(R)) = \psi(\log x) + o(1)$. We choose α at this stage such that $1/\alpha$ be a large number. Hence, $f(P(R))/K \sim 1/\alpha$ and $f(P(R)) \le Kf(P_K)$, i.e., $f(P_K) \ge 1/\alpha$. If we can make the classification of the primes into the K groups so that, with some $x/2 \le N \le x$, and with $\epsilon > 0$ arbitrarily small,

$$\min\{f(N+j) : 1 \le j \le K\} \ge f(P_1) \ge f(P_k) - \epsilon \ge \frac{1}{\alpha} - \epsilon, \quad (5.61)$$

then with α such that $1/\alpha - \epsilon > C$, each $f(N+j) > C$, $1 \le j \le K$, proving $n_{J+1}(C) - n_J(C) \ge K \sim \alpha\psi(\log n_J(C))$, which is an explicit form of the lower bound in (5.54).

It remains to guarantee the inequalities in (5.61). The first inequality is evident because the P_t, $1 \le t \le K$, are relatively prime, and thus there is an integer $x/2 \le N \le x$, x large, and one P_t divides $N+j$, $1 \le j \le K$, $1 \le t \le K$. Since $f(m) > 0$, $f(N+j) \ge f(P_t) \ge f(P_1)$ for all $1 \le j \le K$. To prove that the grouping of the primes $p_1 < p_2 < \cdots < p_R$ is possible into K groups so that $f(P_1) \le f(P_2) \le \cdots \le f(P_K)$ and $f(P_1) \ge f(P_K) - \epsilon$, imagine that we have K boxes, and place p_j into the boxes one by one. The first K primes p_j are placed into the K different boxes (that is, one into each box). Assume that $p_1 < p_2 < \cdots < p_s$ have been placed into the boxes then p_{s+1} is placed into box t if for its members $\sum_{p_j|\text{box } t} f(p_j)$ is the smallest (choose one of the smallest, if there are several possibilities). After this distribution of the primes p_j, we relabel the boxes to satisfy the monotonicity requirement. Clearly, $f(P_1) \geq f(P_K) - f(p^*)$, where p^* is the largest prime in box K. But because $f(P_K) \geq 1/\alpha$, p^* is larger and larger as α decreases. That is, $f(p^*) \leq \epsilon$ if α is small on account of the assumption of $f(p) \to 0$ as $p \to +\infty$.

The theorem is established. ∎

In some special cases, Theorem 5.5 was established in J. Galambos [56] and M. van Rossum-Wijsmuller [133]. The intermediate steps in the proof of Theorem 5.5 are themselves of interest. In particular, the establishment of the limiting distribution of $f(m)$ on the integers $N, N+1, \cdots, N+M(N)$, when $M(N)$ is relatively small, is significant both for the result and for its method for proof.

5.4 Distribution of multiplicative functions: the symmetric case

Let us remain at Bakštys' theorem (Theorem 5.2) for a while. So, we still assume that

$$K = \sum_{g(p)<0} \frac{1}{p} < +\infty, \tag{5.62}$$

and investigate the limiting distribution function $G(x)$ of a multiplicative function $g(n) \neq 0$ under the convergence of the three series in (5.41). The following interesting observation is made in Galambos and Szüsz [68]. Let $g(n)$ be strongly multiplicative. Upon putting

$$\Delta(x) = 1 - G(x) - G(-x), \tag{5.63}$$

and assuming that $g(p) = -1$ whenever $g(p) < 0$, the following inequality

$$\Big|\Delta(x)\Big| \leq e^{-2K} \tag{5.64}$$

holds. In other words, if K is a large finite value, then, upto any number of decimal digits $G(x)$ is as if it were a symmetric distribution function. It is therefore not surprising that if $g(n)$ is (another) multiplicative function taking the values $+1$ and -1 only, and if

$$\sum_{g(p)=-1} \frac{1}{p} = +\infty \tag{5.65}$$

then the limiting distribution function of $g(n)$ is symmetric. This is Wirsing's theorem [156], that is,

Theorem 5.6 *Let $g(n)$ be a multiplicative function taking values $+1$ and -1 only. Then $g(n)$ has a limiting distribution function $F(x)$. Moreover, $F(x)$ is symmetric if, and only if, either $g(2^k) = -1$ for all $k \geq 1$, or (5.65) holds.*

In fact, the conclusion of Wirsing [156] is more general than that of Theorem 5.6 in that he assumes only $|g(n)| \leq 1$, and the conclusion of Theorem 5.6 about the symmetry remains to hold (with divergence of the sum in (5.62) in place of (5.65)).

There are several proofs for Theorem 5.6 and for its extensions; see Jogesh Babu [88], Hildebrand [80], Daboussi and Indlekofer [35], and Indlekofer [83]. None of these proofs come close to the simplicity in proving (5.64) by Galambos and Szüsz [68], which proof now follows.

Recall that $g(n) \neq 0$ is a strongly multiplicative function, that is, $g(p^k) = g(p)$, $k \geq 1$, and the three series in (5.41) converge and (5.62) holds. Thus, by Theorem 5.2,

$$P_N(g(n) \leq x) \to G(x) \quad \text{completely.} \tag{5.66}$$

We assume that $g(p) = -1$ whenever $g(p) < 0$, and denote by $q_1 < q_2 < \cdots$ those primes for which $g(p) = -1$. Define the strongly multiplicative functions $g_o(n) = |g(n)|$, and, for $r \geq 1$,

$$g_r(p) = \begin{cases} g(p) & \text{if } g(p) > 0 \text{ or if } p = q_j,\ j \leq r \\ 1 & \text{if } p = q_j,\ j > r. \end{cases}$$

We show by induction that

$$P_N\Big(g_r(n) \leq x\Big) \to G_r(x) \quad \text{completely,} \tag{5.67}$$

where $G_r(x)$ is a proper distribution function. The claim is a part of Theorem 5.2 for $r = 0$. Note

$$\begin{aligned} &P_N\Big(g_{r+1}(n) \le x\Big) \\ &\quad = P_N\Big(g_r(n) \le x,\, q_{r+1} \nmid n\Big) + P_N\Big(g_{r+1}(n) \le x,\, q_{r+1} | n\Big), \end{aligned} \tag{5.68}$$

where we used $g_r(n) = g_{r+1}(n)$ if $q_{r+1} \nmid n$. Now, if $q_{r+1}|n$, then let $k \ge 1$ be such that $q_{r+1}^k \parallel n$. If we write $n = q_{r+1}^k m$, then $q_{r+1} \nmid m$, and thus

$$g_{r+1}(n) = g_{r+1}(m) g_{r+1}(q_{r+1}^k) = -g_{r+1}(m) = -g_r(n) = g_r(n)/g(q_{r+1})$$

because $g_r(q_{r+1}) = 1$, that is, $g_r(m) = g_r(n)$. Therefore,

$$\begin{aligned} &P_N\Big(g_{r+1}(n) \le x,\, q_{r+1}|n\Big) \\ &\quad = \sum_{k\ge 1} P_N\Big(g_r(n) \ge x/g(q_{r+1}),\, q_{r+1}^k \parallel n\Big). \end{aligned} \tag{5.69}$$

Applying once more that $g_r(q_{r+1}) = 1$, the elementary relation

$$P_N(g_r(n) \le x,\, q_{r+1} \nmid n) = P_N(g_r(n) \le x) - P_N(g_r(n) \le x,\, q_{r+1}|n)$$

becomes

$$\begin{aligned} &P_N(g_r(n) \le x,\, q_{r+1} \nmid n) \\ &\quad P_N(g_r(n) \le x) - \sum_{k\ge 1} P_N(g_r(n) \le x,\, q_{r+1}^k \parallel n). \end{aligned} \tag{5.70}$$

In the sum over k, we actually sum over $q_{r+1}^k \parallel n$, resulting in taking the integers n divisible by q_{r+1}. Hence, the combination of (5.68)-(5.70), and the assumption that (5.67) is valid for r, we get that $P_N(g_{r+1}(n) \le x)$ converges to a function $G_{r+1}(x)$ at continuity points of $G_r(x)$ and the relation

$$G_{r+1}(x) = \left(1 - \frac{1}{q_{r+1}}\right) G_r(x) + \frac{1}{q_{r+1}} \left(1 - G_r\left(\frac{x}{g(q_{r+1})}\right)\right) \tag{5.71}$$

obtains. Since the continuity points of $G_r(x)$ are dense on the real line, $G_{r+1}(x)$ is in effect determined for all x. By rearranging the terms of (5.71) as (since $g(q_{r+1}) = -1$)

$$G_{r+1}(x) - G_r(x) = \frac{1}{q_{r+1}} \{1 - G_r(x) - G_r(-x)\} \tag{5.72}$$

we have

$$0 < G_{r+1}(x) - G_r(x) \leq \frac{1}{q_{r+1}}. \tag{5.73}$$

Because $G_r(x)$, on account of (5.73), is monotonic in r for fixed x,

$$\lim_{r\to+\infty} G_r(x) = G^*(x) \tag{5.74}$$

exists for all x at which each $G_r(x)$ is continuous. But this makes only a denumerable set of exceptional points, $G^*(x)$ is once again uniquely determined. We should add that if $g(n)$ takes only $+1$ and -1, then the only discontinuity points are $+1$ and -1, whatever r.

We claim that $G^*(x)$ of (5.74) is, in fact, $G(x)$ of (5.66). Indeed,

$$\begin{aligned} &|P_N(g(n) \leq x) - P_N(g_r(n) \leq x)| \leq P_N(g(n) \neq g_r(n)) \\ &\quad \leq \sum_{j>r} P_N(q_j|n) \leq \sum_{j>r} \frac{1}{q_j} < \epsilon \end{aligned} \tag{5.75}$$

if r is sufficiently large in view of (5.62). Hence, by letting $N \to +\infty$ and $r \to +\infty$, we get $G^* = G$ at all continuity points (of either of them). But then, setting

$$\Delta_r(x) = 1 - G_r(x) - G_r(-x),$$

(5.72) entails

$$\begin{aligned} \Delta_{r+1}(x) &= \Big(1 - \frac{2}{q_{r+1}}\Big)\Delta_r(x) \\ &= \Big(1 - \frac{2}{q_{r+1}}\Big)\Big(1 - \frac{2}{q_r}\Big)\cdots\Big(1 - \frac{2}{q_1}\Big)\Delta_o(x), \end{aligned}$$

which, by $1 - z \leq e^{-z}$, $z > 0$, yields

$$\Delta_{r+1}(x) \leq e^{-2(K-\epsilon)}$$

with some $\epsilon > 0$ as at (5.75). From (5.75) we also have

$$|\Delta(x) - \Delta_{r+1}(x)| < 2\epsilon,$$

that is,

$$0 \leq \Delta(x) \leq |\Delta_{r+1}(x) + 2\epsilon \leq e^{-2(K-\epsilon)} + 2\epsilon.$$

By letting $r \to +\infty$, and thus $\epsilon \to 0$, (5.64) obtains. ∎

We now turn to the major aim of the present section, that is, to give a new proof for the following theorem of Galambos [53]; see also Timofeev, Tuliaganov and Levin [149], Elliott [43], p. 280, and Jogesh Babu [87]. The method for proof is to reduce the case of symmetric distribution of multiplicative functions to products of independent random variables, relying solely on Wirsing's theorem in the special form of Theorem 5.6. This part of the present section is based on the paper of Galambos and Simonelli [67].

Theorem 5.7 *Let $g(m)$ be a multiplicative function, $g(m) \neq 0$ for all m, and assume*

$$\sum_{p:g(p)<0} \frac{1}{p} = +\infty.$$

Then $g(m)$ has a symmetric limit distribution function $F(x)$, continuous at zero, if, and only if, $|g(m)|$ has a limit distribution function, continuous at zero. Equivalently, $g(m)$ has a symmetric limit distribution function continuous at zero, if, and only if, for arbitrary M, $0 < M < +\infty$, the three series

$$(i) \sum_{p:|ln|g(p)||>M} \frac{1}{p} \quad (ii) \sum_{p:|ln|g(p)||<M} \frac{ln|g(p)|}{p} \quad (iii) \sum_{p:|ln|g(p)||<M} \frac{ln^2|g(p)|}{p}$$

converge. Moreover $g(m)$ has a limit distribution function discontinuous at zero if, and only if, $g(m)$ has a limit distribution function degenerate at zero.

In some abstract probability space let $e_{p_i}(\omega) = e_{p_i}$, $i = 1, 2, \cdots$, be independent random variables, $e_{p_i} = 0, 1, \cdots$,

$$P(e_{p_i} \geq k) = \frac{1}{p_i^k}, \quad k \geq 0,$$

and for arbitrary N define

$$G_N(\omega) = \prod_{p \le N} g(p^{e_p(\omega)}).$$

Let $g(m)$ be a multiplicative function, $g(m) \neq 0$. With minor changes the proof of Theorem 5.2 can be used to show that $|g(m)|$ converges weakly to a random variable continuous at zero if, and only if, $|G_N(\omega)|$ does, and that this holds independently of the convergence or divergence of the sum $\sum_{p:\, g(p)<0} 1/p$. In fact from the proof of Theorem 5.2 one obtains the validity of the following result.

Theorem 5.8 *Let $g(m)$ be a multiplicative function, $g(m) \neq 0$ for all m. Then $|g(m)|$ has a limit distribution function continuous at zero, if, and only if, $|G_N(\omega)|$ does. Equivalently, $|g(m)|$ has a limit distribution function continuous at zero, if, and only if, for arbitrary M, $0 < M < +\infty$, the three series*

$$(i) \sum_{p:|ln|g(p)||>M} \frac{1}{p} \quad (ii) \sum_{p:|ln|g(p)||<M} \frac{ln|g(p)|}{p} \quad (iii) \sum_{p:|ln|g(p)||<M} \frac{ln^2|g(p)|}{p}$$

converge. Moreover $|g(m)|$ has a limit distribution function discontinuous at zero if, and only if, $|g(m)|$ has a limit distribution function degenerate at zero.

To simplify the proof of Theorem 5.7 we first prove the following result.

Lemma 5.1 *Let $g(m)$ be a multiplicative function, $g(m) \neq 0$ for all m. Then for any finite collection of prime integers $q_1, q_2, \cdots, q_t$, and arbitrary nonnegative integers k_i, $i = 1, \cdots, t$,*

$$\lim_{N\to+\infty} P_N\Big(\prod_{\substack{p\le N\\ p\neq q_j}} g(p^{s_p(m)}) < 0,\ s_{q_j}(m) = k_j,\ {\scriptstyle j=1,\cdots,t}\Big) \quad (5.76)$$

$$= \lim_{N\to+\infty} P_N\Big(\prod_{\substack{p\le N\\ p\neq q_j}} g(p^{s_p(m)}) < 0\Big)$$

$$\cdot \lim_{N\to+\infty} P_N\Big(s_{q_j}(m) = k_j,\ {\scriptstyle j=1,\cdots,t}\Big).$$

PROOF. We start our proof by observing that each of the two limits following (5.76) exists. The first limit exists by Theorem 5.6 (see the paragraph after (5.78) for details), whereas the second one exists because for arbitrary nonnegative integers k_i, $1 \le i \le t$, and arbitrary prime integers $p_{j_1}, \cdots, p_{j_t}$,

$$\lim_{N\to+\infty} P_N\Big(s_{p_{j_i}}(m) = k_i,\ {\scriptstyle i=1,\cdots,t}\Big) = P\Big(e_{p_{j_i}} = k_i,\ {\scriptstyle i=1,\cdots,t}\Big).$$

For any fixed N, $q_i \le N$ for $i = 1, \cdots, t$, let $p_1, \cdots, p_l$ denote all prime integers less than or equal to N different from the q_is. By applying the inclusion exclusion principle one can express

$$P_N\Big(\prod_{\substack{p\le N\\ p\ne q_j}} g(p) < 0,\ s_{q_j}(m) = k_j,\ {\scriptstyle j=1,\cdots,t}\Big) \tag{5.77}$$

as sums or differences of terms of the type

$$\frac{[\frac{N}{Q_iPj}]}{N},$$

where $Q_i = q_1^{i_1}\cdots q_t^{i_t}$, $i_j = k_j$ or $k_j + 1$, and $P_j = p_1^{j_1}\cdots p_l^{j_l}$, $j_k \ge 0$. If one writes each of these terms as

$$\frac{[\frac{N}{Q_iPj}]}{N} = \frac{[\frac{N}{Q_i}]}{N}\frac{\left[\frac{[\frac{N}{Q_i}]}{P_j}\right]}{[\frac{N}{Q_i}]},$$

and let $A_{i,N}$ be the collection of all terms in the inclusion exclusion expansion of (5.77) containing Q_i, then it is easy to see that

$$\lim_{N\to+\infty} A_{i,N} = \frac{1}{Q_i}\lim_{N\to+\infty} P_{[\frac{N}{Q_i}]}\Big(\prod_{\substack{p\le[\frac{N}{Q_i}]\\ p\ne q_j}} g(p^{s_p(m)}) < 0\Big). \tag{5.78}$$

Let $\tilde{g}(m)$ be such that $\tilde{g}(p^{s_p(m)}) = g(p^{s_p(m)})$, if $p \ne q_i$, $i = 1, \cdots, t$, and $\tilde{g}(p^{s_p(m)}) = |g(p^{s_p(m)})|$, if $p = q_i$, for some i. Then $\tilde{g}(m)$ is a multiplicative function, $\tilde{g}(m) \ne 0$ for all m, and

$$P_{[\frac{N}{Q_i}]}\Big(\prod_{\substack{p\le[\frac{N}{Q_i}]\\ p\ne q_j}} g(p^{s_p(m)}) < 0\Big) = P_{[\frac{N}{Q_i}]}\Big(\prod_{p\le[\frac{N}{Q_i}]} \tilde{g}(p^{s_p(m)}) < 0\Big).$$

This and Theorem 5.6 imply that

$$\begin{aligned}\lim_{N\to+\infty} A_{i,N} &= \frac{1}{Q_i}\lim_{N\to+\infty} P_N\Big(\prod_{p\le N}\tilde{g}(p^{s_p(m)})<0\Big)\\ &= \frac{1}{Q_i}\lim_{N\to+\infty} P_N\Big(\prod_{\substack{p\le N\\ p\ne q_j}} g(p^{s_p(m)})<0\Big).\end{aligned}$$

If we repeat the above calculations for every Q_i, and then combine the obtained limits, we obtain that

$$\lim_{N\to+\infty} P_N\Big(\prod_{\substack{p\le N\\ p\ne q_j}} g(p^{s_p(m)})<0,\ s_{q_j}(m)=k_j,\ {\scriptstyle j=1,\cdots,t}\Big)$$

$$= \lim_{N\to+\infty} P_N\Big(\prod_{\substack{p\le N\\ p\ne q_j}} g(p^{s_p(m)})<0\Big)\lim_{N\to+\infty} P_N\Big(s_{q_j}(m)=k_j,\ {\scriptstyle j=1,\cdots,t}\Big),$$

thus proving the lemma. ∎

It is interesting to note that Lemma 5.1 implies that for any finite collection of $q_1, q_2, \cdots, q_t$, and arbitrary nonnegative integers k_i, $i=1,\cdots,t$,

$$\lim_{N\to+\infty} P_N\Big(\prod_{\substack{p\le N\\ p\ne q_j}} g(p^{s_p(m)})<0\Big|\ s_{q_j}(m)=k_j,\ {\scriptstyle j=1,\cdots,t}\Big)$$

$$= \lim_{N\to+\infty} P_N\Big(\prod_{\substack{p\le N\\ p\ne q_j}} g(p^{s_p(m)})<0\Big),$$

from which we obtain the following result, which will be used in the proof of Theorem 5.7.

Lemma 5.2 *Let $g(m)$ be a multiplicative function, $g(m)\ne 0$ for all m, and for an arbitrary collection of prime integers $q_1, q_2, \cdots, q_t$, and arbitrary nonnegative integers k_i, $i=1,\cdots,t$, let $B=\{s_{q_j}(m)=k_j,\ {\scriptstyle j=1,\cdots,t}\}$. Then $sign(g(m))|B$ always has a limit distribution* *function, and this limit is symmetric if, and only if, either* $sign(g(2^k))$ $= -$ *for all* $k \geq 1$ *and* $q_i \neq 2$ *for* $i = 1, \cdots, t$, *or*

$$\sum_{p:\, g(p)<0} \frac{1}{p} = +\infty.$$

We now turn to the proof of Theorem 5.7

PROOF OF THEOREM 5.7. Let

$$g_N(m) = \prod_{p \leq N} g(p^{s_p(m)}).$$

Clearly, the weak convergence of $g_N(m)$ and $G_N(\omega)$ imply the weak convergence of $|g_N(m)|$ and $|G_N(\omega)|$, respectively. This and Theorem 5.2 imply that in what follows we can assume that $|g_N(m)|$ and $|G_N(\omega)|$ both converge weakly, and that their limits are continuous at zero. Under this assumption, Theorem 1.26 further gives that $G_N(\omega)$ converges weakly to a random variable X (see also Example 1.18 and Remark 1.1). Let

$$\nu_{i,j} = sign(\prod_{i<p\leq j} g(p^{s_p(m)})), \qquad \nu_j = sign(\prod_{p\leq j} g(p^{s_p(m)})),$$

and for arbitrary continuity point a of X we consider

$$L = \lim_{T\to+\infty} \lim_{N\to+\infty} E_N\Big[1_{\{|g_T(m)|\leq a\}} 1_{\{\nu_N=-\}}\Big].$$

We claim that the above quantity is well defined. That is, each of the above limits exists. Let

$$B_\eta = \left\{ m : 1_{\{|g_T(m)|\leq a\}} 1_{\{\nu_T=\eta\}} \prod_{p\leq T} 1_{\{s_p(m)\leq k\}} \right\},$$

$\eta = +, -$. Then

$$\begin{aligned} L &= \lim_{k\to+\infty} \lim_{T\to+\infty} \lim_{N\to+\infty} E_N\big[1_{\{|g_T(m)|\leq a\}} 1_{\{\nu_N=-\}} \prod_{p\leq T} 1_{\{s_p(m)\leq k\}}\big] \\ &= \lim_{k\to+\infty} \lim_{T\to+\infty} \lim_{N\to+\infty} \Big(E_N\Big[1_{\{\nu_{T,N}=-\}}\Big|B_+\Big] P_N\Big(B_+\Big) \\ &\quad + E_N\Big[1_{\{\nu_{T,N}=+\}}\Big|B_-\Big] P_N\Big(B_-\Big)\Big). \end{aligned}$$

Since

$$\lim_{N\to+\infty} P_N\Big(B_\eta\Big) = E\Big[1_{\{|G_T(\omega)|\le a\}}1_{\{sign(G_T(\omega))=\eta\}}\prod_{p\le T}1_{\{e_p(\omega)\le k\}}\Big],$$

and by Theorem 1.23 the right hand side of the above equation has a limit as T and K go to infinity, then to prove our claim it suffices to show that

$$\lim_{N\to+\infty} E_N\Big[1_{\{\nu_{T,N}=\eta\}}\Big|B_{-\eta}\Big] = \frac{1}{2}.$$

This and Theorem 1.23 will further imply that

$$\prod_{p\le T} g(p^{s_p(m)}) \quad \text{and} \quad \prod_{p\le T} |g(p^{s_p(m)})|\nu_N$$

have the same limiting distribution (as N, T, and K go to infinity, in this order).

By utilizing the prime factorization of the elements in B_η we are going to partition B_η into a finite number of subsets, say $B_{\eta,i}$, $i = 1, 2, \cdots, l$. The sets $B_{\eta,i}$ are defined as follows. Let $p_1 < p_2 < \cdots < p_t$ denote the collection of all prime integers less than or equal to T, and consider the collection of all t-element sequences of the form $\{(p_j, l_j)\}_{j=1}^t$, $0 \le l_j \le k$. For each sequence $I_i = \{(p_j, l_j)\}_{j=1}^t$, $i = 1, 2, \cdots, (k+1)^t$, we define $B_{\eta,i}$ to be the collection of all $m \in B_\eta$ such that $s_{p_j}(m) = l_j$, $1 \le j \le t$. Clearly the collection of the nonempty $B_{\eta,i}$ constitutes a partition of B_η, and hence we only need to show that for any nonempty $B_{\eta,i}$,

$$\lim_{T\to+\infty}\lim_{N\to+\infty} E_N\big[I_{\nu_{T,N}=-1}\big|B_{\eta,i}\big] = \frac{1}{2}.$$

But this immediately follows from Lemma 5.2. Hence our claim holds.

The above discussion and

$$\lim_{T\to+\infty}\limsup_{N\to+\infty} P_N\Big[\big||g_T(m)|1_{\{\nu_N=\eta\}} - |g_N(m)|1_{\{\nu_N\}=\eta}\big| \ge a\Big]$$

$$\le \lim_{T\to+\infty}\limsup_{N\to+\infty} P_N\Big[\big||g_T(m)| - |g_N(m)|\big| \ge a\Big] = 0,$$

which follows from the proof Theorem 5.2, complete the proof of the theorem. ∎

In order to deal with limiting distributions which are continuous at zero, we always assumed so far in the present chapter that multiplicative functions $g(n) \neq 0$ for any n. This however, can easily be changed. First, assume that

$$\sum_{p;\, g(p)=0} \frac{1}{p} = +\infty. \tag{5.79}$$

We use the estimate,

$$\begin{aligned}
\limsup_{N\to+\infty} P_N\Big(g(n)\neq 0\Big) &\le \limsup_{N\to+\infty} P_N\Big(\prod_{\substack{p\le k\\ p^s\|n}} g(p^s)\neq 0\Big)\\
&\le \limsup_{N\to+\infty} P_N\Big(p \not\| n \text{ for } p\le k \text{ with } g(p)=0\Big)\\
&= \prod_{\substack{p\le k\\ g(p)=0}} \Big(1-\frac{1}{p}+\frac{1}{p^2}\Big).
\end{aligned}$$

By letting $k \to +\infty$, the assumed divergence in (5.79) implies that

$$P_N\Big(g(n)=0\Big) \to 1 \qquad \text{as } N\to+\infty.$$

So, in order to get a nondegenerate limiting distribution of $g(n)$, the sum (5.79) should converge. On the other hand, if the sum in question converges, then Theorems 5.2 and 5.8 remain unchanged with the addition that (5.79) converges. We thus have

Theorem 5.9 *Let $g(m)$ be a multiplicative function. Then $|g(m)|$ has a nondegenerate limiting distribution function if, and only if, $|G_N(\cdots)|$ does. Equivalently, $|g(m)|$ has a nondegenerate limiting distribution function if, and only if, the sum at (5.79) converges and, for arbitrary M, $0 < M < +\infty$, the three series*

$$(i) \sum_{p:|ln|g(p)||>M} \frac{1}{p} \quad (ii) \sum_{p:|ln|g(p)||<M} \frac{ln|g(p)|}{p} \quad (iii) \sum_{p:|ln|g(p)||<M} \frac{ln^2|g(p)|}{p}$$

converge.

5.5 Variants of a conjecture of Erdös

In connection with the existence of the limiting distribution

$$F(x) = \lim_{N \to +\infty} P_N\big(f(n) \leq x\big)$$

of an additive function $f(n)$, Erdös [47] asked whether the (necessary and sufficient) conditions of Theorem 5.3 follow from the simple assumption that, for a single pair $a < b$ of numbers,

$$P(a, b) = \lim_{N \to +\infty} P_N\big(a < f(n) \leq b\big)$$

exists and $0 < P(a, b) < 1$. Only partial results are available to answer this question, and our results in Section 2.3 come close to an affirmative answer. For earlier results, see Jogesh Babu [87], Paul [117], Hildebrand [81], and F. Kong and Q. Tang [92].

With our approach to number theoretic functions via sums and products of independent random variables, it turned out to be equally easy to treat additive and multiplicative functions. However, not only at the time of Erdös [48] , but even in the late 1970s (see Elliott [43]) there was a considerable difference between distribution problems for additive and multiplicative functions. Bakštys [14] had to avoid the symmetric case for multiplicative functions, and Galambos [53]'s original proof for the results of the preceding section was based on deep number theoretic analytical tools. Therefore, we separate the cases of additive and multiplicative functions into two theorems, mainly for historical reason. We shall, however, combine their proofs into a single argument. The cited earlier results do not automatically cover the two cases either: This fact, too, justifies to state our results separately.

Recall Section 5.1. An additive function $f(n)$ can be viewed as an infinite sum (see (5.5)) and a multiplicative function $g(n)$ can be approached via infinite products (see (5.8)). In both representations 'the random terms' are associated with the variables $\epsilon_{k,p}(n)$ of (5.6). Note that the summands $f(p^k)\epsilon_{k,p}(n)$, with $p = p_j$, satisfy that $m_j = 0$ is a median for each j. Similarly, for the multiplicative case $g(n)$, the factors $g(p^k)^{\epsilon_{k,p}(n)}$, $p = p_j$, are such that $m_j = 1$ is a median for each j. This is why our model of Section 2.3 is directly applicable to both additive and multiplicative functions. The dependence of the above specified terms have already been overcome by the comparison methods of Sections 5.3 and 5.4, and thus we have the following results by a relatively short argument.

Theorem 5.10 *Let $g(n)$ be a (real valued) multiplicative function for which the density d_0 of $\{n : g(n) = 0\}$ satisfies $d_0 < 1$. Then $g(n)$ has a nondegenerate limiting distribution $G(x)$ if, and only if, there is a real number $a > 0$ such that the density $d_g(a) = lim_{N\to+\infty} P_N(g(n) \le a)$ exists and $d_0 < d_g(a) < 1$.*

Theorem 5.11 *A (real valued) additive function $f(n)$ has a nondegenerate limiting distribution $F(x)$ if, and only if, there is a real number c such that the density $d_f(c) = lim_{N\to+\infty} P_N\big(f(n) \le c\big)$ exists and $0 < d_f(c) < 1$.*

PROOFS OF THEOREMS 5.10 AND 5.11. In both theorems the necessary part is trivial, so the sufficiency only needs proof. Note first that the densities

$$d_0 = \lim_{N\to+\infty} P_N\big(g(n) = 0\big)$$

and

$$d^{(-)} = \lim_{N\to+\infty} P_N\big(g(n) < 0\big)$$

always exist for an arbitrary multiplicative function $g(n)$. We have established earlier (and a direct proof also is immediate) that

$$d_0 < 1 \quad \text{if, and only if,} \quad \sum_{p:\, g(p)=0} \frac{1}{p} < +\infty. \tag{5.80}$$

The statement about $d^{(-)}$ is Wirsing's theorem (Section 5.4) and let us record here, although not needed for the sequel of the proof, that $d^{(-)} = (1/2)(1 - d_0)$ if, and only if either

$$\text{(A)} \sum_{p:\, g(p)<0} 1/p = +\infty, \quad \text{or} \quad \text{(B)} \quad g(2^k) = -1 \ \text{ for each } \ k \ge 1.$$

In other words, if (5.80) holds, then $G(x)$ is symmetric if, and only if, $G(x)$ exists, nondegenerate and either (A) or (B) holds.

The assumption that $d_0 < d_g(a) < 1$ implies that (5.80) must hold. Based on this, we introduce a new multiplicative function $g_1(n)$ in order to bring additive and multiplicative functions into equivalent forms. Let $g_1(p^k) = g(p^k)$ if $g(p^k)$ is not zero, and $g_1(p^k) = 1$ otherwise. Then $g_1(n) \neq 0$, so we can define $f_1(n) = \log |g_1(n)|$, which is an additive function. By Sections 5.3 and 5.4, the existence of a nondegenerate limiting distribution for $g(n)$, $g_1(n)$, $f_1(n)$ and $f(n)$ follows for any one from the others. We can therefore concentrate on $g_1(n)$. But because of the reduction both of the existence and of the particular form of a nondegenerate limiting distribution of $g_1(n)$ to the case of products of independent random variables, by Theorem 2.6 we have that the Three Series Theorem applies to $f_1(n)$, and thus the limiting distributions of both $f(n)$ and $g(n)$ exist, and they are nondegenerate. The proof is complete. ∎

One can similarly transform Theorem 2.7 to additive and multiplicative functions, whose details we leave to the reader.

5.6 Sequences of prime divisors

For a given n, we denote the sequence of prime divisors of n by $q_1(n) < q_2(n) < \cdots < q_{\omega(n)}(n)$. We have shown in Section 5.2 that the number $\omega(n)$ of the prime divisors of n is asymptotically $\log\log n$ for almost all integers (precise statement in Corollary 5.1). In the present section we further analyze the sequence $q_j(n)$ and the additive function $\omega(n)$. A classical theorem on $\omega(n)$ is the refinement of Corollary 5.1 by Erdös and Kac [48].

Theorem 5.12 *As $N \to +\infty$,*

$$\lim P_N\left(\omega(n) \leq \log\log n + x\sqrt{\log\log n}\right) = N(x),$$

the standard normal distribution.

PROOF. We only outline a proof. First note that if we truncate $\omega(n)$ by introducing

$$\omega_y(n) = \sum_{\substack{p|n \\ p \leq y}} 1$$

then, with $y = y(N) = N^{1/\log\log N}$, say, following the introduction 5.1,

$$E_N(w_y(n)) \approx \sum_{p \le y} \frac{1}{p} \approx \log\log y = \log\log N - \log\log\log N$$

and, similarly,

$$V_N(w_y(n)) \sim V_N(\omega_y(n)) \sim \log\log N.$$

Since $\omega(n) - \omega_y(n)$ is an additive function, by the Kubilius-Turán inequality

$$P_N\left(\left|\frac{\omega(n) - \log\log n}{\sqrt{\log\log n}} - \frac{\omega_y(n) - \log\log n}{\sqrt{\log\log n}}\right| \ge \epsilon\right) \to 0$$

as $N \to +\infty$. Hence, in order to prove Theorem 5.12, it suffices to prove the asymptotic normality of $(\omega_y(n) - \log\log n)/\sqrt{\log\log n}$. By the method of moments, this latter complete convergence can be achieved by first observing that, for fixed $k \ge 1$, the binomial moments

$$E_N\left(\binom{\omega_y(n)}{k}\right) = \sum_{1 \le p_{i_1} < p_{i_2} < \cdots < p_{i_k} \le y} P_N\left(p_{i_1} p_{i_2} \cdots p_{i_k} | n\right)$$

$$= \sum_{1 \le p_{i_1} < \cdots < p_{i_k} \le y} \frac{1}{N}\left[\frac{N}{p_{i_1} p_{i_2} \cdots p_{i_k}}\right] = \sum \frac{1}{p_{i_1} \cdots p_{i_k}} + O\left(\frac{1}{N}\binom{\pi(y)}{k}\right)$$

where $\pi(y)$ is the number of primes upto y. Now, by the choice of y,

$$\frac{1}{N}\binom{\pi(y)}{k} \le \frac{y^k}{k!N} \to 0$$

for all fixed k. Therefore, the moments above are asymptotically the same as if $\omega_y(n)$ were the sum of independent random variables $(p \le y)$

$$\epsilon_p = \begin{cases} 1 & \text{with probability } 1/p \\ 0 & \text{with probability } 1 - 1/p. \end{cases}$$

For independent random variables, $\sum_{p \le y} \epsilon_p$ does converge to the normal law, when normalized, and the moments of such normalized sum also converge to the moments of the normal distribution. But, as just observed, the moments $\omega_y(n)$ are asymptotically equal to those of the normalized sums of independent variables, we get the asymptotic normal law first for $\omega_y(n)$, and then for $\omega(n)$. This completes the proof of Theorem 5.12. ∎

The proof of Theorem 5.12 by the method of moments was originally carried out by Halberstam [76], without the truncation and reduction to independent random variables. The truncation method for general additive functions is due to Kubilius, for which we refer to his monograph Kubilius [96]. The need for the truncation is due to the fact that, with large prime divisors q_ω or $q_{\omega-j}$, j fixed, ϵ_q are not independent, and indeed, their influence on an additive function can become considerable. Another implication of the dependence of the preceding ϵ_q is that if we consider the distribution

$$P_N\Big(\log q_\omega(n) \le u \log n\Big) \to M(u), \tag{5.81}$$

which is known to exist, it does not fit into an extreme value distribution for the maximum of independent random variables. We do not go into the details of the theory leading to (5.81), we refer to De Koninck and Ivič [38], instead.

What we want to discuss is the sequence of intermediate prime divisors $q_j(n)$, $j = j(N)$, $q_j \to +\infty$ with j (or N) and $\log q_j(n)/\log N \to 0$ as $N \to +\infty$. We now present a result of De Koninck and Galambos [39]

Theorem 5.13 *Let $j = j(N)$ be a sequence of integers such that the properties of $q_j(n)$ in the preceding paragraph hold for all n with perhaps the exception of a set of integers n of density zero. Then the points $\log\log q_{j+k}$, $k \ge 1$, form a Poisson process in limit, as $N \to +\infty$.*

PROOF. Let $j = j(N)$ be a sequence satisfying the conditions of the theorem. Fix such a sequence, and $\log\log q_j(n)$ will be the starting point of our point process

$$a_k = \log\log q_{j+k}(n), \quad k \ge 1.$$

Consider a fixed number d of disjoint intervals $(t_{m,N}, t^*_{m,N})$, $1 \le m \le d$, and assume that both $t_{m,N} \to t_m$ and $t^*_{m,N} \to t^*_m$, as $N \to +\infty$, and the intervals (t_m, t^*_m) remain disjoint. Form the intervals, $1 \le m \le d$,

$$I(m,N) = \Big(\log\log q_j(n) + t_{m,N}, \log\log q_j(n) + t^*_{m,N}\Big) \tag{5.82}$$

and let $\alpha_N(d;t,t^*)$ be the number of points a_k in the union of $I(m,N)$, $1 \le m \le d$. In other words, since the intervals $I(m,N)$ are disjoint,

$$\alpha_N(d;t,t^*) = \sum_{m=1}^{d} \alpha_{m,N}(t,t^*), \tag{5.83}$$

where $\alpha_{m,N}(t,t^*)$ is the number of points a_k satisfying

$$\{p|n;\ \log\log p \in I(m,N)\} = \{p|n;\ q_j(n)^{e^{t_{m,n}}} \le p \le q_j(n)^{e^{t^*_{m,n}}}\}.$$

Let us calculate the binomial moments $S_{k,m} = S_{k,m,N}$ of $\alpha_{m,N}(t,t^*)$, $t \ge 1$. Instead of computing the actual moments $S_{k,n}$, we establish a recursive formula for them. First,

$$S_{1,N} = \sum P_N\Big(p|n,\ u_{j,m} \le p \le u^*_{j,m}\Big) = \sum_p{}^{*} \frac{1}{p} + o(1) \tag{5.84}$$

where

$$u_{j,m} = u_{j,m,N} = q_j(n)^{e^{t_{m,N}}}, \qquad u^*_{j,m} = q_j(n)^{e^{t^*_{m,N}}}$$

and $\sum^*$ signifies summation between $u_{j,m}$ and $u^*_{j,m}$. The $o(1)$ follows from the assumption that $\log q_j(n)/\log N \to 0$. By the formula (see (5.36))

$$\sum_{p \le T} \frac{1}{p} = \log\log T + C^{**} + o(1), \quad T \to +\infty,$$

we have

$$S_{1,m} = \log\log u^*_{j,m} - \log\log u_{j,m} + o(1) \to t^*_m - t_m. \tag{5.85}$$

Next, in $S_{k,m}$ we sum the terms

$$P_N\Big(p_{i_1}p_{i_2}\cdots p_{i_k}\big|n,\ \text{ each } p_{i_s} \text{ satisfies } u_{j,m} \le p_{i_s} \le u^*_{j,m}\Big).$$

Once again, due to $\log q_j(n)/\log N \to 0$ as $N \to +\infty$, that is, any power $q_j^k(n)/N \to 0$, if we replace the terms

$$\frac{1}{N}\left[\frac{N}{p_{i_1}\cdots p_{i_k}}\right] \quad \text{by} \quad \frac{1}{p_{i_1}\cdots p_{i_k}},$$

even the aggregate errors is $o(1)$ for every k of $S_{k,m}$. With this in mind, we get

$$S_{1,m}S_{k-1,m} = kS_{k,m} + O\Big(\frac{1}{q_j(n)}S_{k-1,m}\Big) + o\Big(S_{k-1,m}\Big) + o(1)$$

from which, by induction, based on (5.86),

$$S_{k,m} \to \frac{(t_m^* - t_m)^k}{k!}, \; N \to +\infty. \tag{5.86}$$

By an appeal to the method of inclusion and exclusion

$$P_N\Big(\alpha_{m,N}(t,t^*) = r\Big) \to \frac{(t_m^* - t_m)^r e^{-(t_m^* - t_m)}}{r!}, \tag{5.87}$$

as $N \to +\infty$. Since the intervals $(t_{m,N}, t_{m,N}^*)$ and (t_m, t_m^*) are disjoint, $\alpha_N(d;t,t^*)$ of (5.83) has an asymptotically Poisson distribution. Now the following lemma, due to Rényi [130], completes the proof of the theorem, that is, that the points a_k form a Poisson process as $N \to +\infty$.

Lemma 5.3 *Let J denote the set of unions of a finite number of intervals of the $[a_j, b_j)$. Let $X(E)$ be a set function of $E \in J$, assuming that $X(E_1 \cup E_2) = X(E_1) + X(E_2)$ for disjoint members E_1 and E_2 of J. Furthermore, if $X([a,b))$ is Poisson distributed with finite nonatomic measure $\lambda(E)$ for $E \in J$, then, for disjoint $E_j \in J$, $X(E_j)$ are independent, and $X(E)$ is a Poisson process.*

The proof can be found in our basic reference [58], p. 361. ∎

Note that, what is true for Poisson processes, Theorem 5.13 implies that the differences

$$\log\log q_{j+k}(n) - \log\log q_{j+k-1}(n)$$

are asymptotically exponentially distributed (under the conditions on $q_j(n)$), and, for $1 \le k \le m$ for any fixed m, the above differences are asymptotically independent. It was proved in Galambos [55] that, for a single k, the above difference is exponential. Far reaching generalizations of Theorem 5.13 are established in De Koninck, Kátai, and Mercier [40], which, however, should belong to books on number theory.

Chapter 6

Miscellaneous results

This chapter contains supplements to previous chapters, but its sections are not necessarily closely related.

6.1 Binary variables: the continuity of the limiting distribution

Let X_j, $j \geq 1$, be independent random variables, each taking two values only: $P(X_j = \alpha_j) = p_j$ and $P(X_j = \beta_j) = 1 - p_j$. We assume that neither α_j nor β_j equals zero. We define $Y_j = \log|X_j|$. Then, by the Three Series Theorem, the infinite sum $S = \sum_j Y_j$, or equivalently, the infinite product $\prod |X_j|$, converges almost surely, if, and only if, each of the series

$$\sum E(Y_{j,M}), \quad \sum V(Y_{j,M}), \quad \text{and} \quad \sum P(|Y_j| \geq M) \tag{6.1}$$

converges for any value of M, where $Y_{j,M}$ is Y_j, truncated at M.

If $|\alpha_j\beta_j| = 1$ and $p_j = 1/2$, that is, if each Y_j is symmetric about zero, then Y has a continuous distribution function. In fact, by a theorem of Lévy [98],the distribution function $F(x)$ of Y is continuous if, and only if,

$$\sum_{\substack{j \geq 1 \\ |\alpha_j| \neq |\beta_j|}} \min(p_j, 1 - p_j) = +\infty. \tag{6.2}$$

Evidently, (6.2) is satisfied if $p_j = 1/2$ for each j, and this case is relatively easy to prove (Lukacs [105], p. 65). Therefore, the emphasis is on the general case and we prove that (6.2) implies that $F(x)$ is continuous for all x. The proof follows a simple argument of Galambos and Kátai [60].

Note that the assumption of (almost sure) convergence of $\sum Y_j$ implies that $|\alpha_j| \neq |\beta_j|$ beyond some finite value J of j. Indeed, if a common value $|\alpha_j| = |\beta_j| \geq e^M$ (recall that Y_j is the logarithm of $|X_j|$), then the third series in (6.1) and the Borel-Cantelli lemma ([58], p. 10) entail that, with some J, $j \leq$ J. On the other hand, if $|\alpha_j| = |\beta_j| < e^M$ then the first series of (6.1) is applicable to conclude that this can happen only a finite number of times. So, $|\alpha_j| \neq |\beta_j|$ can actually be dropped from (6.2).

Rename the values α_j and β_j, if necessary, in such a manner that $|\alpha_j| > |\beta_j|$ for $j \geq$ J, J finite. For simplicity, we put $a_j = \log|\alpha_j| > b_j = \log|\beta_j|$, and $c_j = a_j - b_j \geq 0$.

Let S_1 and S_2 be two independent copies of $S = \sum Y_j$. Now, if $F(x)$ is discontinuous at a point x_o, that is, $P(S = x_o) > 0$ then $P(S_1 - S_2 = 0) > 0$. Consequently, if we show that $S_1 - S_2$ is continuous at 0 then we have that $F(x)$ is continuous everywhere.

Let $\varphi(t)$ be the characteristic function of $F(x)$. That is,

$$\varphi(t) = \prod_{j=1}^{+\infty} \left(e^{ita_j} p_j + e^{itb_j}(1 - p_j)\right) = \prod_{j=1}^{+\infty} e^{itb_j}\left[1 + p_j(e^{itc_j} - 1)\right]. \quad (6.3)$$

We also have that the characteristic function $\phi(t)$ of $S_1 - S_2$ equals $|\varphi(t)|^2$. We show that

$$\lim_{u \to +\infty} \frac{1}{2u} \int_{-u}^{u} |\varphi(t)|^2 \, dt = 0 \qquad (6.4)$$

implies that the distribution function $H(x)$ of $S_1 - S_2$ is continuous at zero. Indeed, by a typical inversion formula type of argument (interchanging integration)

$$\begin{aligned} \frac{1}{2u} \int_{-u}^{u} |\varphi(t)|^2 \, dt &= \frac{1}{2u} \int_{-u}^{u} \left(\int_{-\infty}^{+\infty} e^{itx} \, dH(x) \right) dt \\ &= \int_{-\infty}^{+\infty} \frac{\sin ux}{ux} \, dH(x). \end{aligned}$$

Choose a number $\epsilon > 0$. Then

$$\int_{-\infty}^{+\infty} \frac{\sin ux}{ux} \, dH(x) = \int_{|x| \leq \epsilon} \frac{\sin ux}{ux} \, dH(x) + \int_{|x| > \epsilon} \frac{\sin ux}{ux} \, dH(x),$$

and the second integral goes to zero as $u \to +\infty$ by the estimate

$$\frac{1}{\epsilon T}\int_{|x|>\epsilon} dH(x) \le \frac{1}{\epsilon T}.$$

In the first integral we utilize that $(\sin ux)/(ux) \to 1$ as $x \to 0$, and thus $c(H(\epsilon) - H(-\epsilon))$ with some $c > 0$ as an upper estimate. That is, $H(\epsilon) - H(-\epsilon)$ must converge to zero as $\epsilon \to 0$ if (6.4) is valid.

We now prove that (6.4) holds for $\varphi(t)$ of (6.3). For some j in the product of (6.3) we use the estimate

$$\left|e^{itb_j}\left(1 + p_j(e^{itc_j} - 1)\right)\right|^2 \le 1, \tag{6.5}$$

valid for any characteristic function (each factor in (6.3) is a characteristic function). On the other hand,

$$\begin{aligned}\left|e^{itb_j}\left(1 + p_j(e^{itc_j} - 1)\right)\right|^2 &= \left[1 + p_j(\cos tc_j - 1)\right]^2 + p_j^2 \sin^2 tc_j \\ &= 1 - 2p_j(1 - p_j)(1 - \cos tc_j) \\ &\le e^{-[\min(p_j, 1-p_j)](1-\cos tc_j)} \end{aligned} \tag{6.6}$$

where, in the last inequality, we used the elementary estimates

$$\min(p, 1-p) \le 2p(1-p) \quad \text{for all } 0 \le p \le 1, \quad \text{and} \quad 1 - x \le e^{-x}.$$

Let us combine (6.3), (6.5), and (6.6). We have

$$|\varphi(t)|^2 \le e^{-\sum_{j\in T}[\min(p_j, 1-p_j)](1-\cos tc_j)}, \tag{6.7}$$

where T is an arbitrary subset of the positive integers. We choose a sequence $M_s \to +\infty$ with s and an associated sequence $T_s = T(M_s)$, such that

$$\sum_{j\in T_s} \min(p_j, 1-p_j) = M_s$$

and $c_j > 0$ for all $j \in T_s$. We rewrite (6.7) by appealing to Jensen's inequality (take the unconditional form of [58], p. 317)

$$|\varphi(t)|^2 \le \frac{1}{M_s}\sum_{j\in T_s}[\min(p_j, 1-p_j)]e^{-M_s(1-\cos tc_j)}. \tag{6.8}$$

Now, going back to (6.4), integration in t, and the limit as $u \to +\infty$,

$$\frac{1}{2u}\int_{-u}^{u} e^{-M_s(1-\cos tc_j)}\, dt = \frac{1}{uc_j}\int_{0}^{uc_j} e^{-M_s(1-\cos u)}\, du$$

does not depend on c_j, and the limit is zero. Hence, by (6.8), (6.4) is valid. This completes the proof that, under (6.1), (6.2) suffices for the continuity of $F(x)$ for all x.

That (6.2) is also necessary for $F(x)$ to be continuous for all x is evident. Namely, if (6.2) were finite, then both $\sum a_j$ and $\sum b_j$ would be taken with positive probability.

Let us now investigate further. In the conclusion about (6.2) and the continuity of $F(x)$, the *sign* of α_j and β_j played no role. Therefore, if one of α_j and β_j is positive and the other one is negative (but $|\alpha_j| > |\beta_j|$ for all $j \geq$ J with some finite J), then

$$\sum_{j\geq 1} \min(p_j, 1-p_j) = +\infty \tag{6.9}$$

is a necessary and sufficient condition for $\prod X_j$ to be symmetric, assuming that $X = \prod X_j$ is almost sure convergent, and $|\alpha_j| \neq |\beta_j|$ if $p_j \neq 1/2$. If we eliminate the cases of $p_j = 1/2$ and $|\alpha_j| = |\beta_j|$ from consideration, we actually proved

Theorem 6.1 *Let X_j be independent random variables, taking the values α_j and β_j with $P(X_j = \alpha_j) = p_j$ and $P(X_j = \beta_j) = 1 - p_j$. Assume that one of α_j and β_j is negative and the other one is positive, $|\alpha_j| > |\beta_j|$, and $p_j \neq 1/2$ for any j. Then $X = \prod_{j=1}^{+\infty} X_j$ is almost surely convergent and has a symmetric distribution function $G(x)$ if, and only if, (6.1) holds and $G(x)$ is continuous. On the other hand, $G(x)$ is continuous if, and only if, (6.2) holds. If (6.2) holds, the restriction $p_j \neq 1/2$ can be dropped.*

The association of continuity with symmetry is new here, and the symmetry part follows from Simonelli's theorem [139] (see Theorem 1.21).

Theorem 6.1 can be reformulated for arithmetical functions. This is made possible by the transformations in Chapter 5.

Theorem 6.2 *Let $g(m)$ be a multiplicative function and assume that $g(p^k) < 0$ for all prime p and integer $k \geq 1$. Assume that $g(2^k) \neq$*

-1 *for all* $k \geq 1$. *Assume that* $g(m)$ *has a nondegenerate limiting distribution* $G(x)$. *Then* $G(x)$ *is symmetric if, and only if, it is continuous.*

One can also combine the results of Section 2.3 and Theorem 6.1 for relaxing the conditions in (6.1) for the existence of $G(x)$. A similar simplification of Theorem 6.2 obtains by considering the results of Section 5.6. The details are left to the reader.

6.2 The sign of a product and Markov chains

For independent random variables X_j, $j \geq 1$, the products $T_n = \prod_{j=1}^{n} X_j$, $n \geq 1$, form a Markov chain. We make this connection only for the case $Y_j = sign X_j$, which take only two values +1 and -1. This has the relation to the general case via investigating the *sign* of T_n. That is, assuming that $P(X_j = 0) = 0$, $j \geq 1$,

$$Y_j = \begin{cases} 1 & \text{if } sign X_j > 0 \\ -1 & \text{if } sign X_j < 0. \end{cases} \tag{6.10}$$

We put $S_n = \prod_{j=1}^{n} Y_j$. Clearly, for every $n \geq 2$, and for $s_j = 1$ or -1, $1 \leq j \leq n$, for every $1 \leq u \leq n-1$,

$$\begin{aligned} &P\Big(S_n = s_n \Big| S_{n-1} = s_{n-1}, \cdots, S_u = s_u\Big) \\ &\quad = P\Big(S_n = s_n \Big| S_{n-1} = s_{n-1}\Big) \end{aligned} \tag{6.11}$$

and

$$\begin{aligned} &P\Big(S_n = 1 \Big| S_{n-1} = 1\Big) = P\Big(Y_n = 1\Big) \\ &\quad = P\Big(X_n > 0\Big) = P\Big(S_n = -1 \Big| S_{n-1} = -1\Big) \end{aligned} \tag{6.12}$$

$$\begin{aligned} &P\Big(S_n = 1 \Big| S_{n-1} = -1\Big) = P\Big(Y_n = -1\Big) \\ &\quad = P\Big(X_n < 0\Big) = P\Big(S_n = -1 \Big| S_{n-1} = 1\Big) \end{aligned} \tag{6.13}$$

The Markov property at (6.11) and the transition probabilities (6.12) and (6.13), together with the initial condition $P(X_1 > 0) = p_1$

and $P(X_1 < 0) = 1 - p_1$, uniquely determine the distribution of S_n as well as the k-th step transitions

$$P\Big(S_{n+k} = 1\Big|S_n = 1\Big) = 1 - P\Big(S_{n+k} = -1\Big|S_n = 1\Big) = p_{n,k} \quad (6.14)$$

for all $k \geq 1$. To see that $p_{n,k}$ in (6.14) is uniquely determined by the claimed values, we need an extended form of the total probability rule.

Lemma 6.1 *Let A be an arbitrary event. Let C_j, $j \geq 1$, form a complete system, i.e., the C_j are mutually exclusive and $P(\cup_j C_j) = 1$. Then for an event B such that $P(B \cap C_j) > 0$ for every j whenever $P(C_j) > 0$,*

$$P(A|B) = \sum_j P(A|B \cap C_j)P(C_j|B).$$

PROOF. Note that $P(B) > 0$. Indeed, $P(B) = \sum P(B \cap C_j)$ and $P(B \cap C_j) > 0$ if $P(C_j) > 0$. But $P(C_j)$ cannot be zero for all j by assumption. The proof now goes as for the classical total probability rule. Since

$$P(A|B \cap C_j)P(C_j|B) = \frac{P(A \cap B \cap C_j)}{P(B \cap C_j)} \frac{P(B \cap C_j)}{P(B)} = P(A \cap C_j|B),$$

summation over j gives $P(A|B)$. ∎

We apply Lemma 6.1 for a variety of identities. For simplifying writing we put $p_{n,k}^{(+)} = p_{n,k}$ and $p_{n,k}^{(-)} = 1 - p_{n,k}$. We first establish the relations

$$P\Big(S_{n+1} = -1\Big) = p_{1,n}^{(+)} P(\Big(S_1 = -1) + p_{1,n}^{(-)} P\Big(S_1 = +1\Big), \quad (6.15)$$

and

$$P\Big(S_{n+1} = 1\Big) = p_{1,n}^{(-)} P(\Big(S_1 = -1) + p_{1,n}^{(+)} P\Big(S_1 = +1\Big). \quad (6.16)$$

Simply, choose $B = \Omega$, $A = \{S_{n+1} = -1\}$ and the system $C_1 = \{S_1 = -1\}$ and $C_2 = \{S_1 = 1\}$, yielding (6.15) by Lemma 6.1. The proof of (6.16) is similar. Next, we prove for $1 \leq t < k$,

$$p_{n,k}^{(+)} = p_{n,t}^{(+)} p_{n+t,k-t}^{(+)} + p_{n,t}^{(-)} p_{n+t,k-t}^{(-)}. \quad (6.17)$$

Implicit in (6.17) is the fact that

$$\begin{aligned} P\Big(S_{n+1} - 1\Big|S_n = -1\Big) &= p_{n,k}^{(+)} \quad \text{and} \\ P\Big(S_{n+k} = 1\Big|S_n = -1\Big) &= p_{n,k}^{(-)}. \end{aligned} \tag{6.18}$$

We first prove (6.18). If we choose $A = \{S_{n+k} = -1\}$, $B = \{S_n = -1\}$, and the system C_j, $j \geq 1$, as the collection $\{S_{n+1} = s_{n+1}, \cdots, S_{n+k-1} = s_{n+k-1}\}$, where each s_t is either $+1$ or -1, then

$$P(A|B) = P(S_{n+k} = -1|S_n = -1)$$

and a typical term in the lemma is of the form

$$\begin{aligned} &P\Big(S_{n+k} = -1\Big|S_{n+k-1} = s_{n+k-1}, \cdots, S_{n+1} = s_{n+1}, S_n = -1\Big) \\ &\quad \cdot P\Big(S_{n+k-1} = s_{n+k-1}, \cdots, S_{n+1} = s_{n+1}\Big|S_n = -1\Big) \\ &= P\Big(S_{n+k} = -1\Big|S_{n+k-1} = s_{n+k-1}\Big) \qquad (6.19) \\ &\quad \cdot P\Big(S_{n+k-1} = s_{n+k-1}, \cdots, S_{n+1} = s_{n+1}\Big|S_n = -1\Big). \qquad (6.20) \end{aligned}$$

Now, by an obvious change in the definition of A and B for computing $P(S_{n+k} = 1|S_n = 1)$ by the lemma, with the same system for C_j as above, a general term formally looks as (6.19)-(6.20). In both cases we sum over all $+1$ and -1 for $(s_{n+1}, s_{n+2}, \cdots, s_{n+k-1})$. However, if we switch the terms $(s_{n+1}, s_{n+2}, \cdots, s_{n+k-1})$ when computing $P(S_{n+1} = -1|S_n = -1)$ to $(-s_{n+1}, -s_{n+2}, \cdots, -s_{n+k-1})$ when computing $P(S_{n+k} = 1|S_n = 1)$ and then upon using the multiplication rule

$$\begin{aligned} &P\Big(S_{n+k-1} = s_{n+k-1}, \cdots, S_{n+1} = s_{n+1}\Big|S_n = s_n\Big) \\ &\quad = \prod_{t=1}^{k-1} P\Big(S_{n+t} = s_{n+t}\Big|S_{n+t-1} = s_{n+t-1}\Big) \end{aligned}$$

the same probabilities obtain in the light of (6.12) and (6.13). This proves (6.18). For (6.17), write,

$$\begin{aligned} P\Big(S_{n+k} = 1\Big|S_n = 1\Big) &= P\Big(S_{n+k} = 1,\ S_{n+t} = 1\Big|S_n = 1\Big) \\ &\quad + P\Big(S_{n+k} = 1,\ S_{n+t} = -1\Big|S_n = 1\Big) \end{aligned}$$

and use the multiplication rule as previously. The decomposition for $p_{n,k}^{(-)}$ is, by a similar argument as for (6.17),

$$p_{n,k}^{(-)} = p_{n,t}^{(-)} p_{n+t,k-t}^{(+)} + p_{n,t}^{(+)} p_{n+t,k-t}^{(-)}. \tag{6.21}$$

If we write the transition probabilities in the following matrix form (we combine (6.14) and (6.18))

$$\begin{aligned} \mathbf{P}_{n,k} &= \begin{pmatrix} P(S_{n+k} = 1 | S_n = 1) & P(S_{n+k} = -1 | S_n = 1) \\ P(S_{n+k} = 1 | S_n = -1) & P(S_{n+k} = -1 | S_n = -1) \end{pmatrix} \\ &= \begin{pmatrix} p_{n,k} & 1 - p_{n,k} \\ 1 - p_{n,k} & p_{n,k} \end{pmatrix} \end{aligned} \tag{6.22}$$

then (6.17) and (6.19)-(6.20) entail

$$\mathbf{P}_{n,k} = \mathbf{P}_{n,t}\mathbf{P}_{n+t,k-t} \tag{6.23}$$

and

$$\mathbf{P}_{n,1} = \mathbf{P}_{2,1}\mathbf{P}_{3,2} \cdots \mathbf{P}_{n,n-1}. \tag{6.24}$$

By an appeal to (6.12) and (6.13), we have

$$\mathbf{P}_{n,1} = \prod_{j=2}^{n} \begin{pmatrix} P(X_j > 0) & P(X_j < 0) \\ P(X_j < 0) & P(X_j > 0) \end{pmatrix}. \tag{6.25}$$

The combination of (6.15) and (6.16) with (6.25) gives a formula for $P(T_n < 0) = P(S_n = -1)$ and $P(T_n > 0) = P(S_n = 1)$, which, of course, coincides with results of Chapter 1. In particular, if, for a single j, $P(X_j > 0) = 1/2$, $\mathbf{P}_{n,1}$ of (6.25) is of the form

$$\begin{pmatrix} \frac{1}{2} & \frac{1}{2} \\ \frac{1}{2} & \frac{1}{2} \end{pmatrix}, \tag{6.26}$$

and (6.15)-(6.16) yield $P(T_n < 0) = P(T_n > 0) = 1/2$ for every n. Furthermore, if the entries of $\mathbf{P}_{n,1}$ satisfy $p_{n,1}^{(+)} - p_{n,1}^{(-)} \to 0$ as $n \to +\infty$, we have (6.26) as the limit of $\mathbf{P}_{n,1}$. This now gives $\lim P(T_n < 0) = 1/2$ as $n \to +\infty$ by appealing to (6.15)-(6.16). Since

$$\begin{aligned} p_{n,1}^{(+)} - p_{n,1}^{(-)} &= \prod_{j=1}^{n} \Big[P(X_j > 0) - P(X_j < 0) \Big] \\ &= \prod_{j=2}^{n} \Big| 1 - 2P(X_j < 0) \Big| \\ &= \prod_{j=2}^{n} \Big(1 - 2\min\big(P(X_j > 0), P(X_j < 0)\big) \end{aligned}$$

which is the formula on page 33, when we combine the above product with (6.15)-(6.16).

6.3 Sequences of the prime numbers

We avoided references to number theory as much as possible in Chapter 5 in order to see the extent to which number theoretic functions can be treated by purely probabilistic tools. We now add some number theoretic results. We do not go into the most sophisticated techniques and results of number theory, but want to give some flavor of the subject matter as the prime numbers are concerned.

Just as in Chapter 5, $2 = p_1 < p_2 < \cdots$ will denote the consecutive prime numbers, and p always stands in this section for a generic prime number. The starting point in this section is to connect the formula for all primes $\pi(N)$ which do not exceed N, and Wirsing's theorem on the mean value of multiplicative functions $g(m)$, taking the values +1 and -1, only.

Just as in Chapter 5, $\Omega_N = \{1, 2, \cdots, N\}$, and $P_N(\cdots)$ is the uniform distribution on all subsets of Ω_N. That is, if $\nu_N(\cdots)$ denotes the number of integers $n \leq N$ which satisfy the property in the dotted space, then $P_N(\cdots) = \nu_N(\cdots)/N$. Let A_p be the set of all integers n which are divisible by p. For any set B of integers, B^c denotes the complement of B. Since both $\nu_N(\cdots)$ and $P_N(\cdots)$ limit any set B to $n \in B$ such that $n \leq N$, we can always define a set initially for all integers. We put $S_{0,T} = 1$, and, for $k \geq 1$,

$$S_{k,T} = \sum_{1 \leq i_1 < i_2 < \cdots < i_k \leq T} P_N(A_{p_{i_1}} \cap A_{p_{i_2}} \cap \cdots \cap A_{p_{i_k}}). \tag{6.27}$$

Let $m(T)$ be the number of those A_p, $p \leq T$, which occur, that is, $m(T)$ counts the number of prime divisors p of an integer $n \leq N$ such that $p \leq T$. Then

$$P_N\Big(m(\sqrt{N}) = 0\Big) = \frac{\pi(N) - \pi(\sqrt{N})}{N}. \tag{6.28}$$

Now, if $g(m)$ is a strongly multiplicative function with $g(p) = -1$ for each p, then

$$P_N(m(N) \text{ odd }) = P_N(g(m) < 0), \tag{6.29}$$

from which one can, of course, compute the arithmetical mean of $g(m)$. For the left-hand sides of (6.28) and (6.29) we have the classical method of inclusion and exclusion, and the new Bonferroni-type formulas of Theorem 1.25, respectively. These yield (by (6.27) and (6.28))

$$\pi(N) - \pi(\sqrt{N}) = N \cdot \sum_{k=0}^{\sqrt{N}} (-1)^{k+1} S_{k,\sqrt{N}} \tag{6.30}$$

and (by (6.27) and (6.29))

$$\nu_N(n \leq N : g(n) = -1) = N \cdot \sum_{k=1}^{N} (-2)^{k+1} S_{k,N}. \tag{6.31}$$

First, if one argues with asymptotics,

$$P_N\big(A_{p_{i_1}} \cap A_{p_{i_2}} \cap \cdots \cap A_{p_{i_k}}\big) = \frac{1}{p_{i_1} p_{i_2} \cdots p_{i_k}} + O\Big(\frac{1}{N}\Big), \tag{6.32}$$

by ignoring error terms, one gets

$$\pi(N) \approx \pi(\sqrt{N}) + N \prod_{p \leq \sqrt{N}} (1 - \frac{1}{p}) \tag{6.33}$$

and

$$\nu_N\big(n \leq N : g(n) = -1\big) \approx N\Big[\frac{1}{2} - \prod_{p \leq N} (1 - \frac{2}{p})\Big]. \tag{6.34}$$

The next step is to argue that if one has a proof for (6.34) by any specific method, then, by the similarity of terms, (6.33) must also be valid, at least its order of magnitude. This would give that the 'real' order of magnitude of

$$\pi(N) \sim N e^{\sum_{p \le \sqrt{N}} \log(1-1/p)} + O(\sqrt{N}) \tag{6.35}$$

and

$$\nu_N(n \le N;\, g(n) = -1) \sim \frac{N}{2} - N e^{\sum_{p \le N} \log(1-1/p)}. \tag{6.36}$$

The results in (6.35) and (6.36) are quite close to the real situations. Should one improve on developing asymptotics for one of (6.30) and (6.31), then the other formula will hopefully improve as well. That is, one can deduce much information from one of these formulas for the other.

For making the preceding argument accurate, and thus obtaining proper asymptotics for the left hand side of (6.35) and (6.36), one should make the following steps: (i) replace the identities (6.30) and (6.31) by inequalities; (ii) restrict the number of terms in $S_{k,T}$ of (6.27) in such a way that the inequalities of (i) remain valid, the remaining main terms lead to the same upper and lower bounds, and the aggregate error terms of (6.32) become negligible. Such a method is known in number theory as Brun's method, and a similar technique appears in details in Galambos and Simonelli [62] when dealing with the restrictive sieve method such as the graph-sieves. We do not go into further details, but quote the actual results for $\pi(N)$:

$$\pi(N) = (1 + o(1))\frac{N}{\log N}, \quad N \to +\infty. \tag{6.37}$$

Then an easy summation formula yields

$$\sum_{p \le N} \frac{1}{p} = \sum_{n=2}^{N} \frac{\pi(n) - \pi(n-1)}{n} = \sum_{n=2}^{N-1} \pi(n)\left(\frac{1}{n} - \frac{1}{n+1}\right) + \frac{\pi(N)}{N}$$
$$\sim \sum_{2 \le n \le N} \frac{1}{n \log n} \sim \log\log N.$$

In order to see that (6.35) provides the magnitude $N/\log N$ for $\pi(N)$ but not the accurate coefficient as at (6.37), write

$$\sum_{p\le\sqrt{N}}\left[\log(1-\frac{1}{p})+\frac{1}{p}\right]-\sum_{p\le\sqrt{N}}\frac{1}{p}=\sum_{p\le\sqrt{N}}\left[\log(1-\frac{1}{p})\right]$$

and observe that the first sum on the left hand side is convergent but not leading to the coefficient 'one' in (6.37).

6.4 Hilbert spaces

In Section 2.4 we used some fundamental properties of the L_2-space of random variables, whose proofs we postponed to the present section. We now prove those properties in a somewhat more general set-up than needed in Section 2.4. First, some definitions and notations.

Let R be the set of real numbers. Let H be a nonempty set of elements for which the operation, which we call sum, denoted $+$, is defined with the properties that $+$ is commutative, associative, and $x+z=y$, for every x and $y\in$ H has a solution with $z\in$ H. The solution z of $x+z=x$ is the zero member of H (not to be confused with $0\in R$, though we may use the same notation). For every $a\in R$ and $x\in$ H , ax is defined and has the properties ($a\in R, x, y\in$ H)

$$a(x+y)=ax+ay,\quad (a+b)x=ax+bx,\quad a(bx)=(ab)x \tag{6.38}$$

and $1\cdot x=x$. When the above properties are valid, H is called a linear space.

A function $\| x \|$ on a linear space H is called a norm if, for every $x\in$ H, $\| x \|\ge 0$ real, $\| ax \|=|a|\,\| x \|$ for all $a\in R$, satisfies the triangular inequality

$$\| x+y \|\le\| x \|+\| y \|,\quad x,\, y\in \text{H} \tag{6.39}$$

and $\| x \|=0$ if, and only if, $x=0$ (recall the double meaning of 0, $0\in R$ and $0\in$ H).

We take as an example H $=L_2$, the set of all random variables on a given probability space $(\Omega,\mathcal{A},P)$ whose variance is finite. Obviously, aX, and $X+Y$, are defined for all $X,Y\in L_2$ and $a\in R$, and satisfy the linearity properties. The norm $\| X \|$ is defined as

$$\| X \|=\| X \|_{L_2}=\Big(\int_\Omega X^2dP\Big)^{1/2}. \tag{6.40}$$

In L_2, $X = Y$ if $P(X = Y) = 1$. Then $\| X \| = 0$ if, and only if, $X = 0$ a.s., and all other properties required for a norm are evidently satisfied. By Chebyshev's inequality,

$$X_n \to X \quad \text{in probability, if} \quad \| X_n - X \| \to 0.$$

We return to the general discussion of a linear space H with a norm $\| x \|$, which we call a normed linear space. Assume that $\| x \|$ is such that, for $x, y \in$ H,

$$\| x + y \|^2 + \| x - y \|^2 = 2 \| x \|^2 + 2 \| y \|^2, \tag{6.41}$$

which we interpret as the sum of squares of the diagonals of a parallelogram equals the sum of squares of its sides, then we call

$$(x, y) = \frac{1}{4}\Big(\| x + y \|^2 - \| x - y \|^2 \Big) \tag{6.42}$$

the inner product of x and y. Clearly,

$$(x, y) = (y, x), \qquad (x, x) = \| x \|^2 \geq 0 \tag{6.43}$$

with

$$(x, y) = 0 \quad \text{if, and only if,} \quad x = 0. \tag{6.44}$$

Furthermore, for a fixed $y \in$ H, $A(x) = (x, y)$ is a linear functional on H. That is, for all $z \in$ H, $a, b \in R$, $A(ax + bz) = aA(x) + bA(z)$. We can reverse the order of definitions of a norm and an inner product. Simply, upon defining (x, y) with the above listed properties, then $\sqrt{(x, x)} = \| x \|$ is a norm. Linearity also yields (6.41).

In L_2, the inner product

$$\Big(X, Y\Big) = \int_\Omega (XY) dP. \tag{6.45}$$

Then $(X, X) = \| X \|^2$ as in (6.40).

We now prove the abstract form of the Cauchy-Schwarz inequality:

Lemma 6.2 *If H is a linear space with the inner product (x, y) and norm $\| x \| = \sqrt{(x, x)}$, $x, y \in H$, then*

$$\Big|(x, y)\Big| \leq \| x \| \| y \| . \tag{6.46}$$

PROOF. Let $a \in R$. Then

$$\begin{aligned} 0 \leq \| x + ay \|^2 &= (x + ay, x + ay) = (x + ay, x) + (x + ay, ay) \\ &= (x, x) + a(y, x) + a(x, y) + |a|^2 (y, y). \end{aligned}$$

Now, if $y = 0$, then (6.46) is evidently valid. Hence, we choose $y \neq 0$, that is, $(y, y) > 0$, and let $a = -(x, y)/(y, y)$. We have

$$0 \leq (x, x) - 2\frac{(x, y)^2}{(y, y)} + \frac{(x, y)^2}{(y, y)^2}(y, y)$$

which is indeed (6.46). The proof is complete. ∎

The Cauchy-Schwarz inequality (6.46) entails that if (x, y) is an inner product, then $\| x \| = \sqrt{(x, x)}$ satisfies the triangular inequality (6.39), which now justifies our earlier claim (following (6.44)) that $\sqrt{(x, x)}$ is a norm. Indeed, by the linearity of the inner product,

$$\| x + y \|^2 = (x + y, x + y) = \| x \|^2 + \| y \|^2 + 2(x, y) \tag{6.47}$$

which, by (6.46), is increased by

$$\| x \|^2 + \| y \|^2 + 2 \| x \| \| y \| = (\| x \| + \| y \|)^2.$$

In a linear space H with inner product (x, y), we say that x and y are perpendicular or orthogonal if $(x, y) = 0$. From (6.47) we have that if x and y are orthogonal then

$$\| x + y \|^2 = \| x \|^2 + \| y \|^2,$$

which extends to any finite number of members $x_j \in$ H such that, for all $i \neq j$, $(x_i, x_j) = 0$. We refer to such an identity as the Pythagorean relation.

Theorem 6.3 *Let H be a linear space with an inner product (x, y). Let $x_j \in H$, $1 \leq j \leq n$, be orthogonal for every pair $i \neq j$, and assume that $\| x_j \| = 1$ for all $1 \leq j \leq n$. Then, for $x \in H$, and $a_j \in R$,*

$$\min \| x - \sum_{j=1}^{n} a_j x_j \| \qquad \textit{is reached if } a_j = (x, x_j).$$

PROOF. By the linearity of the inner product,

$$\| x - \sum_{j=1}^{n} a_j x_j \|^2 = \| x \|^2 - 2\sum_{j=1}^{n} a_j (x, x_j) + \sum_{j=1}^{n} a_j^2,$$

where we used the Pythagorean relation in the last term. Since

$$-2a_j(x, x_j) + a_j^2 = (a_j - (x, x_j))^2 - (x, x_j)^2,$$

we have

$$\begin{aligned} 0 &\leq \| x - \sum_{j=1}^{n} a_j x_j \|^2 \\ &= \| x \|^2 - \sum_{j=1}^{n} (x, x_j)^2 + \sum_{j=1}^{n} (a_j - (x, x_j))^2, \end{aligned} \tag{6.48}$$

which is minimized if we take $a_j = (x, x_j)$. ∎

The last inequality immediately yields Bessel's inequality: In a linear space H if $H_1 \subset$ H is such that, for every $u, y \in H_1$, $(u, y) = 0$ and $\| y \| = 1$, for any $x \in$ H, and for $y_j \in H_1$,

$$\| x \|^2 \geq \sum_{y_j \in H_1} (x, y_j)^2. \tag{6.49}$$

As a matter of fact, for any finite number of $y_j \in H_1$, (6.48) entails (6.49), and then letting $n \to +\infty$, (6.49) remains to hold.

We say that a normed linear space is a Banach space if every Cauchy sequence has a limit in the space. That is, if $x_j \in$ H, and

$$\| x_n - x_m \| \to 0 \quad \text{as} \quad n, m \to +\infty,$$

then there is an $x \in$ H such that $\| x_n - x \| \to 0$. If the norm $\| y \|$ can be defined through an inner product (x, y), then a Banach space is a Hilbert space. In other words, if every Cauchy sequence $\{x_n\}$ of a linear normed space H has a limit in H, which we refer to as H is complete, and if (6.41) holds, so (6.42) is an inner product, then H is a Hilbert space.

The space L_2 is a Hilbert space. Only the completeness of L_2 remains to be proved. We present this as a theorem, thus establishing Lemma 2.1.

Theorem 6.4 *The space L_2 is complete. That is, for every sequence of random variables X_j on a probability space $(\Omega, \mathcal{A}, P)$ such that $X_j \in L_2$, and*

$$\| X_n - X_m \|^2 = \int_\Omega |X_n - X_m|^2 \, dP \to 0, \quad n, m \to +\infty,$$

there is a random variable Y such that $Y \in L_2$, and

$$\| X_n - Y \|^2 = \int_\Omega |X_n - Y|^2 \, dP \to 0, \quad n \to +\infty.$$

PROOF. Let N_1 be such that, for all $n, m \geq N_1$, $\| X_n - X_m \| < 1/4$, and let $Y_1 = X_{N_1}$. Assume that $N_1 < N_2 < \cdots < N_k$ and $Y_1, Y_2, \cdots, Y_k$, have been determined. Let now N_{k+1} be such that $\| X_n - X_m \| < (1/4)^{k+1}$ for all $n, m \geq N_{k+1}$, and let $Y_{k+1} = X_{N_{k+1}}$. By Chebyshev's inequalty,

$$P\Big(|Y_k - Y_{k+1}| \geq 2^{-k}\Big) \leq 2^k \| Y_k - Y_{k+1} \|^2 \leq 2^{-k}.$$

Upon summing over k, the Borel-Cantelli lemma entails, that the sequence Y_k, $k \geq 1$, is a Cauchy sequence of real numbers a.s., that is, $Y_k \to Y$ a.s.. Choose a subscript N_o such that, for all $n, m \geq N_o$, $\| X_n - X_m \| < \epsilon$. Fix $\epsilon > 0$ and n. Consider the subsequence $Y_k = X_{N_k}$, and let $k \to +\infty$. We have

$$\begin{aligned} \epsilon &\geq \liminf \| X_n - Y_k \|^2 = \liminf \int_\Omega |X_n - Y_k|^2 \, dP \\ &\geq \int_\Omega \liminf |X_n - Y_k|^2 \, dP = \| X_n - Y \|^2, \end{aligned}$$

where in the last but one step, we used Fatou's lemma (see [58], p. 56). Thus, $\| X_n - Y \| \to 0$ as $n \to +\infty$. Since

$$\| Y \| = \| Y - X_n + X_n \| \leq \| Y - X_n \| + \| X_n \| \leq \epsilon + \| X_n \|,$$

$X_n \in L_2$ implies that $Y \in L_2$, which completes the proof. ∎

We now prove a theorem for a Hilbert space from which we deduce Lemma 2.2.

Theorem 6.5 *Let H be a Hilbert space. Let $H^* \subset H$ such that H^* is closed under the two operations of multiplying an element by a real number and the basic operation of addition. Furthermore, H^* is closed under limits in the norm of H. Then every element $h \in H$ has a unique representation $h = x + y$ where $x \in H^*$ and y is orthogonal to every member of H^*.*

PROOF. The proof, in a sense, is constructive. We show that to every $h \in$ H there is a unique $x \in H^*$ such that

$$\| h - x \| = \inf\{\| h - u \| : u \in H^*\}. \tag{6.50}$$

Then we show that $y = h - x$ is orthogonal to every member of H^*. Finally, we show that the representation in the form of the theorem is unique.

For proving (6.50), set $\alpha = \inf\{\| h - u \| : u \in H^*\}$ for a fixed $h \in$ H. Choose a sequence $u_j \in H^*$, $j \geq 1$, such that $\| h - u_j \| \to \alpha$ as $j \to +\infty$. We show that $u_j \to$ some $x \in H^*$ in the norm $\| \cdots \|$. Because of the completeness of H, and the assumed completeness of H^*, we have to show that u_j is a Cauchy sequence. But by (6.41), applied with $u_n - h$ and $u_m - h$,

$$\| u_n + u_m - 2h \|^2 + \| u_n - u_m \|^2 = 2 \| u_n - h \|^2 + 2 \| u_m - h \|^2,$$

where the right hand side converges to $4\alpha^2$ by the choice of the sequence u_j, and

$$\| u_n + u_m - 2h \|^2 = 4 \| \frac{1}{2}u_n + \frac{1}{2}u_m - h \|^2 \geq 4\alpha^2$$

by the definition of α and by the closedness of H^* (implying that $u_n/2 + u_m/2 \in H^*$). Consequently, $\| u_n - u_m \|$ must converge to zero.

The fact that x is unique in (6.50) easily follows from (6.41), when applied with $h - x$ and $h - x_1$ where both x and x_1 are assumed to satisfy (6.50). In other words, since $\| h - x \| = \| h - x_1 \| = \alpha$, and $\| x/2 + x_1/2 - h \| \geq \alpha$, $\| x - x_1 \| = 0$ must be true in view of (6.41).

Next, we show that $y = h - x$, with x uniquely determined by (6.50), is orthogonal to every $z \in H^*$. That is, $(y, z) = 0$ if $z \in H^*$. Write, with an arbitrary constant a (note that $x + az \in H^*$),

$$\| h - x - az \|^2 = \| h - x \|^2 + a^2 \| z \|^2 - 2(h - x, az) \geq \alpha^2$$

and $\| h - x \|^2 = \alpha^2$, so $a^2 \| z \|^2 - 2(h - x, az) \geq 0$. However, if we choose $a = b(h - x, z)$ with $b > 0$, then $(h - x, az) = b(h - x, z)^2$, and $a^2 \| z \|^2 - 2(h - x, az) = [b^2 \| z \|^2 - 2b] \| h - x, z \|^2$ becomes negative as $b \to 0$. Consequently, $(h - x, z) = 0$. This proves that $h = x + y$ is a representation of the required form. If it were not unique, then $h = x + y = x_1 + y_1$ with $x, x_1 \in H^*$ and, by the closedness of H^*, $x - x_1 \in H^*$ as well. But

$$z = x - x_1 = (h - y) - (h - y_1) = y_1 - y \quad \text{and} \quad (y, z) = (y_1, z) = 0,$$

$z \in H^*$, we get $(z, z) = \| z \|^2 = (y_1 - y, z) = (y_1, z) - (y, z) = 0$, that is, $z = 0$, $x = x_1$, and the theorem is proved. ∎

We return to the space L_2 and prove Lemma 2.2. Let $X_n \in L_2$, $n \geq 1$, and let $X_o \equiv 1$. Assume that for every fixed $k \geq 0$,

$$\int_\Omega X_n X_k \, dP \to 0 \quad \text{as } n \to +\infty \tag{6.51}$$

and that $\| X_n \|$ is bounded in n. We want to prove that, for any $Y \in L_2$,

$$\int_\Omega X_n Y \, dP \to 0 \quad \text{as } n \to +\infty. \tag{6.52}$$

Let H^* be the smallest subset of L_2 such that it contains each X_n, $n \geq 0$, and is closed as specified in Theorem 6.5. If $Y = \sum_{k=0}^t c_k X_k$ with a fixed t, c_k real, then (6.51) evidently implies (6.52). Next, let $Y \in H^*$. Then, to every $\epsilon > 0$, there is a linear combination $\sum_{k=0}^t c_k X_k$ such that

$$\| Y - \sum_{k=0}^{t} c_k X_k \| < \epsilon.$$

Hence, by the boundedness of $\| X_k \| \leq M$ and by the Cauchy-

Schwarz inequality (6.46),

$$\begin{aligned}
|(Y, X_n)| &= |(Y - \sum_{k=0}^{t} c_k X_k + \sum_{k=0}^{t} c_k X_k, X_n)| \\
&= |((Y - \sum_{k=0}^{t} c_k X_k), X_n) + \sum_{k=0}^{t} c_k (X_k, X_n)| \\
&\leq \| Y - \sum_{k=0}^{t} c_k X_k \| \cdot \| X_n \| + |\sum_{k=0}^{t} c_k (X_k, X_n)| \\
&\leq \epsilon M + |\sum_{k=0}^{t} c_k (X_k, X_n)| \to \epsilon M \quad \text{as } n \to +\infty.
\end{aligned}$$

Since $\epsilon > 0$ was arbitrary, we have $(Y, X_n) \to 0$ as $n \to +\infty$. Finally, if $Y \in L_2$, then there is a (unique) representation $Y = X + Z$, $X \in H^*$ and Z is orthogonal to every member of H^* (Theorem 6.5). Hence, $(Y, X_n) = (X, X_n)$. Since we have proved (6.52) for members of H^*, (6.52) is valid for Y, which completes the proof of Lemma 2.2.

Since only fundamental results from the theory of Hilbert spaces are presented here, and all statements are fully proved, the reader is not required to know or to read additional material.

6.5 Applied models with products

In this section we list, and in some cases we give details for, statistical applications and some applied models which appeared in the literature and which use some aspects of products of random variables. We emphasize that our aim is not to give a detailed description of these topics, but to show the importance of products of random variables in statistical applications.

6.5.1 Models related to income

An early, and somewhat bold, model for random growth of person's income was postulated by Gibrat [71]. He assumed that the income $X(t)$ of an individual at time t is a stochastic process which, at random times, changes by random factors Y_j, $j \geq 1$, where the Y_j, even after incorporating the random times for change, are independent random variables. Thus

$$X(t) = X(0) \prod_{j=1}^{M} Y_j, \tag{6.53}$$

where $X(0)$ is the initial income, $M = M(t)$ is the (random) number of times of change in the income level of the individual upto time t. Because $X(0)$ and the Y_j are positive, a reference to the central limit theorem for $logX(t)$ is utilized to conclude that $X(t)$ is asymptotically lognormal. There are several flows in such a conclusion of lognormality which have not been pointed out even in subsequent critical reviews of this model (see Ord [116] and Arnold [8], Section 2.2). Namely, the central limit theorem with a random sample size $M(t)$ was not heard of in 1931, furthermore, as we now know, even with very nicely behaving $M(t)$ (such as the independence of $M(t)$ and the Y_j, and the existence of a limiting distribution of $M(t)$) would not necessarily keep lognormality of $X(t)$, when normalized. On the other hand, without normalization, $X(t)$ cannot be logarithmically normal, because the Y_j are certainly bounded, and each Y_j has a nonnegligible contribution to $X(t)$. Now, if we ignore the practical meaning of Y_j, and thus allow each Y_j to have distributions for which $logY_j$ is unlimited, the following example, based on the paper Galambos [54], is very instructive. Its conclusion is that $X(t)$ of (6.53) does become logarithmically normal after normalization and passage to the limit, but it is solely due to $M(t)$, the number of changes in $X(t)$, rather than the actual changes expressed by Y_j.

The example is as follows. Let U_j, $j \geq 1$, be independent logarithmically exponential variables with $E(\log U_j) = 1$. Define the sequence of records $U_{L(j)}$ as follows. Let $L(1) = 1$. After $L(n-1)$ has been defined, we set

$$U_{L(n)} = \min\{k : k > L(n-1),\, U_k > U_{L(n-1)}\}.$$

Since inequalities remain to hold if we take the logarithm of the variables, $\log U_{L(n)}$ is the n-th record in the sequence $\log U_j$, $j \geq 1$. We know from a result of Tata [148] that the variables $\log U_{L(j)} - logU_{L(j-1)}$ also are independent exponential variables with unit expectation. Hence, if we put $Y_1 = U_1$, and $Y_j = U_{L(j)}/U_{L(j-1)}$, $j \geq 2$, and $M(n) = L(n)$, defined by

$$M(n) = \max\{L(j) : L(j) \leq n\},$$

then the product $X(n)$ of (6.53) becomes

$$X(n) = X(0)U_{L(n)} = X(0)\max\{U_1, U_2, \cdots, U_n\}.$$

Since

$$\log X(n) = \log X(0) + max\{\log U_1, \log U_2, \cdots, \log U_n\},$$

by the exponentiality of the $\log U_j$ (and by choosing $X(0) = 1$), we have the limiting distribution

$$\begin{aligned} P\Big(\log X(n) - \log n \leq z\Big) &= (1 - e^{-z-logn})^n \\ &= (1 - e^{-z}/n)^n \to e^{-e^{-z}}. \quad (6.54) \end{aligned}$$

On the other hand,

$$\log X(n) = \sum_{j=1}^{M(t)} \log Y_j,$$

which, by Anscombe's theorem (see Theorem 2.11) entails the limiting distribution

$$P\Big(\frac{\log X(n) - M(n)}{\sqrt{M(n)}} \leq x\Big) \to N(x), \qquad (6.55)$$

where $N(x)$ is the standard normal distribution function. The seemingly contardictory limits in (6.54) and (6.55) are due to the fact that $M(n)/\log n \to 1$ in probability (see Galambos [57], p. 346). Indeed, in view of (6.55), the asymptotic normality of

$$\frac{\log X(n) - M(n)}{\sqrt{M(n)}} = \frac{\log X(n) - \log n}{\sqrt{M(n)}} - \frac{M(n) - \log n}{\sqrt{\log n}} \frac{\sqrt{\log n}}{\sqrt{M(n)}}$$

yields that the last term on the right hand side must be asymptotically normal , while the first term on the right hand side is degenerated at zero in light of (6.54). Hence, the asymptotic lognormality of $X(t)$ has nothing to do with the Y_j.

There are, of course, general situations when(6.53) is asymptotically logarithmically normal, in which the Y_j have dominant role, as demonstrated in Theorem 2.11.

Another model related to income is about under-reporting to the tax offices. Let X be the actual income and U the amount reported to the tax office. The distribution of X is determined by random sampling (a large number of people go through thorough audit). If U has a different distribution than X, then it is easily recognized that cheating goes on. So, assume that $U = VX$, $0 < V < 1$, V is independent of X, and U has the same distribution as X. Consider the difference

$$D = X - VX = (1 - V)X,$$

the amount not taxed. Note that D has a similar form to U. Now, by the assumed independence,

$$E(D|X = x) = x\,E(1 - V) = ax, \;\; x \leq \sup\{y : F(y) < 1\} = \omega_F.$$

We further assume that the conditional expectation $E(D|X > x) = Ax + B$, also linear. Then, denoting by $F(x) = P(X \leq x)$,

$$E(D|X > x) = \frac{\int_x^{\omega_F} ay\,dF(y)}{1 - F(x)} = Ax + B.$$

In the middle term we recognize the expected residual life function $R_F(x)$ of $F(x)$, a term central in extreme value theory, which uniquely determines $F(x)$. We should add that, by the nature of the practical side of the model, ω_F is finite, and people are taxed only over some income level $\alpha = \alpha_F$, hence all formulas above are for $\alpha \leq x \leq \omega$. Since we established that $R(x) = R_F(x) = Ax + B$, and since, with some constant $c > 0$,

$$1 - F(x) = \frac{c}{R(y)} e^{-\int_\alpha^x \frac{1}{R(u)}\,du}$$

is the general relation between expected residual life R and F itself (just use differentiation - in general, with respect to Lebesgue measure, but here $R = Ax + B$, so the respective F is differentiable), we have for $\alpha \leq y \leq \omega$,

$$1 - F(y) = \frac{c^*}{Ay + B} e^{-(a/A)\log(Ay+B)} = \frac{c^*}{(Ay + B)^\gamma}$$

with some $\gamma > 1$. That is, $F(y)$ is Pareto (α, ω), where (α, ω) refers to the restrictive domain above. Recall that $F(y)$ is Pareto (α, ω) if taxpayers can make it hard for the tax office to discover underreporting of income. The distribution of V has no role in the model. The model is due to Revankar et al [131].

6.5.2 Multiplying observations for smoothing

If observations X_j come from a distribution $F(x)$ with

$$\alpha_F = \inf\{x : F(x) > 0\}, \qquad \text{finite,}$$

then the normalized minimum of the X_j is frequently asymptotically Weibull (see Chapter 2 in Galambos [57]). The shape parameter γ of this asymptotic Weibull law $(1-e^{-x^\gamma})$ is determined by the property

$$\lim_{t\to+\infty} \frac{F(\alpha(F)+\frac{1}{tx})}{F(\alpha(F)+\frac{1}{t})} = x^{-\gamma} \tag{6.56}$$

which is both necessary and sufficient for F to be in the domain of attraction of the Weibull law for the minimum.

The authors of Cousineau et al [33] argue that in some psychological data sets the observations are not identically distributed, and thus γ in the Weibull approximation is not a fixed constant. To smooth out this variation, the authors suggest to 'generate new observations' by multiplying the original observations from different populations, thus creating new observations which behave as if they came from a single population distribution. That is, assume that the observations can be identified as $X_1, X_2, \cdots$, and $Y_1, Y_2, \cdots$ where the X_j come from distribution $F(x)$ and the Y_j from distribution $G(x)$. It is assumed that F and/or G satisfy (6.56). Then 'the new observations' X_jY_j are distributed with H, also satisfying (6.56), and the new shape parameter γ, computable by F and G, will be the same for all products X_jY_j as observations. Several cases apply, out of which we reproduce here the case when $\alpha(F) = 0$ and $\alpha(G) = 1$. Note that $\alpha(G) = 1$ is essentially refers to $\alpha(G) \neq 0$, because by writing

$$\min Y_j = \alpha(G) \min(Y_j/\alpha(G)),$$

the modified variables $Y_j/\alpha(G)$ now have $\alpha = 1$ for their distribution. That is, we show that if F satisfies (6.56) with $\alpha(F) = 0$ and $Y_j \geq$

1 (no further assumption is made) then the distribution function $H(x) = P(X_i Y_j \leq x)$ also satisfies (6.56) with the same γ as one coming from F.

Let us drop the subscripts. We have

$$P\Big(XY \leq x\Big) \int_1^{+\infty} P(X \leq x/y) dG(y) = \int_1^{+\infty} F\Big(\frac{x}{y}\Big)\, dG(y).$$

Choose a subsequence x_n above. Then

$$\frac{H(x_n)}{F(x_n)} = \int_1^{+\infty} \frac{F(x_n/y)}{F(x_n)}\, dG(y) \to \int_1^{+\infty} y^{-\gamma}\, dG(y)$$

where we applied the dominated convergence theorem and (6.56). Since the integral on the right hand side is a finite number k, we have that (6.56) applies to $H(x)$ as well. We also have that $\alpha(H) = 0$ and $\gamma_H = \gamma_F$.

For other cases, we refer the reader to the original paper. However, we mention that the conclusions are far from obvious. For example, if both $\alpha(F) = 0$ and $\alpha(G) = 0$, and the tails of both F and G at close to zero are exact power functions: as $x \longrightarrow 0$,

$$\frac{F(x)}{x^\gamma} \longrightarrow \text{const.} \quad \text{and} \quad \frac{G(x)}{x^{\gamma^*}} \longrightarrow \text{const.}$$

then

$$\frac{H(x)}{x^{\min(\gamma,\gamma^*)}} \longrightarrow \text{const.} \quad \text{if } \gamma \neq \gamma^*$$

but

$$\frac{H(x)}{x^\gamma \log(1/x)} \longrightarrow \text{const.} \quad \text{if } \gamma = \gamma^*.$$

The latter case still expresses regular variation of $H(x)$ at 0, so both cases imply that (6.56) applies with $\gamma_H = \min(\gamma, \gamma^*)$ (which is, of course, γ if $\gamma = \gamma^*$).

When one has more than two mixtures in the data set, the idea of generating new sample set by multiplication still works when one extends the number of factors in the production of 'new observations'.

6.5.3 Estimation for the Pareto distribution

A random variable X is said to have a Pareto distribution function if its density function is given by

$$f(x) = \begin{cases} ab^a x^{-a-1} & x > b \\ 0 & \text{otherwise}, \end{cases} \tag{6.57}$$

where a and b are positive real numbers.

In the previous section we have described the importance of the Pareto distribution in applications dealing with the statistical distribution of certain socio-economic quantities. In these applications it is important to be able to estimate the values, or to test the hypothesis concerning the values, of the parameters involved. Since the arithmetic mean, as well as other standard statistics, are not defined for all values of a, other statistics have been investigated.

In what follows we derive the density function of the geometric mean of a sample from a Pareto distribution, and show that the sample geometric mean is a sufficient statistics for a when the value of b is known. This result is due to Malik [107] and [108].

Let $X, X_1, X_2, \cdots, X_n$ be i.i.d. random variables with common density function given by (6.57), and let $U^{1/n} = (X_1 X_2 \cdots X_n)^{1/n}$. To compute the Mellin transform of $U^{1/n}$ we first compute the Mellin transform of $X^{1/n}$,

$$\mathcal{M}_{X^{1/n}}(s) = \int_b^{+\infty} x^{s/n}\, ab^a x^{-a-1}\, dx = -\frac{nab^{s/n}}{s - na},$$

$Re(s) > na$. Then the independence of the X_i immediately gives

$$\mathcal{M}_{U^{1/n}}(s) = \left(\mathcal{M}_{X^{1/n}}(s)\right)^n = (-1)^n \frac{n^n a^n b^s}{(s - na)^n}.$$

The density function of $U^{1/n}$ can now be recovered by computing its inverse Mellin transform, and for $u > a$ we obtain

$$f_{U^{1/n}}(u) \quad = \quad \frac{n^n a^n}{2\pi i}(-1)^n u^{-1} \int_{c-i\infty}^{c+i\infty} \frac{u^{-s} b^s}{(s - na)^n}\, ds.$$

The above integral can be evaluated by applying standard complex variable techniques (i.e., changing the contour of integration and then applying the residue theorem, see [111]), and obtain

$$\begin{aligned} f_{U^{1/n}}(u) &= \frac{n^n a^n}{(n-1)!}(-1)^n u^{-1} H^{(n-1)}(na) \\ &= \frac{n^n a^n u^{-1}}{\Gamma(n)} \log^{n-1}(\frac{u}{b}) b^{na} u^{-na} \\ &= \frac{n^n a^n b^{na}}{\Gamma(n)} \log^{n-1}\left(\frac{u}{b}\right) u^{-(na+1)}, \qquad (6.58) \end{aligned}$$

where

$$H^{(n-1)}(s) = \frac{d^{n-1}}{ds^{n-1}} e^{s(\log b - \log u)}.$$

From (6.57) and (6.58) it is easy to see that $U^{1/n}$ is sufficient for a when b is known (see [26], p. 274), and this completes our discussion on the Pareto distribution.

6.5.4 Applications and properties of products of independent beta random variables

Let $a > 0$, $b > 0$,

$$B(a,b) = \int_0^1 x^{a-1}(1-x)^{b-1}\, dx,$$

and

$$\Gamma(a) = \int_0^{+\infty} x^{a-1} e^{-x}\, dx.$$

We say that a random variable X has a beta distribution function with parameters (a, b) ($X \sim$ Beta(a, b)) if the density function of X is given by

$$f(x) = \begin{cases} \frac{1}{B(a,b)} x^{a-1}(1-x)^{b-1} & 0 < x < 1 \\ 0 & \text{otherwise.} \end{cases}$$

Let X and Y be independent gamma random variables with parameters (a_1, b_1) and (a_2, b_2), respectively. It is well known (see Rao

[124], for example) that the density function of $W = X/(X+Y)$ is given by

$$f_W(x) = \begin{cases} \frac{\Gamma(a_1+a_2)}{\Gamma(a_1)\Gamma(a_2)} x^{a_1-1}(1-x)^{a_2-1} & 0 < x < 1 \\ 0 & \text{otherwise.} \end{cases}$$

That is, $W \sim$ Beta(a_1, a_2), and

$$B(a_1, a_2) = \frac{\Gamma(a_1)\Gamma(a_2)}{\Gamma(a_1+a_2)}$$

must hold.

Let $X \sim$ Beta(a, b). The Mellin transform of X is given by

$$\begin{aligned} \mathcal{M}_X(s) &= \frac{1}{B(a,b)} \int_0^1 x^{s+a-1}(1-x)^{b-1}\, dx \\ &= \frac{\Gamma(a+b)\Gamma(s+a)}{\Gamma(a)\Gamma(s+a+b)}. \end{aligned} \tag{6.59}$$

Then for any positive integer k,

$$\begin{aligned} E(X^k) &= \mathcal{M}_X(k) \\ &= \frac{\Gamma(a+k)\Gamma(a+b)}{\Gamma(a+k+b)\Gamma(a)} = \frac{(a)_k}{(a+b)_k}, \end{aligned} \tag{6.60}$$

where for arbitrary real number α, $(\alpha)_k = \alpha(\alpha-1)\cdots(\alpha-k+1)$. Therefore if $X_1, X_2, \cdots, X_n$ are independent random variables, $X_i \sim$ Beta(a_i, b_i), $i = 1, 2, \cdots, n$, and $Y \stackrel{d}{=} X_1 X_2 \cdots X_n$, then

$$E(Y^k) = \int_0^1 y^k f_Y(y)\, dy = \prod_{i=1}^{n} \frac{(a_i)_k}{(a_i+b_i)_k}. \tag{6.61}$$

There are many statistical multivariate test criteria U whose moments $E(U^k)$ can be expressed as a product of moments of independent beta random variables $X_1, X_2, \cdots, X_t$. In the case U is of bounded support, which holds in many statistical applications, this further implies that $X \stackrel{d}{=} X_1 X_2 \cdots X_t$. For example, let $x_1, x_2, \cdots, x_n$ be a collection of vector observations with accompanying fixed vectors $z_1, z_2, \cdots, z_n$, where x_i is an observation from $N(\beta_{z_i}, \sum)$. Suppose the matrix β is partitioned as $\beta = (\beta_1 \beta_2)$, where β_1 has q_1

columns, and β_2 has q_2 columns. We consider the null hypothesis $H_o : \beta_1 = \beta^*$, for some given matrix β^*. Wilks [153] considered the likelihood ratio criterion

$$U = \frac{|\Sigma_1|}{|\Sigma_2|},$$

where Σ_1 and Σ_2 are the maximum likelihood estimates of Σ over the full range and over the restricted range under H_o, respectively. Let $p \times p$ denote the domension of $\sum$, and put $m = q_1$, $l = n - q_1 - q_2$, were we assume $n > q_1 + q_2$. Anderson [6] showed that under the assumption H_o is true,

$$E(U^k) = \prod_{j=1}^{p} \frac{\Gamma((n+1-j)/2+k)\Gamma((n+m+1-i)/2)}{\Gamma((n+1-i)/2)\Gamma(n+m+1-i)/2+k)}, \quad (6.62)$$

and $U \stackrel{d}{=} X_1 X_2 \cdots X_p$, where the X_i are independent, $X_i \sim$ Beta$((n+1-i)/2, m/2))$.

Several other likelihood ratio test criteria that can be expressed as products of independent beta random variables can be found in Anderson [6] (page references are given by Tang and Gupta [147]).

The importance in applications of the distribution function of products of independent random variables has generated much effort in finding both an exact form and an approximation of its distribution function (Wilks [153] and [154], Tukey and Wilks [150], and more references below). Different approaches have been used, some of which we describe next.

Let $U = X_1 X_2 \cdots X_n$, where the X_i are independent, $X_i \sim$ Beta(a_i, b_i), and let $f_U(u)$ denote the density function of U. By using inverse Mellin transform Consul [31] was able to derive a closed form for $f_U(u)$, where U is as in (6.62), for special values of the parameters involved. Dennis [41] also considered the inverse Mellin transform of $f_U(u)$, but his approach was to derive a recursive procedure through which one could recover $f_U(u)$. Tang and Gupta [147] derived a structural form for the density function of Y, and provided a simple recursive relation for computing the coefficients of this representation. We describe their result below.

Let $Y_{1,n} = X_1$, and for $2 \leq k < n$, let $Y_{k+1,n} = Y_{k,n} X_{k+1}$. Clearly, $Y_{n,n} = Y$. Tang and Gupta [147] showed that for $1 \leq k \leq n$, the density function of $Y_{k,n}$ can be written as

$$
\begin{aligned}
g_{k,n}(y) &= \left(\prod_{j=1}^{k} n \frac{\Gamma(a_j + b_j)}{\Gamma(a_j)} y^{a_k - 1} (1-y)^{\sum_{j=1}^{k} b_j - 1} \right) \\
&\quad \cdot \sum_{r=0}^{+\infty} \sigma_r^{(k)} (1-y)^r.
\end{aligned} \tag{6.63}
$$

For $k = 1$, $g_{1,n}(y) = f_{X_1}(y)$, and (6.63) certainly holds with $\sigma_0^{(1)} = 1/\Gamma(b_1)$, and $\sigma_r^{(1)} = 0$, if $r \geq 1$. Assume that (6.63) holds for $1 \leq k < n$, and consider the density function $g_{k+1,n}(y)$ of $Y_{k+1,n}$. Since $Y_{k+1,n} = Y_{k,n} X_{k+1}$, and $Y_{k,n}$ and X_{k+1} are independent, then (1.14) gives that

$$
g_{k+1,n}(y) = \int_0^1 \frac{1}{x} g_{k,n}\left(\frac{y}{x}\right) f_{X_{k+1}}(x)\, dx, \tag{6.64}
$$

which, by using properties of the hypergeometric function, can be written as in (6.63). Moreover for $r \geq 0$,

$$
\sigma_r(k) = \frac{\Gamma(\sum_{j=1}^{k} b_j - 1)}{\Gamma(\sum_{j=1}^{k+1} b_j + r)} \sum_{s=0}^{r} \left[\frac{(a_k + b_k - a_{k-1})_s}{s!} \right] \sigma_{r-s}^{(k-1)}.
$$

An alternative approach which only uses (6.61) is to approximate the distribution function of Y by a simple beta random variable $Z \sim$ Beta(a, b), where the parameters a and b are chosen such that $E(Z) = E(Y)$ and $E(Z^k)$ is as close as possible to $E(Z^k)$ for $2 \leq k \leq T$, for some fixed T. This idea was first introduced by Tukey and Wilks [150]. In the special case $T = 2$, Da-Yin Fan [37] considered the equations

$$
\begin{aligned}
E(Y) &= \prod_{j=1}^{n} \frac{a_j}{a_j + b_j} = \frac{a}{c} \\
E(Y^2) &= \prod_{j=1}^{n} \frac{a_j(a_j+1)}{(a_j+b_j)(a_j+b_j+1)} = \frac{a(a+1)}{c(c+1)},
\end{aligned}
$$

which yield

$$
c = E(Y)a, \quad \text{and} \quad a = \frac{E^2(Y) - E(Y)E(Y^2)}{E(Y^2) - E^2(Y)}.
$$

Hence one can approximate Y by $Z \sim$ Beta$(a, c-a)$, and $E(Z^k) = E(Y^k)$, $k = 1, 2$. Da-Yin Fan [37] provided some numerical considerations on the accuracy of this approximation.

The previous discussion naturally leads to the following question: under which conditions is the distribution function of the product of n independent beta random variables itself beta?

Note that if $X_1, X_2, \cdots, X_n$ are independent, $X_j \sim$ Beta(a_j, b_j), then the Mellin trasform of $Y_n = X_1 X_2 \cdots X_n$ is given by

$$\begin{aligned} \mathcal{M}_{Y_n}(s) &= \prod_{j=1}^{n} \mathcal{M}_{X_j}(s) \\ &= \prod_{j=1}^{n} \frac{\Gamma(a_j + s)\Gamma(a_j + b_j)}{\Gamma(a_j)\Gamma(a_j + b_j + s)}. \end{aligned}$$

From the above equation we immediately have that if there exists a permutation $(j_1, j_2, \cdots, j_n)$ such that for $1 \leq i < n$, $a_{j_i} = a_{j_{i+1}} + b_{j_{i+1}}$, then

$$\mathcal{M}_{Y_n}(s) = \frac{\Gamma(a_{j_n} + s)\Gamma(a_{j_n} + b_1 + b_2 + \cdots + b_n)}{\Gamma(a_{j_n})\Gamma(a_{j_n} + b_1 + b_2 + \cdots + b_n + s)}, \tag{6.65}$$

which gives that $Y_n \sim$ Beta$(a_{j_n}, \sum_{j=1}^{n} b_j)$. The converse is true. That is, the following result holds.

Theorem 6.6 *Let $X_1, X_2, \cdots, X_n$ be independent, $X_j \sim$ Beta(a_j, b_j), and let $Y_n = X_1 X_2 \cdots X_n$. Then $Y_n \sim$ Beta(a, b) if, and only if, there exists a permutation $(j_1, j_2, \cdots, j_n)$ such that for $1 \leq i < n$, $a_{j_i} = a_{j_{i+1}} + b_{j_{i+1}}$. Moreover $a = a_{j_n}$, and $b = b_1 + b_2 + \cdots + b_n$.*

The following result is due to James [86]. The special case $n = 2$ was earlier proved by Ramachandran [123].

PROOF OF THEOREM 6.6. The sufficiency part of the theorem was proved in the discussion leading to Theorem 6.6. For the necessity part, let us assume that $X_1, X_2, \cdots, X_n$ are independent, $X_j \sim$ Beta(a_j, b_j), $j = 1, 2, \cdots, n$, and put $Y_n = X_1 X_2 \cdots X_n$,

$Y_n \sim$ Beta(a, b). Then with $s = u$ real, $u \geq 0$, (6.59) gives

$$\begin{aligned}\mathcal{M}_{Y_n}(u) &= E(Y_n^u) = \frac{\Gamma(a+u)\Gamma(a+b)}{\Gamma(a)\Gamma(a+b+u)} \\ &= \prod_{j=1}^{n} \frac{\Gamma(a_j+u)\Gamma(a_j+b_j)}{\Gamma(a_j)\Gamma(a_j+b_j+u)},\end{aligned}$$

and this and (6.60) give that

$$\frac{E(Y_n^{u+1})}{E(Y_n^u)} = \frac{a+u}{a+b+u} = \prod_{j=1}^{n} \frac{a_j}{a_j+b_j+u}. \tag{6.66}$$

Continuity considerations give that the above equation must hold for all $u \geq 0$ (i.e., if one side is zero, the other side is also zero, and if one side is undefined, so is the other). Since we also have that all the constants involved in (6.66) are greater than zero, the choice of $u = -a$ immediately implies the existence of j, which we label j_n, such that $a_j = a_{j_n} = a$. The choices of $u = -a_j$, $j = 1, 2, \cdots, n$, $j \neq j_n$, further imply the existence of a j^*, which we label j_{n-1}, such that $a_{j^*} = a_{j_{n-1}} = a_{j_n} + b_{j_n}$. Similarly we can find a j^{**}, $j^{**} \neq j_{n-1}, j_n$, which we label j_{n-2}, such that $a_{j^{**}} = a_{j_{n-2}} = a_{j_{n-1}} + b_{j_{n-1}}$. By repeating the same argument $n-3$ times we are able to relabel all the a_j in such a way that for $1 \leq i < n$, $a_{j_i} = a_{j_{i+1}} + b_{j_{i+1}}$. The value $u = -a - b$, and our previous discussion further imply that

$$\begin{aligned}a+b &= a_{j_n} + b = a_{j_1} + b_{j_1} = a_{j_2} + b_{j_2} + b_{j_1} = \cdots \\ &= a_{j_n} + b_1 + b_2 + \cdots + b_n,\end{aligned}$$

thus completing the proof of the theorem. ∎

From the sufficiency part of Theorem 6.6 we can derive a nice property of the beta distribution. That is, any beta random variable can be written as a product of independent beta random variables with an arbitrary number of factors. For example, let $X \sim$ Beta(a, b), and for arbitrary n, let $b_j > 0$ such that $b = \sum_{j=1}^{n} b_j$ (i.e., $b_j = b/n$). Put

$$(a_1, c_1) = (a + b - b_1, b_1),$$

and for $2 \leq j \leq n$,

$$(a_j, c_j) = (a_{j-1} + b_j, b_j).$$

It follows from the definition of a_j and c_j that

$$a_n = a, \quad \sum_{j=1}^{n} c_j = b, \quad \text{and } a_j = a_{j+1} + c_{j+1}, \quad 1 \leq j < n,$$

thus giving that if $X_1, X_2, \cdots, X_n$ are independent, $X_j \sim$ Beta(a_j, c_j), then by Theorem 6.6

$$X \stackrel{d}{=} \prod_{j=1}^{n} X_j.$$

In the case $b = \sum_{j=1}^{+\infty} b_j, b_j > 0$ (i.e., $b_j/2^j$), then

$$X \stackrel{d}{=} \prod_{j=1}^{+\infty} X_j,$$

which implies that $\sum_{j=1}^{+\infty} \log X_j$ converges in distribution, and therefore almost surely to $\log X$.

The above result does not characterize the beta distribution. In fact Kotlarski [94] showed that it is possible to express beta random variables by a product of independent random variables which are not beta distributed. Krysicki [95] extended this result, and showed that any beta random variable can be expressed as a geometric mean of n independent beta random variables with an arbitrary number of factors.

Let $X \sim$ Beta(a, b). By utilizing Gauss' multiplication formula

$$\Gamma(z) = n^{z-1/2}(2\pi)^{-(n-1)/2} \prod_{j=1}^{n} \Gamma\left(\frac{z+j-1}{n}\right), \tag{6.67}$$

Krysicki [95] expressed the Mellin transform of X (6.59) as

$$\begin{aligned}\mathcal{M}_X(s) &= \prod_{j=1}^{n}\frac{\Gamma(\frac{a+s+j-1}{n})\Gamma(\frac{a+b+j-1}{n})}{\Gamma(\frac{a+j-1}{n})\Gamma(\frac{a+b+s+j-1}{n})}\\ &= \prod_{j=1}^{n}\frac{\Gamma(\frac{a+j-1}{n}+\frac{s}{n})\Gamma(\frac{a+b+j-1}{n})}{\Gamma(\frac{a+j-1}{n})\Gamma(\frac{a+b+j-1}{n}+\frac{s}{n})}\\ &= \prod_{j=1}^{n}\mathcal{M}_{Y_j}(\frac{s}{n}) = \prod_{j=1}^{n}\mathcal{M}_{Y_j^{1/n}}(s),\end{aligned}$$

where $Y_j \sim$ Beta$((a+j-1)/n, b/n)$. That is, X can be written as a geometric mean of n independent beta random variables.

Krysicki [95] extended the above result to infinite products, by utilizing the infinite multiplication formula

$$\Gamma(s+1) = 2^{2s}\prod_{j=1}^{+\infty}\frac{\Gamma(\frac{1}{2}+\frac{s}{2^j})}{\Gamma(\frac{1}{2})}$$

obtaining

$$X \stackrel{d}{=} \prod_{j}^{+\infty} V_j^{1/2^j},$$

where the V_j are independent, and $V_j \sim$ Beta$(1/2+(a-1)/2^j, b/2^j)$, $j \geq 1$. Even in this case this implies that $\sum_{j=1}^{+\infty}\log V_j^{1/2^j}$ converges almost surely to $\log X$.

We end this section by mentioning a connection between a beta random variables and beta random variables of the second kind.

A random variable Y is said to have a beta distribution function of the second kind with parameters (a, b) ($Y \sim$ Beta$_2(a, b)$) if X has density function

$$f(x) = \begin{cases} \frac{1}{B(a,b)}x^{a-1}(1+x)^{-a-b} & 0 < x < +\infty \\ 0 & \text{otherwise.}\end{cases}$$

Let $X \sim$ Beta(a, b). It is easy to see that if $Y = (1-X)/X$, then $Y \sim$ Beta$_2(b, a)$. Similarly, if $Y \sim$ Beta$_2(a, b)$, and $X = 1/(1+Y)$, then $X \sim$ Beta(b, a). Jambunathan [85] used this property to prove the following result: Let $U = (1+Y)/(1+X)$, and suppose that $U \sim$ Beta$(b-d, d)$, $Y \sim$ Beta$_2(a, b)$, X and Y independent. Then $X \sim$ Beta$_2(a+d, b-d)$.

If one deals with products, rather than with quotients, then we have the following result, due to James [86]: Let $Y \sim$ Beta$_2(a, b)$, X a nondegenerate random variable taking values on $(0, 1)$, X and Y independent. Then $U = XY$ is a beta random variable of the second kind if, and only if, $X \sim$ Beta$_2(c, d)$, with $a = c + d$. When the conditions are satisfied, $U \sim$ Beta$_2(c, b)$.

6.5.5 Properties of products of independent gamma random variables

The decomposition property of the beta distribution is not a characteristic property of this distribution. In fact in what follows we show that the gamma random variable has similar properties. The following results are due to Gordon [72] (see also Bondesson [23]).

A random variable X is said to have a gamma distribution function, or simply to be gamma, with shape parameter α and scale parameter β ($X \sim$ Gamma(α, β)), α, $\beta > 0$, if X has density function

$$f(x) = \begin{cases} \frac{1}{\Gamma(\alpha)} \beta^\alpha x^{\alpha-1} e^{-\beta x} & x > 0 \\ 0 & \text{otherwise.} \end{cases}$$

Let $X \sim$ Gamma$(\alpha, 1)$. Then its Mellin transform is given by

$$\begin{aligned} \mathcal{M}_X(s) &= \frac{1}{\Gamma(\alpha)} \int_0^{+\infty} x^s x^{\alpha-1} e^{\beta x}\, dx \\ &= \frac{\Gamma(\alpha+s)}{\Gamma(\alpha)}, \end{aligned}$$

from which it is easy to recover

$$\mathcal{M}_{X^{1/k}}(s) = \frac{\Gamma(\alpha + s/k)}{\Gamma(\alpha)}.$$

Let $X_1, X_2, \cdots, X_k$ be independent random variables, $X_j \sim$ Gamma$(\alpha+$

$j/k, \beta)$, $j = 1, 2, \cdots, k$, and put $U = k(X_1 X_2 \cdots X_k)^{1/k}$. Then,

$$\mathcal{M}_U(s) = k^s \prod_{j=1}^{k} \mathcal{M}_{X_j^{1/k}}(s)$$

$$= k^s \prod_{j=1}^{k} \frac{\Gamma((k\alpha + s + j)/k)}{\Gamma((k\alpha + j)/k)},$$

which after applying (6.67) can be written as

$$\mathcal{M}_U(s) = \frac{\Gamma(k\alpha + 1 + s)}{\Gamma(k\alpha + 1)} = \mathcal{M}_Y(s),$$

$Y \sim$ Gamma$(k\alpha + 1, 1)$. Hence $U \stackrel{d}{=} Y$, and U is gamma.

It is also possible to represent a gamma random variable as an infinite product. In fact Gordon [72] showed that if $X \sim$ Gamma$(\alpha, 1)$, then

$$X \stackrel{d}{=} e^{-\gamma} \prod_{j=1}^{+\infty} e^{(\frac{1}{j+1} - \frac{X_j}{j+\alpha})},$$

where γ is the Euler-Mascheroni constant, and the X_j are independent random variables, $X_j \sim$ Gamma$(1, 1)$, $j \geq 1$.

We end this section by describing a connection between gamma and beta random variable.

Let X and Y be independent, $X \sim$ Gamma$(\alpha, 1)$, $Y \sim$ Beta(a, b), $a + b = \alpha$. Then

$$\mathcal{M}_{XY}(s) = \mathcal{M}_X(s)\mathcal{M}_Y(s) = \frac{\Gamma(a + s)}{\Gamma(a)} = \mathcal{M}_V(s),$$

where $V \sim$ Gamma$(a, 1)$. That is, the product of beta and a gamma random variables with the above properties is a gamma random variables. Stuart [143] considered the above result in the special case $X \stackrel{d}{=} V_1^2/(2B)$, and $Y \stackrel{d}{=} V_2^2/A$, $A, B > 0$, and by letting $a = 1/2$ obtained the interesting result that under the above conditions $V_1 V_2$ is normal with mean 0 and variance AB.

6.5.6 Random discriminants

Another instance in which the form of products of independent variables is very important is when one deals with random discriminants. For example, let X be a random variable with moments $m_0 = 1$, $m_1 = E(X)$, $m_2 = E(X^2)$, $\cdots$. The matrix

$$M_n = \begin{pmatrix} 1 & m_1 & m_2 & \cdots & m_n \\ m_1 & m_2 & m_3 & \cdots & m_{n+1} \\ \cdot & \cdot & \cdot & \cdots & \cdot \\ \cdot & \cdot & \cdot & \cdots & \cdot \\ \cdot & \cdot & \cdot & \cdots & \cdot \\ m_n & m_{n+1} & m_{n+2} & \cdots & m_{2n} \end{pmatrix}$$

is called the moment matrix of X. The importance of the moment matrix of a random variable X in applications was described by Lindsay [101], who showed that if $X, X_0, X_1, \cdots, X_n$ are iid, then

$$\det M_n = \frac{1}{(n+1)!} E\Big[\prod_{0 \le j < k \le n} (X_k - X_j)^2 \Big]. \tag{6.68}$$

Motivated by (6.68), Lu and Richards [104] considered the random quantity

$$\Delta_n(X) = \Delta(X_1, X_2, \cdots, X_n) = \prod_{1 \le j < k \le n} (X_k - X_j)^2,$$

and in special cases (i.e., normal, gamma, and beta) they computed its moments and its distribution. We give the details of their results for $n = 2$, and with X_1 and X_2 having a beta distribution function. We follow their computation, however we argue with Mellin transforms rather than with the moments of $\Delta_n(X)$.

Let $X, X_1, X_2, \cdots, X_n$ be iid with common distribution function Beta(a, b). Then the Mellin transform of $\Delta_n(X)$ can be written as

$$\begin{aligned} &\mathcal{M}_{\Delta_n(X)}(s) \\ &= \frac{1}{B^n(a,b)} \int_0^1 \cdots \int_0^1 \Delta^s(x_1, x_2, \cdots, x_n) \prod_{j=1}^n x_j^{a-1}(1-x_j)^{b-1}\, dx_j \\ &= \frac{1}{B^n(a,b)} \prod_{j=1}^n \frac{\Gamma(1+js)\Gamma(a+(j-1)s)\Gamma(b+(j-1)s)}{\Gamma(1+s)\Gamma(a+b+(n_j-2)s)}, \end{aligned}$$

where the last equation follows from a result of Selberg [136]. Let $n = 2$. After some simplification we get

$$\mathcal{M}_{\Delta_n(X)}(s) = \frac{\Gamma^2(a,b)\Gamma(a+s)\Gamma(b+s)\Gamma(2s+1)}{\Gamma(a)\Gamma(b)\Gamma(a+b+2s)\Gamma(a+b+s)\Gamma(s+1)}.$$

If one rewrites $\Gamma(2s+1)$ and $\Gamma(a+b+2s)$ by utilizing (6.67) with $k = 2$, and cancel like terms, then $\mathcal{M}_{\Delta_n(X)}(s)$ equals

$$\frac{\Gamma^2(a+b)\Gamma(a+s)\Gamma(b+s)\Gamma(s+1/2)}{2^{a+b-1}\Gamma(a)\Gamma(b)\Gamma(a+b+s)\Gamma(s+(a+b)/2)\Gamma(s+(a+b+1)/2)}.$$

Let $c = \max\{a,b\}$, and $d = \min\{a,b\}$. If $c \neq d$, by applying once more (6.67) with $k = 2$ to one of $\Gamma(a+b)$, and by using the identity $\sqrt{\pi} = \Gamma(1/2)$, after cancelling like term we have

$$\begin{aligned}\mathcal{M}_{\Delta_n(X)}(s) &= \frac{\Gamma(c+d)\Gamma(c+s)}{\Gamma(c)\Gamma(c+d+s)} \frac{\Gamma((1+c+d)/2)\Gamma(s+1/2)}{\Gamma(1/2)\Gamma((c+d+1)/2+s)} \\ &\quad \cdot \frac{\Gamma((c+d)/2)\Gamma(d+s)}{\Gamma(d)\Gamma((c+d)/2+s)} \\ &= \mathcal{M}_{V_1}(s)\,\mathcal{M}_{V_2}(s)\,\mathcal{M}_{V_3}(s),\end{aligned}$$

where $V_1 \sim$ Beta(c,d), $V_2 \sim$ Beta$(1/2,(c+d)/2)$, $V_3 \sim$ Beta$(d,(c-d)/2)$. Hence from the uniqueness of the Mellin transform, we have that $X \stackrel{d}{=} V_1V_2V_3$. If $c = d$, that is $a = b$, we rewrites one of the two $\Gamma(2a)$ terms in $\mathcal{M}_{\Delta_n(X)}(s)$ by utilizing (6.67) with $z = 2a$ and $k = 2$, and after cancelling like terms we obtain

$$\mathcal{M}_{\Delta_n(X)}(s) = \frac{\Gamma(a+b)\Gamma(a+s)}{\Gamma(a)\Gamma(a+b+s)} \frac{\Gamma((1+a+b)/2)\Gamma(s+1/2)}{\Gamma(1/2)\Gamma((a+b+1)/2+s)},$$

which by arguing as before gives that in the case $a = b$, $X \stackrel{d}{=} V_1V_2$, where V_1 and V_2 are as before.

Lu and Richards [104] also discussed the asymptotic of $\Delta_n(X)$, and proved the following result.

Theorem 6.7 *Let $X, X_1, X_2, \cdots$ be i.i.d., absolutely continuous, with $E(|X_1)|) < +\infty$, $0 < Var(X_i) < +\infty$, $Var(\log(X_1 - X_2)^2) < +\infty$. Then as $n \to +\infty$,*

$$\frac{\log \Delta_n(X) - E(\log \Delta_n(X))}{\sqrt{Var(\log \Delta_n(X))}} \to Z \sim N(0,1)$$

in distribution.

6.5.7 Normalized products of sums

Let $X_1, X_2, \cdots$ be a sequence of i.i.d. random variables, and for $k = 1, 2, \cdots$, let

$$S_k = X_1 + X_2 + \cdots + X_k.$$

The study of limit distributions for normalized product of sums has recently come up in the literature. For example, while studying the limit distribution of sums of records, Arnold and Villasenõr [11] obtained that in the case X_j has a unit exponential distribution function, then as $n \to +\infty$,

$$\sum_{k=1}^{n} \frac{\log(S_k) - n \log n}{\sqrt{2n}} \to N$$

in distribution, where N is a standard normal random variable. It is easy to see that the above results is equivalent to

$$\left(\prod_{k=1}^{n} \frac{S_k}{k} \right)^{1/\sqrt{n}} \to e^{\sqrt{2}N}$$

in distribution, as $n \to +\infty$. Recently Rempała and Wesołowski [126] extended the above result to square integrable random variables. That is, they proved

Theorem 6.8 *Let $X_1, X_2, \cdots$ be a sequence of i.i.d. positive square integrable random variables with $\mu = E(X_j) > 0$ and $\sigma^2 = Var(X_j) > 0$. Then as $n \to +\infty$,*

$$\left(\frac{\prod_{k=1}^{n} S_k}{n! \mu^n} \right)^{\frac{\mu}{\sigma\sqrt{n}}} \to e^{\sqrt{2}N}$$

in distribution.

The above authors were able to extend the above result in the more general case S_k is replaced by a large class of statistics, the U-statistics, first introduced by Hoeffding [82].

Let $X_1, X_2, \cdots, X_n$ be independent random variables, and let $h(x_1, x_2, \cdots, x_m)$ be a real valued function of $m \leq n$ variables. A statistics of the form

$$U_n = m! \frac{\sum h(X_{i_1} X_{i_2} \cdots X_{i_m})}{\binom{n}{m}} \tag{6.69}$$

where $\sum$ is over all permutations $1 \leq i_1 < i_2 < \cdots < i_m \leq n$ is called a U statistics.

Rempała and Wesołowski assumed that $h(x_1, x_2, \cdots, x_m)$ is a symmetric function of the m variables $x_1, x_2, \cdots, x_m$, and proved the following extension of Theorem 6.8.

Theorem 6.9 *Let $X_1, X_2, \cdots$ be a sequence of i.i.d. random variables, and let U_n be as in (6.69). Assume that*

$$E[(h^2(X_1, X_2, \cdots, X_m)] < +\infty, \quad P(h(X_1, X_2, \cdots, X_m) > 0) = 1,$$

and

$$\mu = E[h(X_1, X_2, \cdots, X_m)] > 0, \quad \sigma^2 = Var[h(X_1, X_2, \cdots, X_m)] > 0.$$

Then as $n \to +\infty$,

$$\left(\frac{\prod_{k=m}^{n} U_k}{m!\mu}\right)^{\frac{\mu}{m\sigma\sqrt{n}}} \to e^{\sqrt{2}N}$$

in distribution.

It is natural to ask whether with suitable constants Theorem 6.8 continues to hold if N is replaced by a random variable S with an arbitrary stable distribution function $F(x)$ (see Section 2.6), and $X_1, X_2, \cdots$ are replaced by i.i.d. in the domain of attraction of $F(x)$. It is worthwhile to notice that for the X_j in Theorem 6.8 the normalizing constants A_n and B_n which assures that, as $n \to +\infty$,

$$\frac{1}{B_n}(X_1 + X_2 + \cdots + X_n) - A_n \to N$$

in distribution, can be chosen to be $B_n = \sigma\sqrt{n}$, and $A_n = (n\mu)/B_n = (\mu\sqrt{n})/\sigma$, which suggests what the normalizing constants should be in the case one considers stable distributions with finite expected value.

Let $Y, Y_1, Y_2, \cdots$ be a sequence of iid with common distribution function $F_Y(x)$. We further assume that $F_Y(x)$ is in the domain of attraction of a positive stable distribution $G(x)$, with $\alpha \in (1, 2]$. That is, if S is a random variable with distribution function $G(x)$, there exist constants A_n and B_n such that, as $n \to +\infty$,

$$\frac{1}{B_n}(Y_1 + Y_2 + \cdots + Y_n) - A_n \to S$$

in distribution. By classical results (see Loève [103]), A_n and B_n can be chosen such that $A_n = (n\mu)/B_n$, and

$$B_n = \inf\left\{x : \frac{1}{1 - F_Y(x)} \geq n\right\}, \qquad \text{if } 1 < \alpha < 2, \tag{6.70}$$

and

$$B_n = \inf\left\{x : \frac{x^2}{E\left[(Y - E(Y))^2 1_{\{|Y - E(Y)| \leq x\}}\right]} \geq n\right\}. \tag{6.71}$$

Qi [119] proved the following theorem.

Theorem 6.10 *Let $X_1, X_2, \cdots$ be a sequence of positive i.i.d. random variables with common distribution $F(x)$, which we assume to be in the domain of attraction of a stable distribution function $G(x)$ with $\alpha \in (1,2]$. Let $\mu = E(X_1)$, and suppose that S is a random variable with distribution function $G(x)$, B_n as in (6.70) and (6.71). Then, as $n \to +\infty$,*

$$\left(\frac{\prod_{k=1}^n S_k}{n!\mu^k}\right)^{\mu/B_n} \to e^{(\Gamma(\alpha+1))^{1/\alpha} S}$$

in distribution.

6.5.8 Distribution of products of order statistics

A useful technique that one can use to recover the density function of the product or quotient of two dependent random variables is based on utilizing a two-dimensional Mellin transform. We only consider nonnegative random variables, but the method easily generalizes to the general case by proceeding as Epstein [44] has done in the one-dimensional case.

Let (X, Y) be a random vector with nonnegative components, and joint density function $f(x, y)$. The Mellin transform of (X, Y) is defined as

$$\mathcal{M}_{(X,Y)}(s_1, s_2) = \int_0^{+\infty} \int_0^{+\infty} x^{s_1} y^{s_2} f(x, y)\, dx\, dy, \tag{6.72}$$

with the inverse

$$f(x, y) = \frac{1}{(2\pi i)^2} \int_{h-i\infty}^{h+i\infty} \int_{t-i\infty}^{t+i\infty} \mathcal{M}_{(X,Y)}(s_1, s_2) x^{-s_1-1} y^{-s_2-1}\, ds_1\, ds_2 \tag{6.73}$$

Conditions under which (6.72) and (6.73) are valid were given by Reed [125] and Fox [50]. Let us assume that (6.72) and (6.73) are valid, and let $V = XY$. Then one immediately gets that

$$\mathcal{M}_V(s) = \mathcal{M}_{(X,Y)}(s, s),$$

and the density function of V can be recovered by computing the one-dimensional inverse Mellin transform introduced in Chapter 1. We illustrate this procedure with the following example, which is due to Subrahmaniam [144].

Let $X_1, X_2, \cdots, X_n$ be iid random variables with common uniform distribution function on (0,1). For $1 \leq j < k \leq n$, let $X_{j:n}$ and $X_{k:n}$ be the j-th and the k-th order statistics of the X_i, respectively (see Section 2.7). We consider the joint density function of $(X_{j:n}, X_{k:n})$ (see (2.79) with $F(x) = x$ and $F(y) = y$)

$$f(x, y) = \frac{n!}{(j-1)!(k-j-1)!(n-k)!}\, x^{j-1}(1-y)^{n-k}(y-x)^{k-j-1},$$

if $0 < x < y < 1$, and $f(x, y) = 0$ otherwise. Then if $X = X_{j:n}$ and $Y = X_{k:n}$,

$$\begin{aligned}
\mathcal{M}_{(X,Y)}(s_1, s_2) &= \int_0^{+\infty}\int_0^{+\infty} x^{s_1} y^{s_2} f(x, y)\, dx\, dy \\
&= \frac{n!}{\Gamma(j)\Gamma(k-j)\Gamma(n-k+1)} \qquad (6.74)\\
&\quad \cdot \int_0^1 \int_0^y x^{j+s_1-1} y^{s_2}(1-y)^{n-k}(y-x)^{k-j-1}\, dx\, dy \\
&= \frac{n!\,\Gamma(j+s_1)}{\Gamma(j)\Gamma(n-k+1)\Gamma(s_1+k)} \\
&\quad \cdot \int_0^1 y^{s_2+s_1+k-1}(1-y)^{n-k}\, dy \\
&= \frac{n!\,\Gamma(j+s_1)\Gamma(s_1+s_2+k)}{\Gamma(j)\Gamma(s_1+k)\Gamma(s_1+s_2+n+1)}. \qquad (6.75)
\end{aligned}$$

From the Mellin transform of (X, Y) it is now simple to recover the Mellin transform of $V = XY$. That is, if we let $s_1 = s_2 = s$ in (6.75),

then

$$\begin{aligned}\mathcal{M}_{(X,Y)}(s,s) &= \mathcal{M}_V(s) \\ &= \frac{n!\,\Gamma(j+s)\Gamma(2s+k)}{\Gamma(j)\Gamma(s+k)\Gamma(n+2s+1)},\end{aligned}$$

from which, by utilizing the inverse Mellin transform of the above function obtained of Consul [32] (equation (2.2), p. 837), we get that if $0 < x < 1$,

$$f_V(x) = \frac{n!\,x^{(k/2)-1}(1-x)^{k-j}}{2\Gamma(j)\Gamma(n-k+1)\Gamma(k-j+1)}$$

$$\cdot \sum_{r=0}^{n-k} \binom{n-k}{r} (-\sqrt{x})^r\, F\Big(k-j, \frac{k-2+r}{2} - j; k-j+1; 1-x\Big),$$

where $F(a,b;c;x)$ is the hypergeometric function (see [46]).

Next we consider the problem of convergence of extremal products as the number of factors goes to infinity.

Let $X_1, X_2, \cdots, X_n$ be a sequence of i.i.d. variables. We define Z_n and W_n to be the largest and the smallest of the X_j, $j = 1, 2, \cdots, n$, respectively. We call the quantities $R_n = |Z_n W_n|$ and $\sqrt{R_n}$ the sample extremal product and the geometric range, respectively. A characterization of the limit distribution of R_n and $\sqrt{R_n}$ for both deterministic and random sample size was given by Barakat and Nigm [20] and [21]. We present their result for R_n, in the case the sample size is a random variable. They showed that the limit distribution of R_n depends on the limit distributions of Z_n and W_n, obtained by introducing appropriate normalizing constants. The collection of all possible limit distributions for Z_n and W_n is given in the following two theorems.

Theorem 6.11 *If for some sequences of real numbers a_n and b_n, and for some nondegenerate distribution function $H(x)$,*

$$\lim_{n\to+\infty} P\Big(\frac{Z_n - b_n}{a_n} \le x\Big) = H(x),$$

for all continuity point x of $H(x)$, then $H(x)$ must be one of the *following type:*

$$(a)\ H_{1,\gamma} = \begin{cases} e^{-x^{-\gamma}} & if\ x > 0, \\ 0 & otherwise. \end{cases}$$

$$(b)\ H_{2,\gamma} = \begin{cases} 1 & if\ x \geq 0, \\ e^{-(-x)^{\gamma}} & x < 0. \end{cases}$$

$$(c)\ H_{3,0} = e^{-e^{-x}}, \quad -\infty < x < +\infty$$

for some constant $\gamma > 0$.

Theorem 6.12 *If for some sequences of real numbers* c_n *and* d_n, *and for some nondegenerate distribution function* $L(x)$,

$$\lim_{n\to+\infty} P\Big(\frac{W_n - d_n}{c_n} \leq x\Big) = L(x),$$

for all continuity point x *of* $L(x)$, *then* $L(x)$ *must be one of the following type:*

$$(a)\ L_{1,\gamma} = \begin{cases} 1 - e^{-(-x)^{-\gamma}} & if\ x < 0, \\ 1 & if\ x > 0. \end{cases}$$

$$(b)\ L_{2,\gamma} = \begin{cases} 1 - e^{-x^{\gamma}} & if\ x > 0, \\ 0 & if\ x < 0. \end{cases}$$

$$(c)\ L_{3,0} = 1 - e^{-e^{x}}, \quad -\infty < x < +\infty.$$

The proofs of the above theorems, as well as a characterization of the normalizing constants a_n, b_n, c_n, and d_n can be found in Galambos [57], Chapter 2.

Let $X_1, X_2, \cdots$ be i.i.d. random variables with common distribution function $F(x)$. We assume that there exist constants a_n, b_n, c_n, d_n such that, as $n \to +\infty$, the distribution function of $(Z_n - b_n)/a_n$ converges weakly to $H_{j,\gamma}$ and the distribution function of $(W_n - d_n)/c_n$ converges weakly to $L_{i,\beta}$, $j, i \in \{1, 2, 3\}$. We further assume that ν_n is independent of the X_j, $j = 1, 2, \cdots$, and that the distribution function of ν_n/n converges weakly to a distribution function $A(x)$, with $A(0)=0$. Finally we set

$$Q(x;\gamma,\beta) = 1 - \int_0^{+\infty} e^{-y-|x|^{-\gamma}y^{\gamma/\beta}}\,dy\,,$$

$$\omega_F = \sup\{x : F(x) < 1\} \qquad \text{and} \qquad \alpha_F = \inf\{x : F(x) > 0\}.$$

The following result is due to Barakat and Nigm [21].

Theorem 6.13 *There exist constants A_n, B_n, C_n, D_n for which the normalized random extremal product $(R_{\nu_n} - B_n)/A_n$ and the normalized random geometric range $(\sqrt{R_{\nu_n}} - D_n)/C_n$ respectively converge weakly to $P_{i,j;\beta,\gamma}(x)$ and $R_{i,j;\beta,\gamma}(x)$ such that*

(i) $P_{1,1;\beta,\gamma}(x) = \left(1 - \int_0^{+\infty} Q(z^{1/\beta-1/\gamma}\,x;-\beta;\gamma)\,dA(z)\right)1_{[0,+\infty)}(x)$ *and* $R_{1,1;\beta,\gamma}(x) = P_{1,1;\beta,\gamma}(x^2)$.

(ii)

$$\begin{aligned} R_{1,2:\beta,\gamma}(\sqrt{x}) &= P_{1,2:\beta,\gamma}(x) \\ &= \left(1 - \int_0^{+\infty} Q(z^{1/\gamma-1/\beta}\,x;-\beta;-\gamma)\,dA(z)\right)1_{[0,+\infty)}(x) \end{aligned}$$

only if $\omega_F = 0$ *and* $R_{2,1;\beta,\gamma}(\sqrt{x}) = P_{1,2;\beta,\gamma}(x)$ *only if* $\alpha_F = 0$.

(iii)

$$\begin{aligned} R_{2,2;\beta,\gamma}(x) &= P_{2,2;\beta,\gamma}(x) \\ &= \begin{cases} Q^{**}(x;\beta) & -\infty < \alpha_F < 0 < \omega_F < +\infty \\ Q^{*}(x;\beta) & 0 < \alpha_F < 0 < \omega_F < +\infty \\ 1 - Q^{*}(-x,\beta) & -\infty < \alpha_F < \omega_F < 0, \end{cases} \end{aligned}$$

where

$$Q^*(x;\beta) = \int_0^{+\infty}\int_{-\infty}^{+\infty} L_{2,\beta}(z^{1/\beta}\rho^{1/\beta}|\frac{\alpha_F}{\omega_F}|(x-y)\,dH_{2,\beta}(z^{1/\beta}y)d(A(z))$$

and

$$Q^{**}(x;\beta) = \int_0^{+\infty}\int_{-\infty}^{+\infty} H_{2,\beta}(z^{1/\beta}\rho^{1/\beta}|\frac{\alpha_F}{\omega_F}|(x-y)\,dH_{2,\beta}(z^{1/\beta}y)d(A(z))$$

only if $\gamma = \beta$ *and as* $x \downarrow 0$, $\rho = \lim F(\alpha_F + x)/(1 - F(\omega_F - x))$ *exists, with* $0 < \rho < +\infty$.

(iv)

$$R_{3,3;0,0} = P_{3,3;0,0}(x) = \begin{cases} \psi^{**}(x;c) & -\infty \leq \alpha_F < 0 < \omega_F \leq +\infty \\ \psi^{*}(x;c) & 0 \leq \alpha_F < \omega_F \leq +\infty \\ 1 - \psi^{*}(-x;c) & -\infty \leq \alpha_F < \omega_F \leq 0, \end{cases}$$

where

$$\psi^{*}(x;c) = \int_0^{+\infty} \int_{-\infty}^{+\infty} L_{3,0}(c^{-1}(x \log x - y))\, dH_{3,0}(y + \log z)\, dA(z)$$

and

$$\psi^{**}(x;c) = \int_0^{+\infty} \int_{-\infty}^{+\infty} H_{3,0}(c^{-1}(x \log x - y))\, dH_{3,0}(y + \log z)\, dA(z),$$

only if there exists a strictly decreasing differentiable function $P(x)$, *which maps a right neighborhood of* α_F *onto a left neighborhood of* ω_F *such that, as* $x \downarrow 0$, $-c = \lim |P(x)/x|/dP(x)/dx$ *exists, with* $0 < c < +\infty$ *and* $1 - F(P(x)) \sim F(x)$ *as* $x \downarrow \alpha_F$.

6.5.9 Characterization of unimodality via products

We end our examples by discussing an interesting characterization of unimodality via products.

We say that a distribution function $F_X(x)$ of a random variable X is unimodal with mode at $x = a$ if $F_X(x)$ is convex for $x < a$, and concave if $x > a$.

It is well known that all stable distribution are unimodal. The next result, due to Zolotarev [160] shows how products can be utilized to characterize unimodality with mode at $x = 0$.

Theorem 6.14 *A distribution function* $F_X(x)$*of a random variable* X *is unimodal with mode at* $x = 0$ *if and only if,*

$$X \stackrel{d}{=} UY,$$

where U *and* Y *are independent random variables, and the distribution function of* U *is uniform on* $(0, 1)$.

PROOF. We follow the proof of Zolotarev [160]. We start by assuming that $F_X(x)$ is unimodal with mode at $x = 0$. Let U be a random variable with distribution function uniform on $(0, 1)$. By the uniqueness of Mellin transform to prove the theorem we need to show that for some random variable Y, independent of U,

$$\begin{aligned}\mathcal{M}_X(t) &= \mathcal{M}_{X^+}(t) + \gamma\mathcal{M}_{X^-}(t) \\ &= \mathcal{M}_U(t)\mathcal{M}_Y(t) = (1+it)^{-1}(\mathcal{M}_{Y^+}(t) + \gamma\mathcal{M}_{Y^-}(t)) \\ &= (1+it)^{-1}\mathcal{M}_{Y^+}(t) + \gamma(1+it)^{-1}\mathcal{M}_{Y^-}(t) \qquad (6.76)\end{aligned}$$

Since we are assuming that $F_X(x)$ is unimodal with mode at $x = 0$, we have that $F_X(x)$ is absolutely continuous, and the function

$$G(x) = F_X(x) - xF'_X(x)$$

is a distribution function (see Zolotarev [160]). A consequence of this is that

$$|x|F'_X(x) \to 0 \text{ if } x \to 0 \text{ and if } |x| \to +\infty. \qquad (6.77)$$

Let Y be a random variable with distribution function $G(x)$. We claim that the Mellin transform of Y satisfies (6.76). Since the positive and negative parts of the Mellin transform of $G(x)$ are computed by performing identical steps, we only prove the validity of (6.76) for the positive parts. By using the definition of $G(x)$ we obtain

$$\begin{aligned}\mathcal{M}_{Y^+}(t) &= \int_0^{+\infty} x^{it}dG(x) \\ &= -\int_0^{+\infty} x^{it+1}dF'_X(x) \\ &= F'_X(x) = (1+it)\mathcal{M}_{X^+}(t),\end{aligned}$$

where the second equation follows integration by parts and property (6.77). Hence

$$\mathcal{M}_{X^+}(t) = (1+it)^{-1}\mathcal{M}_{Y^+}(t).$$

For the converse, let us assume that $X \stackrel{d}{=} UY$, where U and Y are as in the theorem. We are going to show that $F_X(x)$ is absolutely continuous and that $F'_X(x)$ is nondecreasing if $x < 0$, and nondecreasing if $x > 0$, which implies that $F_X(x)$ is unimodal with mode at $x = 0$. For arbitrary x we consider

$$\begin{aligned} F_X(x) &= P\Big(UY \leq x\Big) = \int_0^1 F_Y(\frac{x}{u})\,du \\ &= \int_1^{+\infty} F_Y(tx)\,\frac{1}{t^2}\,dt = F_Y(x) + x\int_1^{+\infty} F_Y(tx)\,\frac{1}{t}\,dt \\ &= F_Y(x) + x\int_x^{+\infty\, sign(x)} F_Y(u)\,\frac{1}{u}\,du, \end{aligned}$$

from which we obtain

$$dF_X(x) = \int_x^{+\infty\, sign(x)} F_Y(u)\,\frac{1}{u}\,du\,dx.$$

The claimed properties of $F_X(x)$ easily follow from the above equation, thus completing the proof of the theorem. ∎

6.6 Quotients of random variables

We conclude this chapter by considering a quantity which is very much related to products: quotients of random variables. Even for quotients the main problem is the characterization of their distributions, and, as in the case of products, they often lead to computational challenges. We do not go into these difficulties, but rather limit our exposition to examples and theoretical considerations.

Let X and Y be independent random variables, $P(Y = 0) = 0$. By exploiting the independence of X and Y we get that the Mellin transform of $Q = X/Y$ is

$$\mathcal{M}_{X/Y}(s) = \mathcal{M}_X(s)\mathcal{M}_{Y^{-1}}(s) = \mathcal{M}_X(s)\mathcal{M}_Y(-s)$$

Hence, the Mellin transform of X/Y can be recovered from the Mellin transforms of X and Y. As an application we describe the following property of the reciprocal of a stable random variable, which is due to Williams [155] and Zolotarev [160]. This latter author has exploited to a great extent the multiplicative property of the Mellin transform to derive properties of stable distributions. The interested reader can further investigate this in [160].

Let X be a positive stable random variable with characteristic exponent $1/n$, n a positive integer, and Mellin transform given by [160]

$$\mathcal{M}_X(s) = \frac{\Gamma(1-ns)}{\Gamma(1-s))}.$$

We consider X^{-1}, and recover its Mellin transform by evaluating $\mathcal{M}_X(s)$ at $s = -s$. That is,

$$\mathcal{M}_{X^{-1}}(s) = \mathcal{M}_X(-s) = \frac{\Gamma(1+ns)}{\Gamma(1+s)}.$$

If we apply (6.67) with $z = 1 + ns$ and $z = 1$ ($\Gamma(1) = 1$), we obtain

$$\mathcal{M}_{X^{-1}}(s) = n^n \prod_{j=1}^{n} \frac{\Gamma(s+1/n)}{\Gamma(1/n)} = n^n \prod_{j=1}^{n} \mathcal{M}_{X_j}(s),$$

where the X_j are independent, $X_j \sim$ Gamma$(1/n, 1)$. Hence, $X^{-1} \stackrel{d}{=} n^n X_1 X_2 \cdots X_n$. The above result can be extended to stable random variables distributed on the real line but first decomposing these random variables as a product of a positive stable random variable and a Cauchy random variable [160].

In the case the X and Y are not independent, the Mellin transform of $Q = X/Y$ could be obtained by utilizing the bivariate Mellin transform, which we introduced in Section 6.5.7. That is, if we let $s_1 = s$ and $s_2 = 2 - s$ in (6.72), then

$$\mathcal{M}_Q(s) = \mathcal{M}_{(X,Y)}(s, -s).$$

Let us continue our example in Section 6.5.7, and set $Q = X_{j:n}/X_{k:n}$, where $X_{j:n}$ and $X_{k:n}$ are the j-th and the k-th order statistics, respectively, of a sample taken from a distribution function uniform on $(0, 1)$. Then if we make the above substitution in (6.75) we obtain

$$\begin{aligned} \mathcal{M}_Q(s) &= \mathcal{M}_{(X,Y)}(s, -s) \\ &= \frac{\Gamma(j+s)\Gamma(k)}{\Gamma(j)\Gamma(s+k)}. \end{aligned}$$

Since the Mellin transform of $x^{j-1}(1-x)^{k-j-1}$, $0 < x < 1$, is

$$\int_0^1 x^{j-1}x^s(1-x)^{k-j-1}\,dx = \frac{\Gamma(j+s)\Gamma(k-j)}{\Gamma(s+k)},$$

this and the uniqueness of the Mellin transform imply that the density function of Q is (see [85])

$$f_Q(x) = \frac{\Gamma(k)}{\Gamma(j)\Gamma(k-j)}\,x^{j-1}(1-x)^{k-j-1}, \quad 0 < x < 1.$$

As in the case of products, in dealing with the characterization, or derivation of specific properties, of the distribution function of quotients of two random variables, at times it is convenient to rely on standard transformation techniques. We show this with the following example, which deal with random variables with given joint density function.

Let $f_{X,Y}(x,y)$ be the joint density function of (X,Y). We are going to derive explicit formulas for the density function of $Q = X/Y$ in terms of $f_{X,Y}(x,y)$. Let a be any arbitrary real number. Then

$$\begin{aligned} F_Q(a) &= \iint_{x/y\leq a} f_{X,Y}(x,y)dx\,dy \\ &= \int_{-\infty}^{0}\int_{ay}^{+\infty} f_{X,Y}(x,y)dx\,dy + \int_{0}^{+\infty}\int_{-\infty}^{ay} f_{X,Y}(x,y)dx\,dy \\ &= \int_{-\infty}^{0}\int_{a}^{-\infty} f_{X,Y}(uy,y)\,ydu\,dy + \int_{0}^{+\infty}\int_{-\infty}^{a} f_{X,Y}(uy,y)\,ydu\,dy \\ &= -\int_{-\infty}^{a}\int_{-\infty}^{0} f_{X,Y}(uy,y)\,ydy\,du + \int_{-\infty}^{a}\int_{0}^{+\infty} f_{X,Y}(uy,y)\,ydy\,du \\ &= \int_{-\infty}^{a}\int_{-\infty}^{+\infty} |y| f_{X,Y}(uy,y)dy\,du, \end{aligned} \tag{6.78}$$

where the third line follows from the second after making the transformation $u = x/y$, $du = dx/y$. The density function of Q can be derived by taking the derivative with respect of a of both sides of (6.78), and it is given by

$$f_Q(a) = \int_{-\infty}^{+\infty} |y| f_{X,Y}(ay,y)\,dy. \tag{6.79}$$

We now apply the above formula by considering quotients of order statistics. The following example is taken from Gumbel and Keeney [75].

Let X and Y be two independent positive unbounded random variables with density functions $f_X(x)$ and $f_Y(y)$. Since the independence of X and Y gives that $f_{X,Y}(x,y) = f_X(x)f_Y(y)$, then if $Q = X/Y$, (6.79) gives

$$f_Q(a) = \int_0^{+\infty} y f_X(ay) f_Y(y)\, dy. \tag{6.80}$$

In the special case $X \stackrel{d}{=} Y$, $f_X(x) = f(x)$, $F_X(x) = F(x)$, interesting properties of Q can be derived by analyzing its distribution function, given by (see (6.78))

$$\begin{aligned} F_Q(a) &= \int_o^a \int_0^{+\infty} y f(uy) f(y) dy\, du \\ &= \int_0^{+\infty} \int_0^a y f(uy) f(y) du\, dy \\ &= \int_0^{+\infty} F(ay) f(y) dy. \end{aligned} \tag{6.81}$$

By applying integration by parts to the integral in the right hand side of (6.81) we obtain

$$F_Q(a) = 1 - a \int_0^{+\infty} F(y) f(ay) dy,$$

which, after applying the transformation $t = ay$, $dt = ady$, becomes

$$F_Q(a) = 1 - \int_0^{+\infty} F(\frac{t}{a}) f(t)\, dt. \tag{6.82}$$

By replacing a with $1/a$ in (6.81) we obtain

$$F_Q(a) = \int_0^{+\infty} F(ay) f(y) dy,$$

and this and (6.82) imply the relation

$$F_Q(a) = 1 - F_Q(\frac{1}{a}). \tag{6.83}$$

The special choice of $a = 1$ immediately give that $F_Q(1) = 1/2$, i.e., $Q = X/Y$, $X \stackrel{d}{=} Y$ has median 1. Gumbel and Keeney [75] suggested that one could exploit relation (6.83) to construct random variables which are symmetric about zero. That is, let $Z = \log Q$. Then

$$F_Z(x) = F_Q(e^x) = 1 - F_Q(e^{-x}) = 1 - F_Z(-x).$$

Next we consider the quotient between the largest and smallest observations in a sample of size n, and describe the conditions under which the quotient can be approximated by an auto-quotient.

Let $X_1, X_2, \cdots, X_n$ be independent random variables with common distribution function $F(x)$ and density function $f(x)$. We assume $F(x)$ is symmetric about $x = 0$ and has unbounded support. We define Z_n and W_n to be the largest and the smallest of the X_i, and we consider the quotient $V = Z_n/(-W_n)$. The symmetry about zero of $F(x)$ implies that

$$P(Z_n > 0) = P(-W_n > 0) = 1 - \left(\frac{1}{2}\right)^n,$$

which implies that the above probabilities can be made arbitrarily close to one by choosing n to be sufficiently large. The asymptotic independence of Z_n and W_n can be obtained by letting n go to infinity and by imposing further conditions on the distribution function of the underlying population, see Gumbel and Keeney [75] for details. One example in which one obtains this desired asymptotic independence of the distribution of the extrema is when $F(x)$ is a Cauchy distribution function. That is,

$$F(x) = \frac{1}{2} + \frac{1}{\pi}\arctan(x), \quad -\infty < x < +\infty.$$

For such distribution we have that (see Galambos [57])

$$\lim_{n \to +\infty} P\Big(Z_n \leq \frac{xn}{\pi}\Big) = \lim_{n \to +\infty} P\Big(-W_n \leq \frac{xn}{\pi}\Big) = e^{-1/x}, \quad x > 0,$$

which gives that for n sufficiently large $F_{Z_n}(x)$ and $F_{-W_n}(x)$ can be closely approximated by $H(x) = e^{-\pi/(nx)}$. When such approximation is made precise, one can see that even densities can similarly be approximated. Hence from (6.81), and n sufficiently large, we obtain that for $a \geq 0$,

$$F_V(a) \approx \int_0^{+\infty} \frac{\pi}{ny^2} e^{-\frac{\pi}{nay}} e^{-\frac{\pi}{ny}}\, dy.$$

By applying the transformation $x = -\pi/(ny)$, $dx = \pi dy/(ny^2)$, we get

$$F_V(a) \approx \int_{-\infty}^{0} e^{x(1+1/a)}\, dx = \frac{a}{1+a}, \quad a \geq 0.$$

More recently Barakat [19] characterized the limit distribution of extremal quotients. As in the case of extremal products, these limit distributions depend on the limit distributions of the normalized Z_n and W_n. Let $H_{i,\gamma}$ and $L_{j,\beta}$ be as in Theorems 6.11 and 6.12. Barakat [19] proved the following result.

Theorem 6.15 *Let $X_1, X_2, \cdots$ be iid random variables, and assume that there exist real numbers a_n, b_n, c_n, and d_n such that the distribution function of $(Z_n - a_n)/b_n$ and $(W_n - c_n)/d_n$ converge to the distribution functions $H_{i,\gamma}$ and $L_{j,\beta}$, respectively, where $H_{i,\gamma}$ and $L_{j,\beta}$ are as in Theorems 6.11 and 6.12. Then there exist real numbers A_n and B_n such that, as $n \to +\infty$, the distribution function of $(U_n - A_n)/B_n$ converges to a nondegenerate distribution function $Q_{i,j;\beta,\gamma}(x)$ such that*

(i) $Q_{1,1;\beta,\gamma}(x) = Q(x;\beta,\gamma)1_{\{(-\infty,0)\}} + 1_{\{[0,+\infty)\}}$, *and*

$$Q(x;\beta,\gamma) = 1 - \int_0^{+\infty} e^{-y-|x|^{-\gamma}y^{\gamma/\beta}}\, dy.$$

The normalizing constants A_n and B_n can be chosen as $A_n = a_n/d_n$ and $B_n = 0$.

(ii) $Q_{1,2:\beta,\gamma}(x) = Q(x;\beta,-\gamma)1_{[0,+\infty)}(x)$, *if $\omega_F = 0$. The normalizing constants can be chosen as in case (i).*

(iii) $Q_{2,1;\beta,\gamma}(x) = [1 - Q(x;-\beta,\gamma)]1_{[0,+\infty)}(x)$, *if $\omega_F = 0$. The normalizing constants can be chosen as in case (i).*

(iv)

$$Q_{2,2:\beta,\gamma}(x) = \begin{cases} Q^{(1)}_{2,2:\beta,\gamma}(x) = 1 - Q^*_2(-x;\beta) & -\infty < \alpha_F < 0 < \omega_F < +\infty, \\ Q^{(2)}_{2,2:\beta,\gamma}(x) = Q^{**}_2(x;\gamma) & 0 < \alpha_F < \omega_F < +\infty, \\ Q^{(3)}_{2,2:\beta,\gamma}(x) = 1 - Q^{**}_2(-x;\gamma) & -\infty < \alpha_F < \omega_F < 0, \end{cases}$$

where

$$Q_2^*(x;\beta) = \int_{-\infty}^{+\infty} L_{2,\beta}(\rho^{1/\beta}|\alpha_F/\omega_F|(x-z))\, dH_{2,\beta}(z)$$

and

$$Q_2^{**}(x;\gamma) = \int_{-\infty}^{+\infty} H_{2,\gamma}(\rho^{1/\gamma}|\alpha_F/\omega_F|(x-z))\, dH_{2,\gamma}(z)$$

if and only if $\gamma = \beta$ *and the limit*

$$\lim_{x\downarrow 0} \frac{F(\alpha_F + x)}{1 - F(\omega_F - x)}$$

exists, with $0 < \rho < +\infty$. *The normalizing constants for the three types* $Q^{(i)}_{2,2;\beta,\gamma}$, $i = 1, 2, 3$, *can be chosen as* $A_n = a_n/|c_n|$ *and* $B_n = b_n/c_n$.

(v) $Q_{3,3;0,0}(x) = L_3(c^{-1}x) * H_3(x)$, *where* $*$ *denotes the convolution operation,*

$$H_3(x) = \begin{cases} H_{3,0}(x) & \alpha_F \geq 0 \\ L_{3,0}(x) & \alpha_F < 0, \end{cases}$$

and

$$L_3(x) = \begin{cases} L_{3,0}(x) & \alpha_F \leq 0 \\ H_{3,0}(x) & \alpha_F > 0, \end{cases}$$

if and only if there exists a strictly decreasing differentiable function $P(x)$, *which maps a right neighborhood of* α_F *onto a left neighborhood of* ω_F *such that the limit*

$$-c = \lim_{x\downarrow \alpha_F} \frac{|P(x)/x|}{dP(x)/dx}$$

exists, with $0 < c < +\infty$ *and* $1 - F(P(x)) \sim F(x)$ *as* $x \downarrow \alpha_F$.

The class of nondegenerate limit distribution functions of $(U_n - B_n)/A_n$ *contains only the distribution functions defined in cases (i)-(v).*

In the original version of Theorem 6.15, Barakat also gives further characterizations of the function $Q_2^*(x;\beta)$ in the case the original distribution function $F_X(x)$ is symmetric, and of $Q_2^{**}(x,\gamma)$ when $\alpha_F\rho^{1/\beta}/\omega_F = 1$. He also characterized the collection of all possible trivial limits. The case of convergence of extremal quotients with random sample size was discussed by Barakat and Nigm [21].

We end this section on quotients by observing that the lack of uniqueness we noticed in expressing a random variable as a product continues to hold if one deals with quotients. For example, it is well known that if X and Y are i.i.d. normal random variables with mean zero and arbitrary variance, then X/Y has a Cauchy distribution function. The converse is not true. In fact Steck [142] and Laha [97] constructed families of distribution functions with this property. Kotlarski [93] investigated further this non-uniqueness property of quotients, and showed that if X and Y are independent gamma random variables, then if $U = X^r/Y^s$, r and s are nonzero real numbers, then there exist independent random variables V and Z with different distribution functions than X^r and Y^s such that $(V/Z) \stackrel{d}{=} U$. Going back to the case X/Y has a Cauchy distribution function, the normality of X and Y can be obtained by imposing restriction on the moments of X and on the zeros of $E(e^{iz\log|X|})$ (see Laha [97]).

Bibliography

[1] V. A. Abramov. General limit theorems in the M-scheme, Theory Probab. Appl. 30, No. 3 (1986), 501-503.

[2] V. A. Abramov. Stability of decompositions in an M-scheme, Theory Probab. Appl. 31 (1986), 123-124.

[3] V. A. Abramov. Limit theorems in the multiplicative scheme, Theory Probab. Appl. 31, No. 4 (1986), 633-644.

[4] J. Aczél. Lectures on functional equations and their applications, Academic Press, New York 1966.

[5] J. Aczél. Personal communication, 1989.

[6] T. W. Anderson. An introduction to multivariate statistical analysis, Wiley, New York 1958.

[7] F. J. Anscombe. Large sample theory of sequential estimation, Proc. Cambridge Philos. Soc. 48 (1952), 600-607.

[8] B. C. Arnold. Pareto distributions, International Cooperative Publishing House, Fairland, Maryland 1983.

[9] B. C. Arnold, E. Castillo, and J. M. Sarabia. Conditional specification of statistical models, Springer, New York 1999.

[10] B. C. Arnold and D. Strauss. Bivariate distributions with conditionals in prescribed exponential families, J. Royal Statist. Soc., Ser. B, 53 (1991), 365-375.

[11] B. C. Arnold and J. A. Villasenõr. The asymptotic distribution of sums of records, Extremes 1 (1998), 351-363.

[12] R. Arratia. Limiting point processes for rescalings of coalescing and annihilating random walks on Z^d, Ann. prob. 9 (1981), 909-936.

[13] A. A. Azlarov and N. A. Volodin. Characterization problems connected with exponential distribution, Springer, New York 1986 (original Russian appeared in 1982, Tashkent).

[14] A. Bakštys. On limiting distribution laws of multiplicative arithmetical functions, Lietuv. mat. rinkinys (=Litovsk. Mat. Sb.) 8 (1968), (1), 5-20 (in Russian).

[15] A. Bakštys. Weak convergence of distribution of products of independent random variables, Lit. Mat. Sb. 11, 4 (1971), 727-744 (in Russian).

[16] A. Bakštys. Convergence to the logarithmic normal law, Lit. Mat. Sb. 12, 1 (1972), 23-39 (in Russian).

[17] A. Bakštys. Convergence to the logarithmic normal law, Lit. Mat. Sb. 12, 2 (1972), 17-27 (in Russian).

[18] N. Balakrishnan and C. R. Rao (editors). Order statistics: Theory and methods, Handbook of statistics 16, North Holland, Amsterdam 1998.

[19] H. M. Barakat. Weak limit of the sample extremal quotient, Austr. & New Zeland J. Stat. 40 (1998), 83-93.

[20] H. M. Barakat. and E. M. Nigm. Weak limit of sample geometric range, J. Indian Stat. Assoc. 34 (1996), 85-95.

[21] H. M. Barakat. and E. M. Nigm. Convergence of random extremal quotient and product, J. Stat. Planning and Inference 81 (1999), 209-221.

[22] G. Bareikis and J. Šiaulys. Nepriklausomų atsitiktinių dydžių sandaugos, Vilniaus Universiteto Leidykla 1998.

[23] L. Bondesson. On infinite divisibility of powers of a gamma variable, Scand. Actuar. J. 61 (1978), 48-61.

[24] M. Bramson, Ding Wan-ding, and R. Durrett. Annihilating branching processes, Stoc. Proc. and Appl. 37 (1991), 1-17.

[25] M. Bramson and L. Gray. The survival of branching annihilationg random walk, Z. Wahrsch. Verw. Gebiete 68 (1985), 447-460.

[26] G. Casella and R. L. Berger. Statistical inference, 2nd ed. Duxbury, 2002.

[27] E. Castillo and J. Galambos. Modelling and estimation of bivariate distributions with conditional based on their marginals, Conference on weighted distributions, Penn State University 1985 (unpublished).

[28] E. Castillo and J. Galambos. Bivariate distributions with normal conditionals, In: Simulation, Modelling and Development-SMD '87 (ed.: M.H. Hamza), Acta Press, Anaheim 1987, 59-62.

[29] E. Castillo and J. Galambos. Conditional distributions and the bivariate normal distribution, Metrika 36 (1989), 209-214.

[30] Y. S. Chow and H. Robbins. On sums of independent random variables with infinite moments and fair games, Proc. Nat. Acad. Sci. USA 47 (1961), 330-335.

[31] P. C. Consul. On the exact distributions of the likelihood ratio critera for testing linear hypotheses about regression coefficients, Ann. Math. Stat. 37 (1966), 1319-1330.

[32] P. C. Consul. On the exact distributions of Votaw's criteria for testing compound symmetry of a covariance matrix, Ann. Math. Stat. 40 (1969), 836-843.

[33] D. Cousineau, V. W. Goodman, and R. M. Shiffrin. Extending statistics of extremes to distributions varying in position and scale and the implications for race models, J. Math. Psychology 46 (2002), 431-454.

[34] H. Cramer. Über eine Eigensschaft der normalen Verteilungsfunction, Math. Z. 41 (1936), 405-414.

[35] H. Daboussi and K.-H. Indlekofer, Two elementary proofs of Halász's theorem, Math. Z. 209 (1992), 43-52.

[36] H. A. David and H. N. Nagaraja. Order statistics, 3rd ed. Wiley, New York 2003.

[37] Da-Yin Fan. The distribution of the product of independent beta variables, Commun. Statist.-Theory Meth. 20 (1991), 4043-4052.

[38] J.-M. De Koninck and A. Ivič. Topics in arithmetical functions, Notas de Matematica 72, North Holland, Amsterdam 1980.

[39] J.-M. De Koninck and J. Galambos. The intermediate prime divisors of integers, Proc. Amer. Math. Soc. 101 (1987), 213-216.

[40] J.-M. De Koninck, I. Kátai, and A. Mercier. On the normal growth of prime fcators of integers, Canadian J. Math. 44 (1992), 1121-1154.

[41] S. Dennis. On the distribution of products of independent beta variates, Communications in Statistics 23 (1994), 1895-1913.

[42] J.L. Doob. Stochastic processes, Wiley, New York, N.Y. 1953.

[43] P.D.T.A. Elliott. Probabilistic number theory I-II, Springer Verlag, New York, 1979/1980.

[44] B. Epstein. Some applications of the Mellin transform in statistics, Ann. of Math. Stat. 19 (1948), 370-379.

[45] B. Epstein and M. Sobel. Life testing, J. Amer. Statist. Assoc. 48 (1953), 486-502,

[46] A. Erdélyi (Editor). Higher trascendental functions I, McGraw-Hill, New York 1954.

[47] P. Erdös. Some remarks and corrections to one of my papers, Bull. Amer. Math. Soc. 53 (1947), 761-763.

[48] P. Erdös and M. Kac. On the Gaussian law of errors in the theory of additive functions, Proc. Nat. Acad. Sci. USA 25 (1939), 206-207) (see also Amer. Journ. Math. 62 (1940), 738-742).

[49] P. Erdös and A. Wintner. Additive arithmetical functions and statistical independence, Amer. J. Math. 61 (1939), 713-721.

[50] C. Fox. Some applications of Mellin transforms to the theory of bivariate statistical distributions, Proc. Cambridge Philos. Soc. 53 (1957), 620-628.

[51] D. Freedman and P. Diaconis. De Finetti's theorem for symmetric location families, Ann. of Stat. 10, No. 1 (1982), 184-189.

[52] J. Galambos. Limiting distribution of sums of (dependent) random variables with applications to arithmetical functions, Z. Wahrscheinlichkeits. verw. Gebiete 18 (1971), 261-270.

[53] J. Galambos. On the distribution of strongly multiplicative functions, Bull. London Math. Soc. (1971), 3, 307-312.

[54] J. Galambos. A remark on the asymptotic theory of sums with random size, Math. Proc. Cambridge Philos. Soc. 79 (1976), 531-532.

[55] J. Galambos. The sequence of prime divisors of integers, Acta Arith. 31 (1976), 213-218.

[56] J. Galambos. On a conjecture of Kátai concerning weakly composite numbers, Proc. Amer. Math. Soc. 96 (1986), 215-216.

[57] J. Galambos. The asymptotic theory of extreme order statistics, 2nd ed. Krieger, Malabar 1987.

[58] J. Galambos. Advanced probability theory, 2nd ed. Marcel Dekker, New York, N.Y. 1995.

[59] J. Galambos and I. Kátai. The gaps in a particular sequence of integers of positive density, J. London Math. Soc. (2) 36 (1987), 429-435.

[60] J. Galambos and I. Kátai. A simple proof for the continuity of infinite convolution of binary random variables, Stat. & Probab. Letters 7 (1989), 369-370.

[61] J. Galambos and S. Kotz. Characterizations of probability distributions, Lecture Notes in Math. Vol. 675, Springer, Heidelberg 1978.

[62] J. Galambos and I. Simonelli. Bonferroni-type inequalities with applications, Springer's Applied Probability Series, New York (1996).

[63] J. Galambos, and I. Simonelli. A purely probabilistic proof for a theorem of Bakštys on multiplicative functions, Ann. Sci. Univ. Budapest, Sectio Comp. 21 (2002), 177-186.

[64] J. Galambos and I. Simonelli. The odd number of occurrences in a sequence of events, Inter. Math. Journal 3, No. 5 (2003), 497-502.

[65] J. Galambos and I. Simonelli. Comments on a recent limit theorem of Quine, Stat. & Probab. Letters 63 (2003), 89-95.

[66] J. Galambos and I. Simonelli. Characterization of probability distributions by properties of products of random variables. J. Appl. Statist. Science, to appear in 2004.

[67] J. Galambos and I. Simonelli. Distribution of multiplicative functions: the symmetric case, Ann. Univ. Sci. Budapest, Sectio Comp., 2003 (to appear in 2004).

[68] J. Galambos and P. Szüsz. On the distribution of multiplicaive functions, Acta Arith. 47 (1986), 57-62.

[69] M. Gardner. Mathematical games, Scientific Americam 223 (1970), 120-123.

[70] U. Gather, U. Kamps, and N. Schweitzer. Characterizations of distributions via identically distributed functions of order statistics, in: Handbook of Statistica 16 (N, Balakrishnan and C. R. Rao eds.), North Holland, Amsterdam (1998), 257-290.

[71] R. Gibrat. Les Inégalités Economiques, Recueil Sirey, Paris 1931.

[72] L. Gordon. Bounds for the distribution of the generalized variance, Ann. Stat. 17 (1989), 1684-1692.

[73] D. Griffeath. Annihilating and coalescing random walks on Z^d, Z. Wahrscheinlichkeitstheorie verw. Geb. 46 (1978), 55-65.

[74] D. Griffeath. Additive and cancellative interacting particle systems,Lecture Notes in Math 724, Springer-Varlag, New York 1979.

[75] E. J. Gumbel and R. D. Kenney. The extremal quotients, Ann. Math. Stat. 21 (1950), 523-538.

[76] H. Halberstam. On the distribution of additive number-theoretic functions I, J. London Math. Soc. 30 (1955), 43-53.

[77] G. H. Hardy and S. Ramanujan. The normal number of prime factors of a number n, Quart. J. Math. (Oxford) 48 (1917), 125-133.

[78] T. E. Harris. Additive set-valued Markov processes and percolation methods, Ann. Prob. 6 (1978),355-378.

[79] T. E. Harris. On a class of set-valued Markov processes, Ann. prob. 4 (1976), 175-194.

[80] A. Hildebrand. On Wirsing's mean value theorem for multiplicative functions, Bull. Lond. Math. Acad. Soc. 18 (1986), 147-152.

[81] A. Hildebrand. A note on the convergence of sums of independent random variables, Ann. of Prob. 20, No.: 3 (1992), 1204-1212.

[82] W. Hoeffding. A class of statistics with asymptotically normal distribution, Ann. Math. Stat. 19 (1948), 293-325.

[83] K.-H. Indlekofer, Remarks on an elementary proof of Halász's theorem, Liet. Matem. Rink. 33, No. 4 (1993), 417-423.

[84] K.-H. Indlekofer and I. Kátai. Generalized moments of additive functions II, Liet. Matem. Rink. 41, No. 1 (2001), 70-83.

[85] M. V. Jambunathan. Some properties of beta and gamma distributions, Ann. Math. Stat. 25 (1954), 401-405.

[86] I. R. James. Products of independent beta variables with applications to Connor an Mosimann's generalized Dirichlet distribution, J. Amer. Stat. Ass. 67, No. 340 (1972), 910-912.

[87] G. Jogesh Babu. Probabilistic methods in the theory of arithmetic functions, Macmillan, New Delhi (1978).

[88] G. Jogesh Babu. On the distribution of multiplicative functions, Acta Arith. 36 (1980), 331-340.

[89] M. Kac. Une remarque sur les équations fonctionnelles, Comm. Math. Helvetici 9 (1937), 170-171.

[90] A. M. Kagan, Yu. V. Linnik, and C. R. Rao. Characterization problems of mathematical statistics, Wiley, New York 1973.

[91] A. V. Kakosyan, L. B. Klebanov, and J. A. Melamed. Characterization of distribution by the method of intensively monotone operators, Lecture Notes in Math. Vol 1088, Springer, Heidelberg 1984.

[92] F. Kong and Q. Tang. A theorem on the convergence of sums of independent random variables, Acta Math. Scientia 21, Sec. 3 (2001), 331-338.

[93] I. Kotlarski. On pairs of independent random variables whose quotients follow some known distribution, Colloq. Math IX, Fasc. 1(1962), 151-162.

[94] I. Kotlarski. On groups of n independent random variables whose product follows the beta distribution, Colloq. Math IX, Fasc. 2(1962), 325-332.

[95] W. Krysicki. On some new properties of the beta distribution, Stat. & Probab. Letters 42 (1999), 131-137.

[96] J. Kubilius. Probabilistic methods in the theory of numbers, Amer. Math. Soc. Translations of Math. Monographs, No. 11 (1964), Providence, R.I..

[97] R. G. Laha. On the law of Cauchy and Gauss, Ann. Math. Stat. 30 (1959), 1165-1174.

[98] P. Lévy. Sur les séries dont les terms sont des variables éventuelles indépendantes, Studia Math. 3 (1931), 119-155.

[99] T. Liggett. Interacting particle systems, Springer-Verlag, New York 1985.

[100] D. Lind. Applications of ergodic theory and sofic systems to cellular automata, Physica 10 D (1984), 36-44.

[101] B. G. Lindsay. On the determinants of matrices, Ann. Stat. 17 (1989), 711-721.

[102] Yu. V. Linnik. Decomposition of probability distributions. Oliver and Boyd, Edinburg 1964 (original Russian appeared in 1960 in Leningrad).

[103] M. Loève. Probability theory I, 4-th ed. Springer, New York 1977.

[104] I-Li Lu and D. Richards. Random discriminants, Ann. Stat. 21 (1993), 1982-2000.

[105] E. Lukacs. Characteristic functions 2nd ed., Hafner Publ. Company, New York 1970.

[106] E. Lukacs and R.G. Laha. Applications of characteristic functions, No. 14, Griffin's Statistical Monographs, London 1964.

[107] H. J. Malik. Estimation of the parameters of the Pareto distribution, Metrika 15 (1970), 126-132.

[108] H. J. Malik. Distribution of product statistics from a Pareto population, Metrika 15 (1970), 19-22.

[109] S. Malmquist. On a property of order statistics from a arectangular distribution, Skand. Actuarietidskr. 33 (1950), 214-222.

[110] K. V. Mardia. Families of bivariate distributions, Griffin, London 1970.

[111] J. E. Marsden and M. J. Hoffman. Basic complex analysis, W. H. Freeman, New York 1999.

[112] M. Miyamoto. An equilibrium state for one-dimensional life game, J. Math. Kyoto Univ. 19 (1979), 525-540.

[113] J. Mogyoródi. A central limit theorem for the sum of a random number of independent random variables, MTA Mat. Kutató Int. Közl. 7 (1962), 407-424.

[114] P. A. P. Moran. An introduction to probability theory, Clarendon Press, Oxford 1968.

[115] J. K. Ord. Families of frequency distributions, Griffin, London 1972.

[116] J. K. Ord. Statistical models for personal income distributions, in: Statistical Distributions in Scientific Work, Vol. 2 (1975), Eds.: G.P. Patil, S. Kotz and J.K. Ord, Reidel, Dordrecht, 151-158.

[117] E.M. Paul. On the distribution of values of additive arithmetical functions, Sankyā Ser A 40 (1970), 1-9.

[118] G. Pólya. Herleitung des Gauss'schen Fehlergesetzes aus einer Funktionalgleichung. Math. Z. 18 (1938), 91-24 (in Russian).

[119] Y. Qi. Limit distributions for products of sums, Stat. & Probab. Letters 62 (2003), 93-100.

[120] M. P. Quine. A result of Shepp, Appl. Math. Lett. 7 (1994), 33-34.

[121] M. P. Quine. A central limit theorem for self-normalized products of random variables, Stat. & Probab. Letters 43 (1999), 137-143.

[122] D. A. Raikov. On the decomposition of Gauss and Poisson laws, Izv. Akad. Nauk. USSR (Ser. Mat.) 2 (1928), 91-124 (in Russian).

[123] B. Ramachandran. On the product of two independent beta random variables, Ann. Math. Stat. 33 (1962), 1212.

[124] C. R. Rao. Linear statistical inference and its applications Wiley, New York 1965.

[125] I. S. Reed. The Mellin type of double integral, Duke Math. J. 11 (1944), 565-572.

[126] G. Rempała and J. Wesołowski. Asymptotics for products of sums and U-statistics, Elect. Comm. in Probab. 7 (2002), 47-54.

[127] A. Rényi. On the theory of order statistics, Acta Math. Acad. Sci. Hung. 4 (1953), 191-231.

[128] A. Rényi. On mixing sequences of sets, Acta Math. Acad. Sci. Hung. 9 (1958), 215-228.

[129] A. Rényi. On the central limit theorem for the sum of a random number of independent random variables, Acta Marh. Acad. Sci. Hung. 11 (1960), 97-102.

[130] A. Rényi. Remarks on the Poisson process, Studia Sci. Math. Hungar. 2 (1967), 119-123.

[131] N. S. Revankar, M. J. Hartley, and M. Pagano. A characterization of the Pareto distribution, Ann. Stat. 2 91974), 599-601.

[132] J. P. Romano and A. F. Siegel. Counterexamples in probability and statistics, Wadsworth & Brooks/Cole 1986.

[133] M. van Rossum-Wijsmuller. A variant of Kátai's minimax theorem for additive functions, Publ. Math. Decebren 34 (1987), 323-326.

[134] W. Schmeidler. Linear Operatoren in Hilbertschen Raum, Teubner, Stuttgart (1954).

[135] D. L. Schwartz. Applications of duality to a class of Markov process, Ann. prob. 5 (1977), 522-532.

[136] A. Selberg. Bermerkninger om et multipelt integral, Nordisk Mathematisk Tidskrift 26 (1944), 71-78.

[137] L. Shepp. Normal functions of normal random variables, SIAM Rev 6 (1964), 459.

[138] I. Simonelli. A limit theorem for a class of interacting particle systems, Ann. Prob. 23 (1995), 141-156.

[139] I. Simonelli. Convergence and symmetry of infinite products of independent random variables, Stat. & Probab. Letters 55 (2001), 45-52.

[140] I. Simonelli. Products of random variables and asymptotic distribution of extremes, 2003 (to appear).

[141] M. D. Springer. The algebra of random variables, Wiley, New York (1979).

[142] G. P. Steck. A uniqueness property not enjoyed by the normal distribution, Ann. Math, Stat. 29 (1958), 604-606.

[143] A. Stuart. Gamma-distributed products of independent random variables, Biometrika 49 (1962), 564-565.

[144] K. Subrahmaniam. On some applications of Mellin transforms to statistics: dependent random variables, SIAM J. Appl. Math. 19 (1970), 658-662.

[145] A. Sudbury. The branching annihilating process: an interacting particle system, Ann. Prob. 18 (1990), 581-601.

[146] P. V. Sukhatme. Tests of significance for samples of the chi-squared population with two degrees of freedom, Ann. Eugen. 8 (1937), 52-56.

[147] J. Tang and A. K. Gupta. On the distribution of the product of independent bet random variables, Stat. & Probab. Letters 2 (1984), 165-168.

[148] M. M. Tata. On outstanding values in a sequence of random variables, Z. Wahrscheinlichkeitstheorie verw. Ceb. 12 (1969), 9-20.

[149] B.V. Timofeev, S. T. Tuliaganov and B.V. Levin. The distribution of values of multiplicative functions, Lietuvos Mat. Rinkinys 13 (1973), 87-100 (in Russian).

[150] J. W. Tukey and S. S. Wilks. Approximation of the products of beta variables by a single beta variable, Ann. Math. Stat. 17 (1946), 318-324.

[151] P. Turán. On a theorem of Hardy and Ramanujan, J. London Math. Soc. 9 (1934), 274-276.

[152] P. Turán. Über einige Verallgemeinerungen eines Satzes von Hardy and Ramanujan, J. London Math. Soc. 11 (1936), 125-133.

[153] S. S. Wilks. Certain generalizations in the analysis of variance, Biometrika 24 (1932), 471-494.

[154] S. S. Wilks. Sample criteria for testing equality of means, equality of variances, and equality of covariances in a normal multivariate distribution, Ann. Math. Stat. 17 (1946), 257-281.

[155] E. J. Williams. Some representation of stable random variables as products, Biometrika 64 (1977), 167-169.

[156] A. Wirsing. Das asymptotische Verhalten von Summen über multiplikative Funktionen II, Acta Math. Acad. Sci. Hung. 18 (1967), 411-467.

[157] S. Wolfram. Geometry of the binomial coefficients, Amer. Math. Montlhy 91 (1984), 566-571.

[158] V. M. Zolotarev. Mellin-Stieltjes transforms in probability theory, Theory Prob. Appl. 2, No. 4 (1957), 433-460.

[159] V. M. Zolotarev. On a general theory of multiplication of independent random variables, Soviet Math. Dokl. 3 (1962), 166-170.

[160] V. M. Zolotarev. One-dimensional stable distributions, Translation of Mathematical Monographs V. 65 (1986), AMS, Providence, RI.

Author Index

Subject Index